VOLUME SIX HUNDRED AND SIXTY THREE

METHODS IN ENZYMOLOGY

Antimicrobial Peptides

METHODS IN ENZYMOLOGY

Editors-in-Chief

ANNA MARIE PYLE
Departments of Molecular, Cellular and Developmental Biology and Department of Chemistry
Investigator, Howard Hughes Medical Institute
Yale University

DAVID W. CHRISTIANSON
Roy and Diana Vagelos Laboratories
Department of Chemistry
University of Pennsylvania
Philadelphia, PA

Founding Editors

SIDNEY P. COLOWICK and NATHAN O. KAPLAN

VOLUME SIX HUNDRED AND SIXTY THREE

METHODS IN ENZYMOLOGY

Antimicrobial Peptides

Edited by

LESLIE M. HICKS
Department of Chemistry,
University of North Carolina at Chapel Hill,
United States

ELSEVIER

ACADEMIC PRESS
An imprint of Elsevier

Academic Press is an imprint of Elsevier
50 Hampshire Street, 5th Floor, Cambridge, MA 02139, United States
525 B Street, Suite 1650, San Diego, CA 92101, United States
The Boulevard, Langford Lane, Kidlington, Oxford OX5 1GB, United Kingdom
125 London Wall, London, EC2Y 5AS, United Kingdom

First edition 2022

ISBN: 978-0-323-90158-1
ISSN: 0076-6879

For information on all Academic Press publications
visit our website at https://www.elsevier.com/books-and-journals

Publisher: Zoe Kruze
Developmental Editor: Federico Paulo S. Mendoza
Production Project Manager: James Selvam
Cover Designer: Alan Studholme

Typeset by STRAIVE, India

Contents

Contributors

Izabela Abissi
Department of Chemistry and Institute of Materials Science, University of Connecticut, Storrs, CT, United States

Jessie L. Allen
University of South Florida, Tampa, FL, United States

Alfredo M. Angeles-Boza
Department of Chemistry and Institute of Materials Science, University of Connecticut, Storrs, CT, United States

Samantha J. Balboa
Department of Chemistry, University of North Carolina at Chapel Hill, Chapel Hill, NC, United States

Amanda M. Brechbill
Department of Chemistry, University of North Carolina at Chapel Hill, Chapel Hill, NC, United States

Marlon Henrique Cardoso
S-Inova Biotech, Pós-Graduação em Biotecnologia, Universidade Católica Dom Bosco, Campo Grande; Centro de Análises Proteômicas e Bioquímicas, Programa de Pós-Graduação em Ciências Genômicas e Biotecnologia, Universidade Católica de Brasília, Brasília, Brazil

Angela Cesaro
Machine Biology Group, Departments of Psychiatry and Microbiology, Institute for Biomedical Informatics, Institute for Translational Medicine and Therapeutics, Perelman School of Medicine; Departments of Bioengineering and Chemical and Biomolecular Engineering, School of Engineering and Applied Science; Penn Institute for Computational Science, University of Pennsylvania, Philadelphia, PA, United States

James W. Checco
Department of Chemistry; The Nebraska Center for Integrated Biomolecular Communication (NCIBC), University of Nebraska-Lincoln, Lincoln, NE, United States

David J. Craik
Institute for Molecular Bioscience, Australian Research Centre of Excellence for Innovations in Peptide Science, The University of Queensland, Brisbane, QLD, Australia

Kevin D. Culver
Department of Chemistry, University of North Carolina at Chapel Hill, Chapel Hill, NC, United States

Nicolau Brito da Cunha
Centro de Análises Proteômicas e Bioquímicas, Programa de Pós-Graduação em Ciências Genômicas e Biotecnologia, Universidade Católica de Brasília, Brasília, Brazil

Cesar de la Fuente-Nunez
Machine Biology Group, Departments of Psychiatry and Microbiology, Institute for Biomedical Informatics, Institute for Translational Medicine and Therapeutics, Perelman School of Medicine; Departments of Bioengineering and Chemical and Biomolecular Engineering, School of Engineering and Applied Science; Penn Institute for Computational Science, University of Pennsylvania, Philadelphia, PA, United States

Caroline M. Donaghy
Department of Chemistry and Institute of Materials Science, University of Connecticut, Storrs, CT, United States

Octávio Luiz Franco
S-Inova Biotech, Pós-Graduação em Biotecnologia, Universidade Católica Dom Bosco, Campo Grande; Centro de Análises Proteômicas e Bioquímicas, Programa de Pós-Graduação em Ciências Genômicas e Biotecnologia, Universidade Católica de Brasília; Programa de Pós-Graduação em Patologia Molecular, Universidade de Brasília, Brasília, Brazil

Neha Garg
School of Chemistry and Biochemistry; Center for Microbial Dynamics and Infection, Georgia Institute of Technology, Atlanta, GA, United States

Morgan B. Harris
Department of Chemistry and Biochemistry; Laboratories of Molecular Anthropology and Microbiome Research, University of Oklahoma, Norman, OK, United States

Julian D. Hegemann
Helmholtz Institute for Pharmaceutical Research Saarland (HIPS), Helmholtz Centre for Infection Research (HZI), Saarland University Campus, Saarbrücken, Germany

Leslie M. Hicks
Department of Chemistry, University of North Carolina at Chapel Hill, Chapel Hill, NC, United States

Matthew Howell
Department of Chemistry and Institute of Materials Science, University of Connecticut, Storrs, CT, United States

Yen-Hua Huang
Institute for Molecular Bioscience, Australian Research Centre of Excellence for Innovations in Peptide Science, The University of Queensland, Brisbane, QLD, Australia

Meng-Wei Kan
Institute for Molecular Bioscience, Australian Research Centre of Excellence for Innovations in Peptide Science, The University of Queensland, Brisbane, QLD, Australia

Sarah J. Kennedy
University of South Florida, Tampa, FL, United States

Michel Lopes Leite
Departamento de Biologia Molecular, Instituto de Ciências Biológicas, Universidade de Brasília, Brasília, Brazil

Mahbobeh Lesani
Department of Microbiology and Plant Biology; Laboratories of Molecular Anthropology and Microbiome Research, University of Oklahoma, Norman, OK, United States

Lingjun Li
Department of Chemistry; School of Pharmacy, University of Wisconsin, Madison, WI, United States

Zongyuan Liu
Department of Chemistry and Biochemistry; Laboratories of Molecular Anthropology and Microbiome Research, University of Oklahoma, Norman, OK, United States

Mariana Rocha Maximiano
S-Inova Biotech, Pós-Graduação em Biotecnologia, Universidade Católica Dom Bosco, Campo Grande; Centro de Análises Proteômicas e Bioquímicas, Programa de Pós-Graduação em Ciências Genômicas e Biotecnologia, Universidade Católica de Brasília, Brasília, Brazil

Andrew C. McAvoy
School of Chemistry and Biochemistry, Georgia Institute of Technology, Atlanta, GA, United States

Laura-Isobel McCall
Department of Chemistry and Biochemistry; Department of Microbiology and Plant Biology; Laboratories of Molecular Anthropology and Microbiome Research, University of Oklahoma, Norman, OK, United States

Biljana Mojsoska
Department of Science and Environment, Roskilde University, Denmark

Tessa B. Moyer
Department of Chemistry, University of North Carolina at Chapel Hill, Chapel Hill, NC, United States

Marvin D. Naing
Department of Chemistry and Institute of Materials Science, University of Connecticut, Storrs, CT, United States

Nicole C. Parsley
Department of Chemistry, University of North Carolina at Chapel Hill, Chapel Hill, NC, United States

Scott Pierce
Department of Chemistry and Institute of Materials Science, University of Connecticut, Storrs, CT, United States

Állan da Silva Pires
Centro de Análises Proteômicas e Bioquímicas, Programa de Pós-Graduação em Ciências Genômicas e Biotecnologia, Universidade Católica de Brasília, Brasília, Brazil

Samilla Beatriz Rezende
S-Inova Biotech, Pós-Graduação em Biotecnologia, Universidade Católica Dom Bosco, Campo Grande, Brazil

Thuanny Borba Rios
S-Inova Biotech, Pós-Graduação em Biotecnologia, Universidade Católica Dom Bosco, Campo Grande; Centro de Análises Proteômicas e Bioquímicas, Programa de Pós-Graduação em Ciências Genômicas e Biotecnologia, Universidade Católica de Brasília, Brasília, Brazil

Christopher S. Sauer
Department of Chemistry, University of Wisconsin, Madison, WI, United States

Lindsey N. Shaw
University of South Florida, Tampa, FL, United States

Marcelo Der Torossian Torres
Machine Biology Group, Departments of Psychiatry and Microbiology, Institute for Biomedical Informatics, Institute for Translational Medicine and Therapeutics, Perelman School of Medicine; Departments of Bioengineering and Chemical and Biomolecular Engineering, School of Engineering and Applied Science; Penn Institute for Computational Science, University of Pennsylvania, Philadelphia, PA, United States

Liana Costa Pereira Vilas Boas
Centro de Análises Proteômicas e Bioquímicas, Programa de Pós-Graduação em Ciências Genômicas e Biotecnologia, Universidade Católica de Brasília; Programa de Pós-Graduação em Patologia Molecular, Universidade de Brasília, Brasília, Brazil

Guangshun Wang
Department of Pathology and Microbiology, College of Medicine, University of Nebraska Medical Center, 985900 Nebraska Medical Center, Omaha, NE, United States

Devon V. Wasche
Department of Chemistry and Institute of Materials Science, University of Connecticut, Storrs, CT, United States

Antonina K. Wenc
Department of Chemistry and Institute of Materials Science, University of Connecticut, Storrs, CT, United States

Baba M. Yussif
Department of Chemistry, University of Nebraska-Lincoln, Lincoln, NE, United States

CHAPTER ONE

Unifying the classification of antimicrobial peptides in the antimicrobial peptide database

Guangshun Wang*
Department of Pathology and Microbiology, College of Medicine, University of Nebraska Medical Center, 985900 Nebraska Medical Center, Omaha, NE, United States
*Corresponding author: e-mail address: gwang@unmc.edu

Contents

Abstract

Natural products offer an important avenue to novel therapeutics against drug-resistant bacteria, viruses, fungi, parasites, and cancer. However, there are numerous hurdles and challenges in discovering such molecules, including antimicrobial peptides (AMPs). While a thorough characterization of AMPs is limited by the amount of material, existing technology, and researcher's expertise, peptide classification is complicated by incomplete information as well as different methods proposed for AMPs from bacteria, plants, and animals. This article describes unified classification schemes for natural AMPs on a common platform: the Antimicrobial Peptide Database (APD; *https://aps.unmc.edu*). The various criteria for these unified classifications include peptide biological source, biosynthesis machinery, biological activity, amino acid sequence, mechanism of action, and

Methods in Enzymology, Volume 663
ISSN 0076-6879
https://doi.org/10.1016/bs.mie.2021.09.006

three-dimensional structure. To overcome the problem with a limited number of known 3D structures, a universal peptide classification has also been refined and executed in the APD database. This universal method, based on the spatial connection patterns of polypeptide chains, is independent of peptide source, size, activity, 3D structure, or mechanism of action. It facilitates information registration, naming, exchange, decoding, prediction, and design of novel antimicrobial peptides.

1. Introduction

Many drugs are derived from natural products, including peptides and small molecules. After years of exploration of small molecule drugs, there is now an increased interest in peptide therapeutics. One important reason is high specificity, making them important candidates for future precise medicine. Nature produces a variety of peptide scaffolds, ranging from linear to cyclic structures. These peptides vary in half-life, allowing them to play a variety of roles in the biosphere such as growth factor, hormone, and anti-infection. At present, 80 peptides have been approved and hundreds more are under development (Craik & Kan, 2021).

Antimicrobial peptides (AMPs) are a special class of natural products that protect the host from microbial infection. This observation stimulated the development of AMPs into novel antibiotics. However, the isolation of novel peptides remains a challenging task. Frequently, the search is limited by the material available for purification and characterization. As a consequence, many peptides do not have known amino acid sequences and/or antimicrobial activity. Fleming noticed lysozyme from a droplet of his nose, but large-scale production only became possible from egg white. Boman had to find a "larger" insect organism for cecropin isolation (Faye & Lindberg, 2016). Peptide characterization can also be challenging, especially for those with extensive post-translationally modifications. Nisin was initially found in 1928. Its structure was established as a lantibiotic only after decades of work (Gross & Morell, 1971). Gassericin A and reutericin 6 were initially thought to differ in the number of D-amino acids. A follow-up study established that these two peptides are identical (Arakawa et al., 2010). Sublancin 168 was originally reported as an unusual lantibiotic where both lanthionine and disulfide bridges confer stability to the peptide (Paik, Chakicherla, & Hansen, 1998). However, there is no Lan enzyme in the genome to facilitate a thioether bond formation. Rather, a glycosyltransferase is encoded, leading to an S-linked glycopeptide (Stepper et al., 2011). Microcin J25 was first proposed to be a head-to-tail circular structure

(Blond et al., 1999). It was later found to have a lasso structure where the sidechain carboxylic group of Glu8 forms a peptide bond with the N-terminal amine of the peptide, allowing the N-terminal region to enter and lock in this cyclic structure (Rosengren et al., 2003; Wilson et al., 2003).

Peptide classification poses another challenge. On one hand, only a limited number of peptides have been characterized today, leading to incomplete information. On the other hand, there are different classification approaches for AMPs from different life kingdoms (Section 2). A unified classification of AMPs from a variety of sources is needed. This chapter describes a data platform for AMPs that enables a detailed classification of natural AMPs (Section 3). We then present unified classification methods for natural AMPs, including a refined universal peptide classification system (Section 4). We also compare the universal peptide classification with those proposed for bacteria, plants, insects, and amphibians (Section 5). Our systematic classifications of natural AMPs lay the foundation for effective information registration, storage, exchange, and use.

2. The problem of peptide classification

There are different classification schemes for antimicrobial peptides from different kingdoms. In animals, AMPs are separated into families such as defensins, cathelicidins, and histatins. Each family has a recognizable trait. Many defensins comprise three disulfide bonds (Schneider, Unholzer, Schaller, Schäfer-Korting, & Korting, 2005). While the precursor sequences of cathelicidins are highly conserved, the mature AMP portions in this family are rather diverse (Scocchi, Wang, & Zanetti, 1997). In amphibians, different family names were also introduced based on peptide sources and sequence patterns. Examples of some commonly accepted family names are aureins, maganins, dermaseptins, brevinins, esculentins, temporins, ranatuerins, and phylloseptins (Amiche, Ladram, & Nicolas, 2008; Conlon, 2008; Wang, 2020b). Unlike amphibians, insect AMPs are grouped into three categories: disulfide bond stabilized hairpin-like β-sheet or α-helical-β-sheet mixed structures, α-helical amphipathic peptides, and proline or glycine-rich peptides (Bulet, Hetru, Dimarcq, & Hoffmann, 1999). In plant AMPs, disulfide bonds frequently occur and are utilized as a basis to classify them into several families, including defensins, thionins, lipid transfer proteins, hevein-like peptides, knottin-type peptides, glycine-rich peptides, hairpin-like peptides, cyclotides, and snakins (Egorov, Odintsova, Pukhalsky, & Grishin, 2005).

The classification of bacteriocins has been evolving with time. Initially, Klaenhammer (1993) classified them into four classes: (I) Lantibiotics, (II) Small heat-stable non-lanthionine containing membrane-active peptides, (III) large heat-labile proteins, and (IV) complex bacteriocins containing lipid or carbohydrate. Class II contains three subclasses: (IIa) *Listeria*-active peptides with a consensus N-terminal sequence YGNGVXC such as pediocin-like peptides, (IIb) poration complexes consisting of two proteinaceous peptides for activity, and (IIc) thiol-activated peptides. Since thioether bonds are characteristic of bacterial lantibiotics, Cotter, Hill, and Ross (2005) recasted the above classification into two major classes: (I) Lanthionine-containing bacteriocins/lantibiotics and (II) Non-lanthionine-containing bacteriocins. Class II was further classified into four subclasses. Class IIa contains heterogeneous small peptides. Class IIb includes two peptides. While cyclic peptides belong to class IIc, non-pediocin single linear peptides form class IId. Also, class III is assigned to non-bacteriocin lytic proteins such as lysostaphin. In a comment article, Heng and Tagg (2006) proposed a "universal" bacteriocin classification scheme. Class I contains lantibiotics, which can be linear (Ia), globular (Ib), and multi-component (Ic). Class II consists of unmodified peptides (IIa: pediocin-like, IIb: miscellaneous, and IIc: multi-component), whereas class III is also reserved for large proteins (IIIa: bacteriolytic and IIIb: non-lytic). However, cyclic peptides became a new class IV. A more recent article puts bacteriocins into two classes: modified and unmodified (Alvarez-Sieiro, Montalbán-López, Mu, & Kuipers, 2016). Class I (modified) includes lantipeptides (Ia), head-to-tail cyclized peptides (Ib), sactibiotics (Ic), linear azol(*in*)e-containing peptides (LAPs) (Id), glycocins (Ie), and lasso peptides (If). Class II (unmodified) comprises pediocin-like bacteriocins (IIa), two-peptide bacteriocins (IIb), leaderless bacteriocins (IIc), and non-pediocin-like, single-peptide bacteriocins (IId). Class III is again reserved for heat-labile large antimicrobial proteins. It is evident that these classification schemes for different kingdoms are diverse and have not been unified under the same umbrella.

3. Classification platform

The information on antimicrobial peptides from different kingdoms is scattered in the literature. To unify peptide classification, it is necessary to register these peptides from various sources into a common platform. For this purpose, we created the APD (Wang & Wang, 2004). The first version contained 525 AMPs. The peptide number increased to 1280 in the second

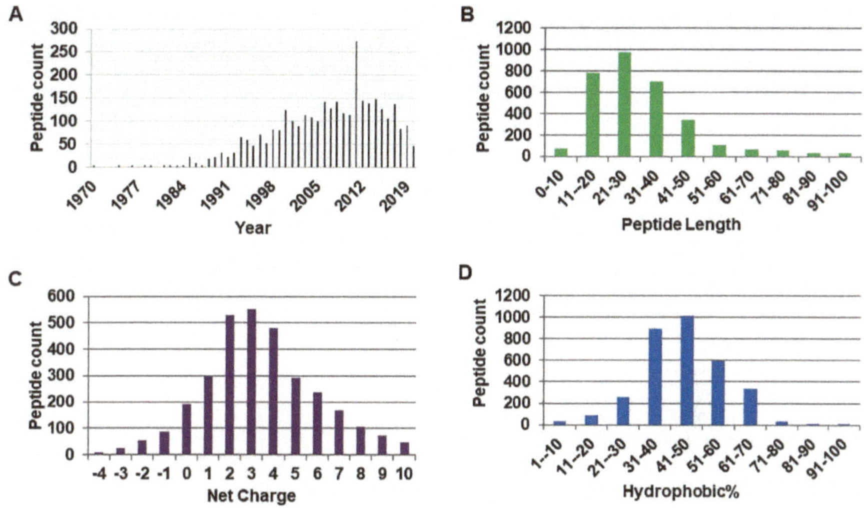

Fig. 1 Antimicrobial peptide discovery and parameter space defined in the antimicrobial peptide database: (A) peptide reported per year; (B) peptide count as a function of length; (C) peptide count as a function of net charge; and (D) Peptide count as a function of peptide hydrophobic content.

version (APD2) (Wang, Li, & Wang, 2009). In the third version with 2619 peptides (Wang, Li, & Wang, 2016), the following set of criteria was applied to data registration: (1) natural peptides, (2) known antimicrobial activity, (3) known amino acid sequences, and (4) a peptide size less than 100 amino acids. Based on these criteria, over 50 such peptides were reported in 1993. The annual total passed 100 AMPs in 2000. A record high of 273 peptides was documented in 2011 (Fig. 1A). Due to continued updates (Wang, 2020a), 3273 such peptides have been registered into the APD as of August 2021 (available at *https://aps.unmc.edu*).

4. Unified classification of antimicrobial peptides

In the following, we describe unified peptide classification methods, each based on a single peptide trait (source, activity, charge, length, and so on).

4.1 Biological sources

AMPs have been discovered in a variety of biological sources, ranging from single-cell prokaryotes to multicellular eukaryotes. The source of each polypeptide is usually unambiguous and can be validated in the genome at the

DNA level. However, there are also cases where the peptide source became complicated. For example, cecropin P1 (AP00134) was originally thought to be isolated from pigs. A revisit uncovered its source as roundworms living within pigs (Andersson, Boman, & Boman, 2003; Sipos, Andersson, & Ehrenberg, 1992). The peptide source classification in the APD3 has been merged into six life kingdoms: bacteria, archaea, protists, fungi, plants, and animals (Wang et al., 2016). Some major sources for natural AMPs are bacteria (369 peptides), plants (361 peptides), and animals (2424 peptides). Most of the animal peptides originate from amphibians (1127 peptides) and insects (325 peptides).

4.2 Peptide machinery

In nature, AMPs are biosynthesized in two mechanisms: ribosomally and non-ribosomally. Such information is annotated in the APD. Ribosomally synthesized peptides dominate (97%). Non-ribosomally made peptides have been discovered in bacteria and fungi. Frequently, they contain non-standard amino acids and special chemical bonds. These natural peptides are excellent templates for chemists to make synthetic or recombinant peptides in laboratories to understand the structure-activity relationships.

4.3 Biological activity

AMPs share a common denominator of being able to inhibit the growth of microbes (viruses, bacteria, fungi, and parasites). They are labeled as such when the minimal inhibitory concentration (MIC) is below 100 μM, or at least inhibition is observed in vitro under laboratory conditions. AMPs may have varying antimicrobial activity spectra. Magainins, defensins, and LL-37 are able to eliminate numerous pathogens (Selsted, Harwig, Ganz, Schilling, & Lehrer, 1985; Wang et al., 2019; Zasloff, 1987). However, plant defensins mainly kill fungi (Egorov et al., 2005), while bacteriocins may only inhibit the growth of competing strains (Cotter et al., 2005).

AMPs may also act on the host cells by binding to cell receptors (usually below MIC) to trigger signal transduction and gene transcription, leading to an alteration in the level of released cytokines or chemokines for instance. Depending on the type of diseases that the peptide has an impact on, AMPs may be anticancer, anti-infectious, anti-diabetics, and wound healing. The APD has annotated over 20 peptide activities/functions (Wang et al., 2016).

4.4 Amino acid sequence

The APD enables the calculations of peptide length, net charge, hydrophobic content, and amino acid composition, which form the basis for peptide classification.

Peptide length: The size distribution of natural AMPs in the APD is shown in Fig. 1B. The shortest peptide comprises two amino acids (aa) plus chemical modifications. The upper limit is defined by the peptides collected into the APD (currently less than 200 amino acids). Based on peptide length, they may be further grouped: ultrashort (2–9 aa), short (10–24 aa), medium (25–50 aa), and long (50–100 aa). The majority of AMPs (88%) are small with less than 50 amino acids (with a peak at 21–30 aa in Fig. 1B). Approximately 10% contain 50–100 amino acids. AMPs with greater than 100 aa are antimicrobial proteins.

Net charge: The classic view is that AMPs are cationic. However, some peptides are anionic (Wang & Wang, 2004). The net charge distribution of the AMPs in the APD is shown in Fig. 1C. 88% of AMPs are cationic (net charge >0) with the maximum at +3. Only 6% are neutral (net charge = 0) and the remaining 6% are anionic (net charge <0). It is important to note that the net charge calculation in the APD has considered the effect of post-translational modifications.

Hydrophobic, hydrophilic, and amphipathic: AMPs are typically amphipathic, consisting both cationic and hydrophobic amino acids. Fig. 1D shows the Pho plot. In the APD, Pho represents the hydrophobic content of a peptide, which is defined as the sum of A, C, F, I, L, M, V, and W divided by the total number of amino acids in the peptide. In Fig. 1D, the majority of AMPs (87%) have hydrophobic contents in the range of 31–70% with the peak at 41–50%. However, there are exceptions to the classic amphipathic concept. Surfactant-associated anionic peptide SAAP lacks hydrophobic amino acids and consists of merely a string of hydrophilic aspartates (Brogden, De Lucca, Bland, & Elliott, 1996). In contrast, two AMPs (e.g., baceridin and lugdunin) are composed of entirely hydrophobic amino acids (Niggemann et al., 2014; Zipperer et al., 2016).

Amino acid composition: Some AMPs are recognized as "amino acid-rich." The classic examples are His-rich, Trp-rich, Pro-rich, Arg-rich, and Gly-rich AMPs. This was judged based on the recurring of a certain amino acid in the sequence. However, there was no consensus definition. The rich amino acid in nearly all the above rich families is greater than 27% except for His-rich (23.3%). The APD defined the word "rich" to be at least >25% (Wang et al., 2016). Such a whole sequence-based definition, however, has a shortcoming since it may miss some AMPs that contain an amino acid-rich motif or domain. Therefore, it may be useful to also define the sequence regions that are rich in certain amino acid for multiple domain peptides (>25%).

4.5 Peptide targets

Many cationic, amphipathic AMPs recognize anionic bacterial membranes. However, the molecular targets of AMPs are not restricted to membranes. While nisins are known to target cell walls (van Heusden, de Kruijff, & Breukink, 2002), Pro-rich peptides can inhibit protein synthesis by targeting ribosomes (Krizsan et al., 2014). Additional molecular mechanisms of AMPs are summarized elsewhere (Wang et al., 2015). However, such mechanistic information is limited, making it impractical to do a full classification of existing AMPs.

4.6 Peptide three-dimensional structure

It is logical to classify AMPs based on 3D structure. Boman classified AMPs into three types: α-helix, β-sheet, and rich (Boman, 2003). This is a mixed scheme that considers both secondary structure (α and β) and amino acid composition (rich or not). However, both His-rich and Trp-rich peptides may form an α-helical structure, causing overlap between classes (Brewer, Hunter, & Lajoie, 1998; Lakshmaiah Narayana et al., 2020). Another classification scheme is also mixed: α-helix, β-sheet, extended structure, and loop (Schneider et al., 2005). The APD proposed and executed a self-consistent structure-based classification scheme: α, β, αβ, and non-αβ (Wang, 2010). While α-helix and β-sheet are self-explanatory, AMPs in the αβ family consist of both α-helix and β-sheet structures. It is worthwhile to point out that the non-αβ family is broad. It can include extended structures, random coiled, turns, spiral structures, and so on. However, structure-based classification suffers from a limited number of known 3D structures determined physically (13.6% in the APD). This is because both multidimensional multinuclear NMR spectroscopy and X-ray crystallography are specialized technologies requiring years of training and experience. Another difficulty is a lack of mechanistic information for many AMPs, making it impractical to determine a meaningful structure in complex with a proper molecular target.

4.7 A universal peptide classification protocol

A universal classification should consider the key aspects of peptides. In our opinion, the peptide connection pattern is a key feature as each chemical bond restricts the folding space of a polypeptide chain substantially, thereby determining both structure and activity. Here we describe a refined universal classification (UC) originally proposed by Wang (2015). In this classification system, peptides are separated into four large classes based on the polypeptide chain bonding types: class O, class P, class S, and class L (Fig. 2). Class O peptides are circular due to a chemical bond between the N-terminal and C-terminal backbone atoms (i.e., UCBB). Class P peptides (also called UCSB) resemble the letter in shape where a chemical bond forms between the sidechain of one amino acid and the backbone of another amino acid in the chain. The third class (Class S) contains a chemical bond between different sidechains (UCSS). Finally, all linear peptides (one or multiple chains) are included in class L (i.e., UCLL). The longer class names (UCBB, UCSB, UCSS, and UCLL) are all unique and searchable in the APD. In the following, we define the classification criteria and priority for this universal classification system.

4.7.1 Key classification criteria

1. Peptide chain connection patterns (UCBB, UCSB, UCSS, and UCLL);
2. Peptide chain number (unlinked and covalently linked);

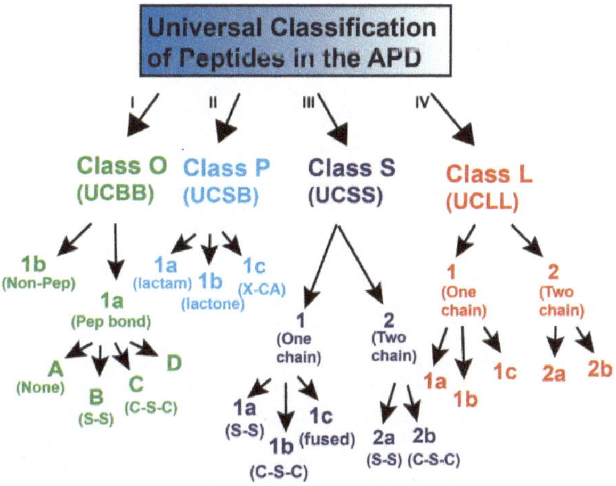

Fig. 2 A universal peptide classification scheme for antimicrobial peptides where the petpides are classified into four classes. For simplicity, a single-letter class name is also introduced. For further details and examples, see the text.

3. Each class can be further separated based on the type and number of additional chemical modifications, which are annotated most comprehensively in the APD.

4.7.2 Classification priority

The first step is to determine the major class (UCBB, UCSB, UCSS, and UCLL) for the peptide based on the chain connection patterns. The classification priority is outlined in Fig. 3 and described below:

1. If the peptide is circular (i.e., head-to-tail connection into a circle), it is assigned as UCBB (e.g., AS-48). This is the highest level (Class I in Fig. 2) in this classification system
2. Next, the peptide is classified as UCSB if there is a chemical bond from sidechain to backbone between different amino acids (e.g., daptomycin). This class has the second priority (Class II in Fig. 2).
3. If the peptide consists of one or more covalent bonds between side chains (e.g., disulfide bonds in defensins and thioether bonds in lantibiotics), they belong to the UCSS class. Such a bond is usually within a single

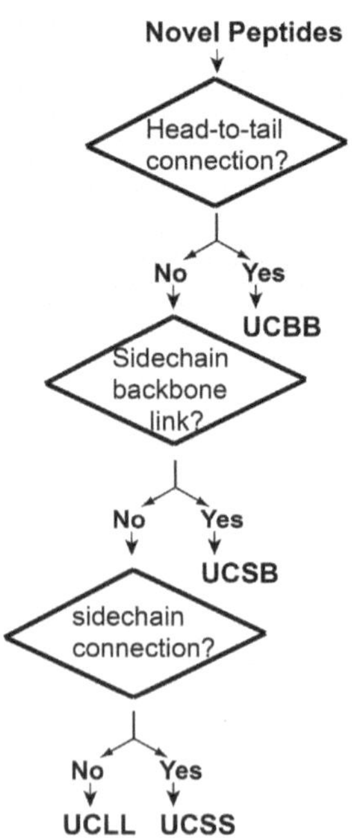

Fig. 3 The classification priority of antimicrobial peptides from cyclic to linear sequences.

peptide chain, but can also occur between two different peptide chains. This class takes the third priority (Class III in Fig. 2).

4. If the peptide is linear and does not have any chemical bond between two different amino acids, it belongs to UCLL (e.g., LL-37 and magainin). This class has the lowest level in this universal classification scheme (Class IV).
5. When the peptide sequence is incomplete, it will be labeled as "unclassified."
6. A peptide may be re-classified once additional information becomes available

4.7.3 Basis for further classification

After assigning the major class for a peptide, one can then further classify it into subclasses based on chain number and additional types of chemical modification (Fig. 2). **UCBB** peptides normally become circular by forming a peptide bond (UCBB1a). This class can be split into four groups. Group A does not have any additional chemical modification (e.g., AS-48). The B group possesses disulfide bridges (S—S). Both plant cyclotides and monkey θ-defensins form three pairs of disulfide bonds that further stabilize the circular structure. The third group C has additional thioether bond (C—S—C) in the ring structure (e.g., subtilosin A). We also define a D group here where D-amino acids have been found (searchable in the APD using UCBB1aD).

UCBB peptides may also become circular via non-peptide bonds (UCBB1b). Lugdunin is a member of the UCBB1b subclass where the two termini cyclization is achieved via a multiple enzyme system (non-ribosomally). The N-terminal amine of L-Cys and C-terminal aldehyde (L-Val) first form a macrocyclic imine/Schiff base. This is followed by a nucleophilic attack of the cysteine thiol group to generate a five-membered thiazolidine heterocycle (Zipperer et al., 2016). This peptide, discovered from human nasal microbiota, is effective against methicillin-resistant *Staphylococcus aureus* (MRSA).

UCSB can also be further classified: (1) UCSB1a with a NH—CO peptide bond (lactam, e.g., lassos MccJ25), (2) UCSB1b with a CO—O ester bond (lactone, e.g., daptomycin), and (3) UCSB1c with a sidechain atom-Cα (X—CA) bond (e.g., thurincin H and tryglysin). Both UCSB1a (17 entries) and UCSB1b (16 entries) peptides are classic members in this class. The UCSB1c peptides (eight members) were first characterized during 2009–10 and most of the members were reported during 2018 and 2021.

These peptides, called sactipeptides (Flühe et al., 2013), contain a thioether bond from the Cys sidechain to peptide backbone Cα of a different amino acid (S—CA). During this study, a new member in the UCSB1c subclass has been reported (Rued et al., 2021). Tryglysin contains a C—Cα (C—CA) bond between aromatic sidechain carbon and Cα. Meanwhile, one additional bond is established between the sidechains of lysine and tryptophan to produce sidechain fusion.

UCSS peptides can be further classified based on the number of chains (UCSS1 and UCSS2 in Fig. 2). Each class can be further classified based on the chemical bond type and number. The known interchain bond types are disulfide bonds S—S (UCSS1a and UCSS2a) and thioether bonds C—S—C between two beta carbons (UCSS1b and UCSS2b). For two-chain peptides, the interchain number of S—S bonds can vary from one (distinctin) to five (rattusin) (Min et al., 2017; Resende et al., 2009). For single-chain peptides, the intrachain S—S bonds in 1092 UCSS1a peptides can vary from one to eight, while the intrachain thioether bonds in 71 UCSS1b lantibiotics can vary from two to seven. We also define a third subclass here reported in 2019 (UCSS1c). Darobactin has an unusual structure with two fused rings formed post-translationally between sidechains. There is an unusual ether bond that links C6 of aromatic W1 and β-carbon of W3. A second carbon-carbon bond connects C7 of aromatic W3 and β-carbon of basic K5 (Imai et al., 2019).

Linear **UCLL** peptides can be further classified based on the peptide chain number. Each can then be further separated based on localized chemical modifications limited to individual amino acid. The UCLL1a subgroup contains linear peptides (e.g., LL-37) without any chemical modifications. In this subclass, some peptides may be rich in certain amino acids (e.g., histatins). An individual sidechain of peptides in the UCLL1b group may be modified (e.g., halogenation, hydroxylation, and glycosylation). This allows a further classification based on the types of modifications. Naturally, any backbone modification of a single amino acid (terminal capping, D-amino acid, dehydration, and heterocycle) generates the third subgroup (UCLL1c). The UCLL2 peptides consist of two independent chains without any interchain bonding. Normally, the two chains need to be combined to be active (UCLL2a). Examples are class 2b bacteriocins such as mutacin IV and carnobacteriocin XY. Under special circumstances, the two peptide chains of gramicidin A can be identical (UCLL2b). As an ionic channel, gramicidin A forms a head-to-head dimer between two single-stranded β-helices (Jadhav et al., 2017).

5. Relationship with source-dependent classifications of antimicrobial peptides

The above-described universal peptide classification serves as a master key to manage the natural antimicrobial peptide information on the same APD platform. It is complementary to the existing classification methods. To illustrate this, Table 1 summarizes the relationships of our universal peptide classification with other kingdom methods. According to this universal classification system, amphibian AMPs fall into two major classes: UCLL and UCSS. The UCLL contains linear peptides that form amphipathic helices or are rich in certain amino acids such as glycine. This class also includes linear AMPs containing D-amino acids (Table 1). The UCSS class includes two major subclasses: those with a Rana box stabilized by intramolecular disulfide bonds and distinctin with an intermolecular disulfide bond that links the two chains. At present, no members from the UCSB or UCBB class have been reported from amphibians.

Like amphibians, insects also make two major classes of AMPs: UCLL and UCSS. The linear UCLL class comprises two subclasses: (1) those that form amphipathic helix in membrane-mimetic environments, and (2) those that are rich in glycines or alanines. Likewise, the UCSS class currently contains two subclasses: (1) those containing intramolecular disulfide bonds (thanatin, drosomycin, and defensin), and (2) those two-chain peptides connected by intermolecular disulfide bonds (delta-myrtoxin-Mp1a) (Table 1).

The plant AMPs consist of three classes: UCLL, UCSS, and UCBB. The linear subclass comprises proline or glycine-rich peptides and "not rich" AMPs. The UCSS class covers a variety of disulfide-bonded defense peptides (Table 1). Numerous circular cyclotides belong to the UCBB class. Finally, bacteria are remarkable organisms that are capable of making all kinds of bacteriocins, covering all the universal classes. The recently proposed bacteriocin families (Alvarez-Sieiro et al., 2016) are well incorporated into this universal classification system (Table 1).

Table 1 also maps the landscape for the currently known antimicrobial peptide scaffolds in various life kingdoms. The linear UCLL class is most widespread and exists in bacteria, plants, and animals. Linear two-chain peptides (UCLL2) are only found in bacteria. However, disulfide bond linked two-chain AMPs (UCSS2a) have been reported in both insects and amphibians, but not in bacteria or plants. Interestingly, the UCSB peptides are only discovered in bacteria. Finally, circular UCBB peptides have examples from

Table 1 Relationship between universal classification and other existing methods.

Universal classification	Subfamily	Bacteria (3 classes)	Plants	Insects	Amphibians
UCLL (Class IV)	UCLL1a	cOB1 (class IId)	(1) Not-rich: Skh-AMP1 (2) Pro-rich: Gly-rich	Gloverin; diptericin	(1) Not-rich XT-2 (2) Rich: Leptoglycin (Gly-rich)
	UCLL1b (sidechain modified)	Delftibactin-A			
	UCLL1c (backbone modified)	Microcin B17 (Class Id)		Antapin	Bombinin H3
	UCLL2	Two-chain peptides (class IIb)			
UCSS (Class III)	UCSS1a	(1) Pediocin-like peptides (class IIa); (2) Glycocins (Ie)	(1) Defensins, (2) thionins; (3) Lipid-transfer proteins; (4) Hevein-like; (5) Knottin-type; (6) Snakin	Defensins (3S—S); Thanatin (1S—S) Drosomycin (4S—S)	Brevinin-1Ea (1S—S)
	UCSS1b (C-S-C)	Lantibiotics (class Ia)			
	UCSS1c (fused Sidechain)	Darobactin			
	UCSS2a (inter)			Delta-Myrtoxin-Mp1a	Distinctin
	UCSS2b (intra)	Lacticin-3147, Smb (class IIb)			

UCSB (Class II)	UCSB1a (CO-NH, lactam)	Microcin J25, polymyxin B (lasso, class If)	
	UCSB1b (CO-O, lactone)	Fusaricidin, daptomycin	
	UCSB1c (Cβ-S-Cα)	Thuricin CD (class Ic: sactibiotic)	
UCBB (Class I)	UCBB1aA (Unmodified)	AS-48 (class Ib, circular)	Cyclosaplin (no S—S)
	UCBB1aB (S—S)	Closticin 574	Circulin A, Kalata B1
	UCBB1aC (3S-CA)	Subtilosin A (class Ic: sactibiotic)	
	UCBB1aD (backbone modified)	Gramicidin S (D-amino acids)	
	UCBB1b	Lugdunin	

bacteria, plants, and non-human primates. Hence, different life kingdoms utilize varying defense chemistries for survival. Because bacteria are able to manufacture a variety of peptide scaffolds (Table 1), they are an important source for identifying future antibiotics. The current interest in microbiota further extends our antimicrobial strategies as well as the search scope.

6. Concluding remarks

Despite numerous challenges, over 3000 thousand antimicrobial peptides have been discovered from nature. In order to improve our understanding, it is important to group them into classes. We advanced peptide classification by achieving unified classifications on the same database platform. Our practice facilitates effective information registration, storage, exchange, and use of antimicrobial peptides. This universal classification can be readily expanded to incorporate new peptide classes. Our systematic classification may enable the development of more effective peptide prediction approaches. Finally, a specific peptide with a defined scaffold and parameters lays out a blueprint for synthetic biologists to engineer future medicine.

Acknowledgment

This study is supported by the NIH grants GM138552, AI137161, and the University of Nebraska Collaborative Initiation Award 2021.

References

Alvarez-Sieiro, P., Montalbán-López, M., Mu, D., & Kuipers, O. P. (2016). Bacteriocins of lactic acid bacteria: Extending the family. *Applied Microbiology and Biotechnology, 100*, 2939–2951.

Amiche, M., Ladram, A., & Nicolas, P. (2008). A consistent nomenclature of antimicrobial peptides isolated from frogs of the subfamily Phyllomedusinae. *Peptides, 29*, 2074–2082.

Andersson, M., Boman, A., & Boman, H. G. (2003). Ascaris nematodes from pig and human make three antibacterial peptides: Isolation of cecropin P1 and two ASABF peptides. *Cellular and Molecular Life Sciences, 60*, 599–606.

Arakawa, K., Kawai, Y., Ito, Y., Nakamura, K., Chujo, T., Nishimura, J., et al. (2010). HPLC purification and re-evaluation of chemical identity of two circular bacteriocins, gassericin A and reutericin 6. *Letters in Applied Microbiology, 50*, 406–411.

Blond, A., Péduzzi, J., Goulard, C., Chiuchiolo, M. J., Barthélémy, M., Prigent, Y., et al. (1999). The cyclic structure of microcin J25, a 21-residue peptide antibiotic from Escherichia coli. *European Journal of Biochemistry, 259*, 747–755.

Boman, H. G. (2003). Antibacterial peptides: Basic facts and emerging concepts. *Journal of Internal Medicine, 254*, 197–215.

Brewer, D., Hunter, H., & Lajoie, G. (1998). NMR studies of the antimicrobial salivary peptides histatin 3 and histatin 5 in aqueous and nonaqueous solutions. *Biochemistry and Cell Biology, 76*, 247–256.

Brogden, K. A., De Lucca, A. J., Bland, J., & Elliott, S. (1996). Isolation of an ovine pulmonary surfactant-associated anionic peptide bactericidal for Pasteurella haemolytica. *Proceedings of the National Academy of Sciences of the United States of America, 93*, 412–416.

Bulet, P., Hetru, C., Dimarcq, J. L., & Hoffmann, D. (1999). Antimicrobial peptides in insects: Structure and function. *Developmental and Comparative Immunology, 23*, 329–344.

Conlon, J. M. (2008). Reflections on a systematic nomenclature for antimicrobial peptides from the skins of frogs of the family Ranidae. *Peptides, 29*, 1815–1819.

Cotter, P. D., Hill, C., & Ross, R. P. (2005). Bacteriocins: Developing innate immunity for food. *Nature Reviews. Microbiology, 3*, 777–788.

Craik, D. J., & Kan, M. W. (2021). How can we improve peptide drug discovery? Learning from the past. *Expert Opinion on Drug Discovery*. https://doi.org/10.1080/17460441.2021.1961740. In press.

Egorov, T. A., Odintsova, T. I., Pukhalsky, V. A., & Grishin, E. V. (2005). Diversity of wheat anti-microbial peptides. *Peptides, 26*, 2064–2073.

Faye, I., & Lindberg, B. G. (2016). Towards a paradigm shift in innate immunity-seminal work by Hans G. Boman and co-workers. *Philosophical Transactions of the Royal Society of London. Series B, Biological Sciences, 371*, 20150303.

Flühe, L., Burghaus, O., Wieckowski, B. M., Giessen, T. W., Linne, U., & Marahiel, M. A. (2013). Two [4Fe-4S] clusters containing radical SAM enzyme SkfB catalyze thioether bond formation during the maturation of the sporulation killing factor. *Journal of the American Chemical Society, 135*, 959–962.

Gross, E., & Morell, J. L. (1971). The structure of nisin. *Journal of the American Chemical Society, 93*, 4634–4635.

Heng, N. C. K., & Tagg, J. R. (2006). What's in a name? Class distinction for bacteriocins. *Nature Reviews. Microbiology, 4*, 160.

Imai, Y., Meyer, K. J., Iinishi, A., Favre-Godal, Q., Green, R., Manuse, S., et al. (2019). A new antibiotic selectively kills Gram-negative pathogens. *Nature, 576*, 459–464.

Jadhav, K. B., Stein, C., Makarewicz, O., Pradel, G., Lichtenecker, R. J., Sack, H., et al. (2017). Bioactivity of topologically confined gramicidin A dimers. *Bioorganic & Medicinal Chemistry, 25*, 261–268.

Klaenhammer, T. R. (1993). Genetics of bacteriocins produced by lactic acid bacteria. *FEMS Microbiology Reviews, 12*, 39–85.

Krizsan, A., Volke, D., Weinert, S., Sträter, N., Knappe, D., & Hoffmann, R. (2014). Insect-derived proline-rich antimicrobial peptides kill bacteria by inhibiting bacterial protein translation at the 70S ribosome. *Angewandte Chemie (International Ed. in English), 53*, 12236–12239.

Lakshmaiah Narayana, J., Mishra, B., Lushnikova, T., Wu, Q., Chhonker, Y. S., Zhang, Y., et al. (2020). Two distinct amphipathic peptide antibiotics with systemic efficacy. *Proceedings of the National Academy of Sciences of the United States of America, 117*, 19446–19454.

Min, H. J., Yun, H., Ji, S., Rajasekaran, G., Kim, J. I., Kim, J. S., et al. (2017). Rattusin structure reveals a novel defensin scaffold formed by intermolecular disulfide exchanges. *Scientific Reports, 7*, 45282.

Niggemann, J., Bozko, P., Bruns, N., Wodtke, A., Gieseler, M. T., Thomas, K., et al. (2014). Baceridin, a cyclic hexapeptide from an epiphytic bacillus strain, inhibits the proteasome. *Chembiochem, 15*, 1021–1029.

Paik, S. H., Chakicherla, A., & Hansen, J. N. (1998). Identification and characterization of the structural and transporter genes for, and the chemical and biological properties of, sublancin 168, a novel lantibiotic produced by Bacillus subtilis 168. *The Journal of Biological Chemistry, 273*, 23134–23142.

Resende, J. M., Moraes, C. M., Munhoz, V. H., Aisenbrey, C., Verly, R. M., Bertani, P., et al. (2009). Membrane structure and conformational changes of the antibiotic

heterodimeric peptide distinctin by solid-state NMR spectroscopy. *Proceedings of the National Academy of Sciences of the United States of America, 106*, 16639–16644.
Rosengren, K. J., Clark, R. J., Daly, N. L., Göransson, U., Jones, A., & Craik, D. J. (2003). Microcin J25 has a threaded sidechain-to-backbone ring structure and not a head-to-tail cyclized backbone. *Journal of the American Chemical Society, 125*, 12464–12474.
Rued, B. E., Covington, B. C., Bushin, L. B., Szewczyk, G., Laczkovich, I., Seyedsayamdost, M. R., et al. (2021). Quorum sensing in *Streptococcus mutans* regulates production of Tryglysin, a novel RaS-RiPP antimicrobial compound. *mBio, 12*, e02688-20.
Schneider, J. J., Unholzer, A., Schaller, M., Schäfer-Korting, M., & Korting, H. C. (2005). Human defensins. *Journal of Molecular Medicine (Berlin, Germany), 83*, 587–595.
Scocchi, M., Wang, S., & Zanetti, M. (1997). Structural organization of the bovine cathelicidin gene family and identification of a novel member. *FEBS Letters, 417*, 311–315.
Selsted, M. E., Harwig, S. S., Ganz, T., Schilling, J. W., & Lehrer, R. I. (1985). Primary structures of three human neutrophil defensins. *The Journal of Clinical Investigation, 76*, 1436–1439.
Sipos, D., Andersson, M., & Ehrenberg, A. (1992). The structure of the mammalian antibacterial peptide cecropin P1 in solution, determined by proton-NMR. *European Journal of Biochemistry, 209*, 163–169.
Stepper, J., Shastri, S., Loo, T. S., Preston, J. C., Novak, P., Man, P., et al. (2011). Cysteine S-glycosylation, a new post-translational modification found in glycopeptide bacteriocins. *FEBS Letters, 585*, 645–650.
van Heusden, H. E., de Kruijff, B., & Breukink, E. (2002). Lipid II induces a transmembrane orientation of the pore-forming peptide lantibiotic nisin. *Biochemistry, 41*, 12171–12178.
Wang, G. (2010). *Antimicrobial peptides: Discovery, design, and novel therapeutic strategies*. England: CABI. version 2, 2017.
Wang, G. (2015). Improved methods for classification, prediction, and design of antimicrobial peptides. *Methods in Molecular Biology, 1268*, 43–66.
Wang, G. (2020a). The antimicrobial peptide database provides a platform for decoding the design principles of naturally occurring antimicrobial peptides. *Protein Science, 29*, 8–18.
Wang, G. (2020b). Bioinformatic analysis of 1000 amphibian antimicrobial peptides uncovers multiple length-dependent correlations for peptide design and prediction. *Antibiotics (Basel), 9*, 491.
Wang, G., Li, X., & Wang, Z. (2009). APD2: The updated antimicrobial peptide database and its application in peptide design. *Nucleic Acids Research, 37*, D933–D937.
Wang, G., Li, X., & Wang, Z. (2016). APD3: The antimicrobial peptide database as a tool for research and education. *Nucleic Acids Research, 44*, D1087–D1093.
Wang, G., Mishra, B., Lau, K., Lushnikova, T., Golla, R., & Wang, X. (2015). Antimicrobial peptides in 2014. *Pharmaceuticals (Basel), 8*, 123–150.
Wang, G., Narayana, J. L., Mishra, B., Zhang, Y., Wang, F., Wang, C., et al. (2019). Design of antimicrobial peptides: Progress made with human cathelicidin LL-37. *Advances in Experimental Medicine and Biology, 1117*, 215–240.
Wang, Z., & Wang, G. (2004). APD: The antimicrobial peptide database. *Nucleic Acids Research, 32*, D590–D592.
Wilson, K. A., Kalkum, M., Ottesen, J., Yuzenkova, J., Chait, B. T., Landick, R., et al. (2003). Structure of microcin J25, a peptide inhibitor of bacterial RNA polymerase, is a lassoed tail. *Journal of the American Chemical Society, 125*, 12475–12483.
Zasloff, M. (1987). Magainins, a class of antimicrobial peptides from Xenopus skin: Isolation, characterization of two active forms, and partial cDNA sequence of a precursor. *Proceedings of the National Academy of Sciences of the United States of America, 84*, 5449–5453.
Zipperer, A., Konnerth, M. C., Laux, C., Berscheid, A., Janek, D., Weidenmaier, C., et al. (2016). Human commensals producing a novel antibiotic impair pathogen colonization. *Nature, 535*, 511–516.

CHAPTER TWO

Protocols for measuring the stability and cytotoxicity of cyclotides

Yen-Hua Huang, Meng-Wei Kan, and David J. Craik*
Institute for Molecular Bioscience, Australian Research Centre of Excellence for Innovations in Peptide Science, The University of Queensland, Brisbane, QLD, Australia
*Corresponding author: e-mail address: d.craik@imb.uq.edu.au

Contents

Abstract

Cyclotides are plant host-defense peptides that have a wide range of biological activities and have diverse potential applications in medicine and agriculture. These 27–37 amino acid peptides have a head-to-tail cyclic backbone and are built around a cystine knot core, which makes them exceptionally stable. This stability and their amenability to sequence modifications has made cyclotides attractive scaffolds in drug design, and many synthetic cyclotides have now been designed and synthesized to test their efficacy as leads for a wide range of diseases, including infectious disease, cancer, pain and multiple sclerosis. Additionally, some natural cyclotides are selectively toxic to certain cancer cell lines, opening their potential as anticancer agents, and others have insecticidal activity, with applications in crop protection. With these applications in mind, there is a need to be able to measure cyclotides in pharmaceutical or

Methods in Enzymology, Volume 663
ISSN 0076-6879
https://doi.org/10.1016/bs.mie.2021.09.007

agrichemical formulations and in biological media such as blood serum, as well as to assess their potential persistence in the environment when used as agrichemical agents. This chapter describes protocols for quantifying cyclotides in biological fluids, measuring their stability, and assessing their relative cytotoxicity on various types of cells.

1. Introduction

Cyclotides are plant-derived peptides with a head-to-tail cyclic backbone and a cystine knot arrangement of their three conserved disulfide bonds (Craik, 2015; Craik, Daly, Bond, & Waine, 1999). The first member of the family, kalata B1, was discovered as the active ingredient in a uterotonic medicinal tea derived from the African plant *Oldenlandia affinis* (Gran, 1970, 1973a, 1973b). Since then, hundreds of cyclotides have been discovered in a wide range of other plants (Göransson, Luijendijk, Johansson, Bohlin, & Claeson, 1999; Gruber et al., 2008). Their prevalence and distribution in the plant kingdom, their structures, biological activities and applications in agriculture and medicine have been extensively described (for recent references, see Camarero & Campbell, 2019; de Veer, Kan, & Craik, 2019; Gerlach & Mondal, 2012; Gründemann, Stenberg, & Gruber, 2019; Kan & Craik, 2018; Koehbach & Gruber, 2015; Weidmann & Craik, 2016).

Cyclotides occur in a wide range of individual species from five major plant families. They are ubiquitous in the Violaceae family, having been found in every single member of that family studied so far, whereas they occur sparsely in the Rubiaceae, Cucurbitaceae, Solanaceae and Fabaceae plant families. Cyclotides occur generally in all tissues within an individual plant, including leaves, stems, flowers and roots, but with a different distribution of cyclotides in different tissues, and with varying contents in different seasons (Trabi et al., 2004; Trabi & Craik, 2004). A single plant can contain anywhere from a few to a few hundred cyclotides.

The natural function of cyclotides is believed to be in host defense. This hypothesis is based on a pioneering feeding trial study in which *Helicoverpa* caterpillar larvae were fed a diet containing cyclotides and either died or were inhibited in their growth compared to controls (Jennings, West, Waine, Craik, & Anderson, 2001). In addition to this insecticidal activity, cyclotides have been reported to have nematocidal (Colgrave et al., 2008; Colgrave et al., 2009; Colgrave, Huang, Craik, & Kotze, 2010; Colgrave, Kotze, Ireland, Wang, & Craik, 2008), antibacterial (Tam, Lu, Yang, & Chiu, 1999), antifungal (Parsley et al., 2018; Slazak et al., 2018; Tam, Lu,

Yang, & Chiu, 1999) and molluscidal activities (Plan, Saska, Cagauan, & Craik, 2008). Numerous studies have revealed a wide range of other biological activities not involving host defense, such as selective toxicity against certain types of cancer cells (Henriques, Huang, Chaousis, Wang, & Craik, 2014; Lindholm et al., 2002; Mehta et al., 2020), anti-HIV activity (Gustafson et al., 1994; Gustafson, McKee, & Bokesch, 2004) and antifouling activity (Göransson, Sjögren, Svangård, Claeson, & Bohlin, 2004). Many of these activities were based on screening assays and do not necessarily reflect the natural functions of cyclotides but are, nevertheless, useful for a wide range of other applications.

Given their diversity of biological activities, cyclotides have found applications in both agriculture and medicine. For example, an extract from the cyclotide-containing plant, *Clitoria ternatea* (butterfly pea), is the basis of a commercial insecticide that is now licensed for the treatment of cotton and vegetable crops in Australia against a wide range of pests. This extract contains more than 70 cyclotides, some (or all) of which are thought to be the bioactive ingredients. Cyclotides also potentially have a wide range of applications in medicine, given their selective cytotoxicity and ability to be grafted to incorporate a desired therapeutic epitope. Such medical applications include cancer, pain, cardiovascular disease, obesity, multiple sclerosis and others. These applications have been extensively reviewed elsewhere (Camarero & Campbell, 2019; de Veer, Kan and Craik, 2019; Mehta et al., 2020; Narayani, Babu, Chadha & Srivastava, 2020; Wang & Craik, 2018; Weidmann & Craik, 2016).

Given the extensive applications of cyclotides in both drug design and agriculture, there is a need to be able to quantify cyclotides and related cyclic peptides in biological samples, e.g., blood (Poth et al., 2021; Poth, Huang, Le, Kan, & Craik, 2019) and in the environment, as well as to determine their stability and potential side effects against non-target cell types. In this article we describe protocols to determine the stability and cytotoxicity of cyclotides. We first provide a short background on cyclotide structures.

2. Structures and syntheses of cyclotides

Cyclotides are typically around 30 amino acids in size with the range of all naturally occurring cyclotides reported so far being 27–37 amino acids. They all contain six conserved cysteine residues and a conserved asparagine or aspartic acid that is essential for the asparaginyl endopeptidase (AEP) mediated biosynthetic cyclization (Du et al., 2020; Harris et al., 2015;

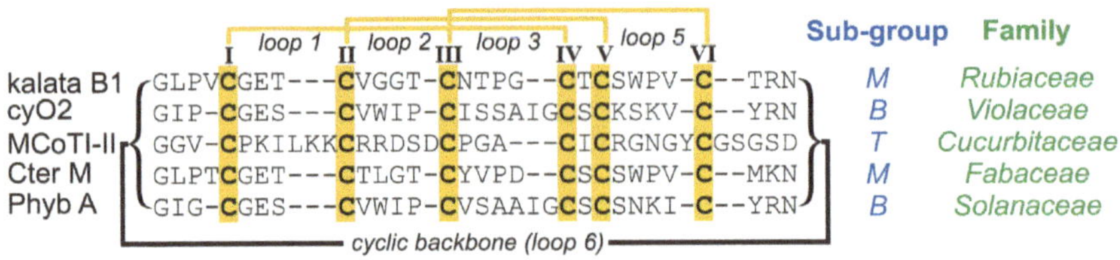

Fig. 1 Sequences of selected cyclotides. Examples include kalata B1, cycloviolacin O2 (cyO2) and *Momordica cochinchinensis* trypsin inhibitor-II (MCoTI-II) as representatives of the Möbius (M), bracelet (B) and trypsin inhibitor (T) sub-groups, respectively. These are the most studied prototypic sequences from these sub-groups, as well as being representative examples from the Rubiaceae, Violaceae and Cucurbitaceae plant families, respectively. Also shown are the sequences of Cter M and Phyb A, as examples of cyclotides from the Fabaceae and Solanaceae plant families. Key features of the sequences include the six conserved Cys residues (numbered I–VI) and the six backbone loops (loops 1–6). A conserved Asn or Asp residue in loop 6 corresponds to the C-terminal residue of the cyclotide domain in the precursor protein and is essential for biosynthetic cyclization of the peptide backbone. *Figure reproduced with permission from de Veer, S. J., Kan, M.-W., & Craik, D. J. (2019). Cyclotides: From structure to function.* Chemical Reviews, 119, *12375–12421.*

Jackson, Nguyen, Gilding, Durek, & Craik, 2020). As illustrated in Fig. 1, the backbone segments between Cys residue are referred to as loops. These are typically exposed on the surface of a cyclotide molecule, protruding from the cystine knot core and contributing to its intermolecular interactions with membranes or receptors. The sequences of currently known cyclotides are documented in the database CyBase (www.cybase.org.au) (Wang, Kaas, Chiche, & Craik, 2008), which contains a wealth of information about their bioactivities and structure as well as a number of tools for their analysis. Mass spectrometry in particular has been pivotal in helping to determine the primary sequences of cyclotides (Kirkpatrick et al., 2017; Parsley et al., 2018).

Cyclotide sequences fall into three sub-groups based on the presence (Möbius sub-group) or absence (bracelet sub-group) of a *cis*-Xaa-proline peptide bond in loop 5, or based on their activity (trypsin inhibitor cyclotides). Fig. 1 shows some prototypic sequences of cyclotides from the various sub-groups and also shows the plant families from which these exemplar cyclotides are derived.

The three-dimensional structures of cyclotides have been determined largely by NMR spectroscopy, with the first structure reported being that of kalata B1 (Saether et al., 1995). The cystine knot which is the core feature of cyclotides comprises two disulfide bonds and their connecting

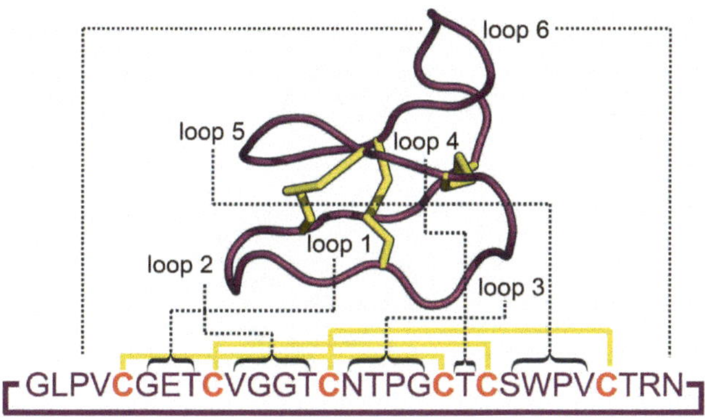

Fig. 2 Sequence and structure of the prototypical cyclotide kalata B1 (PDBID: 1NB1), showing the labeling of backbone loops and the cystine knot motif. *Figure reproduced with permission from Kan, M.-W., & Craik, D. J. (2018). Trends in cyclotide research. In J. Koehnke, J. Naismith, & W. A. van der Donk (Eds.),* Cyclic peptides: From bioorganic synthesis to applications *(Chapter 14, pp. 302–339). Cambridge: The Royal Society of Chemistry.*

backbone segments from loops 1 and 4 forming a ring that is threaded by the third disulfide bond as illustrated in Fig. 2. This motif has now been seen in around 30 cyclotides that have been structurally characterized (Handley, Wang, Harvey, Lawrence, & Craik, 2020) and in its context within a peptide with a cyclic backbone is referred to as a cyclic cystine knot (CCK) motif (Craik, Daly, et al., 1999). The connectivity of the disulfide bonds that form this knot has also been independently characterized using chemical approaches (Göransson & Craik, 2003), validating the NMR conclusions. The CCK structure of cyclotides readily explains their exceptional stability. With no N- or C-termini they are highly resistant to exoprotease degradation, and the highly cross-braced cystine knot structure helps provide resistance to endoprotease digestion.

In early studies structures were only able to be done on native cyclotides, with the first examples being kalata B1 (Saether et al., 1995), circulin A (Daly, Koltay, Gustafson, & Craik, 1999), and kalata B2 (Jennings et al., 2005), along with others that have been reviewed recently (Handley, Wang, Harvey, Lawrence and Craik, 2020). However, the ability to synthesize cyclotides facilitated a range of new possibilities, including preparing cyclotides in large quantities for a wide range of structural and biophysical studies. Synthetic routes also opened up the possibility of making modified or grafted cyclotides. Fig. 3 summarizes the types of modified cyclotides that have been made to explore structure-activity relationships, for example.

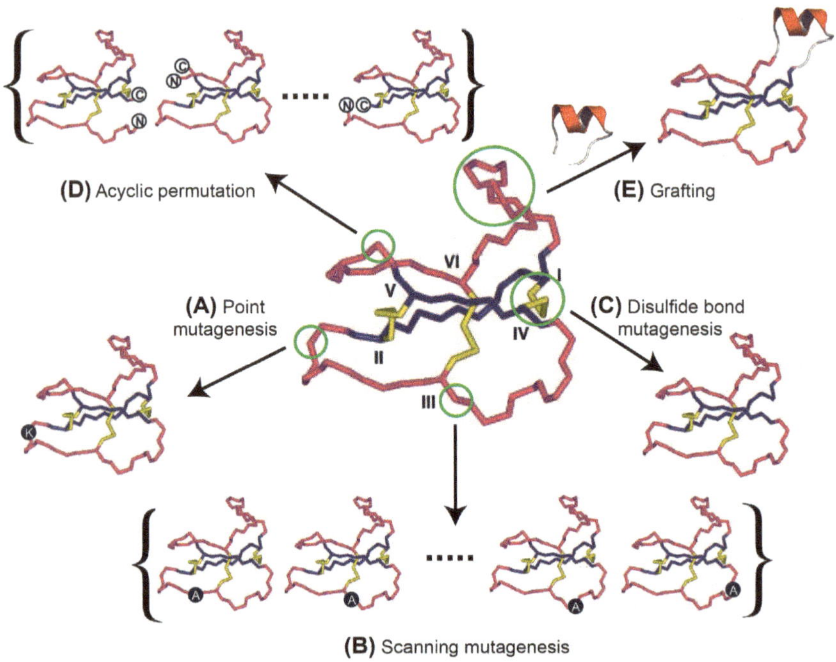

Fig. 3 Overview of structure-activity relationship studies of cyclotides. The prototypical cyclotide kalata B1 is shown in the middle with the types of mutagenesis and structure-activity relationship studies that have been done illustrated schematically. (A) Point mutagenesis – individual residues are replaced with another residue (lysine, for example) to investigate the effect on activity. (B) Scanning mutagenesis – a residue (alanine, for example) is systematically substituted to all positions within the sequence. (C) Disulfide bond mutagenesis – each one of the three disulfide bonds were successively removed to investigate their role in structure and folding. (D) Acyclic permutation – acyclic derivatives are made by breaking single peptide bond in the backbone to investigate the effect of cyclization on structure and folding as described by Daly and Craik (2000). (E) Grafting – bioactive epitopes were grafted into the cyclotide scaffold to investigate the plasticity of the scaffold being used for development of therapeutic and diagnostic agents. *Figure reproduced with permission from a recent book on cyclotides Craik, D. J. (2015).* Advances in botanical research: Plant cyclotides. *Vol. 76. London: Academic Press. Copyright 2015 Elsevier, Ltd.*

In such studies cyclotides are typically made using solid phase peptide chemistry. Initially, Boc sidechain protection chemistry was used, in combination with native chemical ligation (NCL), as described in several early synthetic articles (Daly, Love, Alewood, & Craik, 1999; Tam, Lu, & Yu, 1999), but now Fmoc based methods are becoming increasingly used in the synthesis of cyclotides (Cheneval et al., 2014). In either case the synthesis proceeds via assembly of the peptide chain followed by cyclization and subsequent folding. Oxidative folding remains the most challenging part of cyclotide synthesis, with Möbius and trypsin inhibitor cyclotides folding relatively well but bracelet cyclotides being very difficult to synthesize. The reasons for this difficulty in oxidative folding are not known, but are being explored.

Synthetic chemistry opened a range of other possibilities. For example, being able to make an all-D, (i.e., mirror image) version of cyclotides allowed mechanistic activities of cyclotides to be explored (Henriques, Peacock, Benfield, Wang, & Craik, 2019), as well as facilitating racemic crystallography to enable crystal structures of cyclotides to be determined. It is noteworthy that cyclotides are generally recalcitrant to crystallization, but this can be ameliorated by the use of racemic and quasi-racemic crystallography (Huang et al., 2021; Wang, King, Northfield, Ojeda, & Craik, 2014).

3. Applications

As noted in the introduction, cyclotides have a wide range of applications in drug design and agriculture based on their diverse bioactivities, their amenability to grafting new sequences, and their exceptional stability.

3.1 Applications in agriculture

Given their defense role in plants, it is not surprising that cyclotides have found applications as crop protection agents in agriculture. The first cyclotide-containing product to reach the market is Sero-X developed from an extract of the butterfly pea plant by an Australian company, Innovate Ag (https://innovate-ag.com.au/), and currently licensed in Australia for the protection of cotton, macadamia and vegetable crops. Butterfly pea is a member of the Fabaceae plant family and is readily grown and harvested, providing a convenient source of biomass for the manufacture of Sero-X. Transcriptomics analysis, combined with peptide level proteomics studies on butterfly pea, has established that the suites of cyclotides expressed in different parts of this plant seem to be tailored for the likely pests that will be encountered in different tissues (Gilding et al., 2016). For example, cyclotides with membrane-binding activity relevant to insect pests are expressed primarily in the above ground tissues, and nematocidal-active cyclotides are expressed in roots. Fig. 4 shows the insecticidal activity of Cter M, one of the component peptides in Sero-X (Poth, Colgrave, Lyons, Daly, & Craik, 2011) that is present in leaf tissue of butterly pea.

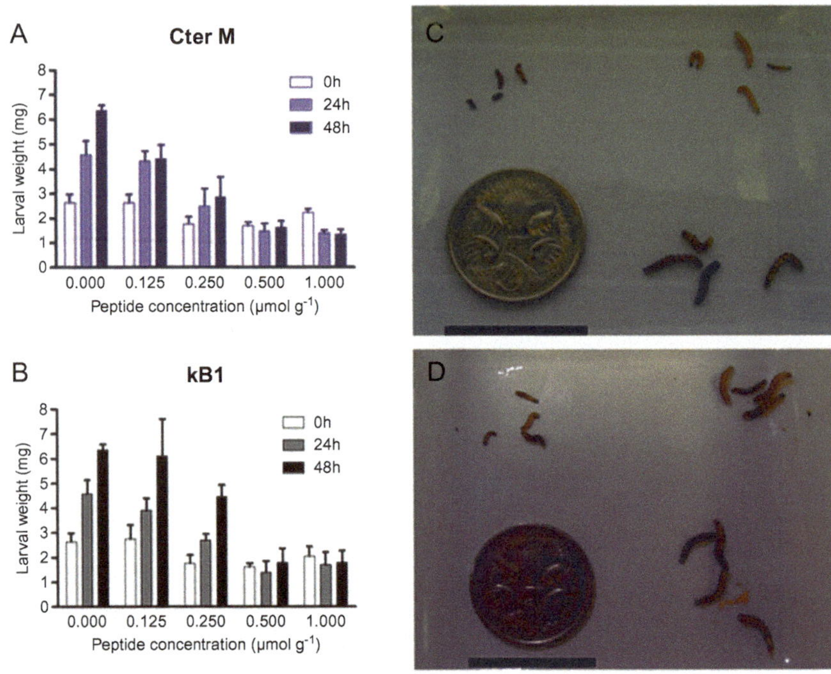

Fig. 4 The effect of Cter M on the growth of *Helicoverpa armigera*. The weight of larvae at 0, 24, and 48 h is plotted versus peptide concentration for Cter M (A) and kalata B1 (kB1) (B) and the size of control larvae alongside larvae fed at medium (0.25 μmol g^{-1} diet) and high (1.0 μmol g^{-1} diet) peptide concentrations at 48 h is depicted for Cter M (C) and kB1 (D). Scale bar = 20 mm. *Figure reproduced with permission from Poth, A. G., Colgrave, M. L., Lyons, R. E., Daly, N. L., & Craik, D. J. (2011). Discovery of an unusual biosynthetic origin for circular proteins in legumes.* Proceedings of the National Academy of Sciences, 108, *10127–10132.*

This feeding trial assay for insecticidal activity was based on an earlier study demonstrating an insecticidal effect for kalata B1 (Jennings et al., 2001). A variety of other assays have been reported for assaying of cyclotides of agricultural interest, including the use of *in vitro* insect Sf9 cells for testing various butterfly pea accessions (Oguis et al., 2020) and for measuring the efficacy of the cyclotide parigidin-br1 against pests of sugar cane (Pinto et al., 2012).

3.2 Pharmaceutical applications

Pharmaceutical applications fall into two categories; the first being the use of natural cyclotides as drug leads based on their intrinsic activities, and the second being the artificial engineering of cyclotides via the insertion of a bioactive epitope into the cyclotide scaffold. For the former activity, there was initial interest in cyclotides as potential antimicrobial agents (Tam, Lu, Yang, & Chiu, 1999). Although some synthetic cyclotides displayed moderate activity against a range of pathogenic bacteria, that activity was typically lost in the presence of physiological concentrations of salts and so cyclotides are not generally regarded as excellent leads for antimicrobial therapies. This is not to say there are no examples of bacteria, presumably bacteria relevant for the infection of plants, that cyclotides may be active against, but so far this has not been reported.

Commonly reported activities of natural cyclotides from a therapeutic perspective pertain to their selective cytotoxicity against certain types of cancer cells (Henriques et al., 2014; Lindholm et al., 2002; Mehta et al., 2020). Again, the selectivity and potency of these activities is not at a level sufficient to consider human trials at this stage. Similarly, earlier studies reported anti-HIV activity of cyclotides (Gustafson et al., 1994; Ireland, Wang, Wilson, Gustafson, & Craik, 2008). One factor responsible for the lack of progression of native cyclotides into the clinic is that they are mildly hemolytic (Barry, Daly, Clark, Sando, & Craik, 2003), which is not a favored property for a systemic therapeutic.

Most activity in the pharmaceutical area for cyclotides has been based on the concept of grafting, i.e., the insertion of a bioactive epitope into a cyclotide scaffold, thereby stabilizing the bioactive epitope and, hopefully, maintaining its natural biological activity (Craik & Du, 2017; de Veer, Kan & Craik, 2019; Poth, Chan, & Craik, 2013). There are more than two dozen such grafting studies that have been reported for Möbius or trypsin inhibitor cyclotides by various groups; these have been extensively reviewed and will not be discussed here, except to note that the activities range from multiple sclerosis (Wang et al., 2014) to blocking protein–protein interactions inside cells (Ji et al., 2013).

With these various applications of cyclotides being explored, there is a need to quantify and measure their stability and cytotoxicity as described in the following protocols.

4. Protocols

4.1 Quantification of cyclotides

Like many peptides cyclotides are typically quantified using UV-based methods, HPLC or LCMS. Various methods for quantifying cyclotides and other cyclic peptides have been reported (Conibear, Daly, & Craik, 2012). The following protocol is a convenient and efficient approach for measuring cyclotides in solution.

1.1 Equipment
UV–Vis spectrophotometer
Nanodrop instrument
Reverse-Phase High Performance Liquid Chromatography (RP-HPLC)
C18 analytical column (300 Å, 150 × 2.0 mm)

1.2 Prepare the following solutions prior to the assay:
Low binding microcentrifuge tubes (1.5 mL)
Ultrapure water
Phosphate-buffered saline (PBS)
A standard peptide solution quantified by amino acid analysis
Solvent A: 0.05% trifluoroacetic acid
Solvent B: 90% acetonitrile/0.05% trifluoroacetic acid

1.3 The quantification method routinely used to measure the concentration of tyrosine- and/or tryptophan-containing cyclotides is UV absorption. Weigh out cyclotides in clean low binding microcentrifuge tubes and add ultrapure water to reach concentrations at 2–3 mg/mL. Alternatively, cyclotides can be dissolved in PBS or other physiological buffers for cell-based or *in vitro* assays. Vortex or sonicate the tubes for 15 s to ensure fully solubilize the peptides. Adjust wavelength of UV–Vis spectrophotometer to 280 nm, calibrate the machine to zero with the aqueous solution, transfer the cyclotide solutions to a clean quartz cuvette with 1 cm optical path, and measure the absorbance at 280 nm. The absorbance at 280 nm can also be obtained using a Nanodrop instrument, where much less amount of peptide solutions is required (usually 2 μL).

1.4 Calculate the peptide concentrations using the Beer-Lambert Law: $A = \varepsilon lc$, where A is the absorbance at 280 nm, ε is the extinction coefficient of the peptide, l is the optical path length in cm, and c is the molar concentration. The extinction coefficient of cyclotides is calculated using the following formula: $\varepsilon_{280} = (n_W \times 5500) + (n_Y \times 1490) + (n_{S\text{-}S} \times 125)$, where n_W is the number of tryptophan residues, n_Y is the number of tyrosines, and $n_{S\text{-}S}$ is the number of disulfide bonds, which is three for all cyclotides. For example, the extinction coefficient of kalata B1 is $(1 \times 5500) + (0 \times 1490) + (3 \times 125)$, which equals to 5875 $M^{-1} cm^{-1}$. If the absorbance at 280 nm of a kalata B1 stock solution is 4 and the optical path length is 1 cm, the molar concentration calculated using Beer-Lamber Law would be c = 4/5875/1 M, i.e., 6.81×10^{-4} M

1.5 Cyclotide quantification can also be achieved by comparing analytical RP-HPLC chromatography traces of peptide analytes with that of a standard solution quantified by amino acid analysis. To build a standard curve, prepare solutions from the peptide standard at concentrations of 600, 400, 300, 100, 50, and 25 μM, and then further dilute them by factors of 0.05 for the top three concentrations and 0.1 for the bottom three concentrations. The final concentrations of the diluted standards are 30, 20, 15, 10, 5, 2.5 μM. Inject 10 μL of each diluted solutions onto an analytical C18 column in triplicate and carry out 2%/min linear gradient of solvent B against solvent A at a flow rate of 0.3 mL/min on RP-HPLC. The peptide peaks are detected at 214 nm and the area of the peaks are divided by the dilution factor and then plotted against the concentrations to construct the calibration curve

1.6 To quantify the cyclotide sample using the analytical RP-HPLC, dilute the sample prepared in subsection 1.3 by a factor of 0.02 and analyze the peptide peak as described in subsection 1.5. Fit the peak area of the peptide sample to the calibration curve to give a concentration relative to the peptide standard determined using amino acid analysis

4.2 Stability in biofluids

The following protocol is convenient for measuring the stability of cyclotides in a range of biofluids, as described in a recent report on the design of anticancer cyclotides (Huang et al., 2015).

2.1 Equipment
RP-HPLC
C18 analytical column (300 Å, 150 × 2.0 mm)

2.2 Prepare the following solutions prior to the assay:
Cyclotide stock solutions prepared in PBS at 300 μM (10 × stock)
A linear peptide with random sequence in PBS at 300 μM (10 × stock)
Human male AB serum (Sigma)
20% trichloroacetic acid (Sigma)
6 M urea (Sigma)
Solvent A: 0.05% trifluoroacetic acid
Solvent B: 90% acetonitrile/0.05% trifluoroacetic acid

2.3 Dissolve cyclotides in sterile PBS, measure the concentration of the peptide samples using the protocol described in section 4.1. Adjust the concentration of the stock solutions to 300 μM with sterile PBS

2.4 Thaw human male AB serum (purchased from Sigma) completely, make 1.2 mL aliquots and store the aliquots at −20 °C until used. On the day of the assay, defrost a 1.2 mL aliquot of human serum and spin it at 17,000 × *g* for 15 min to separate lipid content from the serum. Carefully transfer 360 μL of human serum into three 1.5 mL Eppendorf tubes individually as technical replicates. Pre-warm the serum tubes at 37 °C for 15 min

2.5 Add 40 μL of cyclotide stock into each batch of the temperature-equilibrated human serum to reach a final peptide concentration of 30 μM. Mix well, incubate the reaction solutions at the 37 °C incubator and record the starting time of the assay. The serum stability of cyclotides and their variants is generally monitored over a period of 24 h and the advised time intervals include 0, 1, 2, 4, 6, 8, 12, and 24 h

2.6 Two sets of controls can be included in the cyclotide stability assay: 1. Incubate equivalent amount of cyclotides in PBS and process the PBS-treated samples in parallel with the serum-treated samples at all time points as negative controls, and 2. Incubate equivalent amount of a linear peptide with random sequence in serum in parallel with the serum-treated cyclotides at all time points as controls for the proteolytic activity of serum

2.7 Take 40 μL aliquot from each the reaction solution at designated times and mix with 40 μL of 6 M urea to denature the serum proteins. Leave the samples on ice for 10 min before adding 40 μL of 20% trichloroacetic acid to precipitate the serum proteins. The sample solutions should look cloudy after being treated with trichloroacetic acid. Incubate the cloudy samples on ice for another 10 min and then spin at 17,000 × *g* for 10 min to pellet the serum proteins. Collect 110 μL of the supernatants and transfer into disposable inserts (11 mm) of chromatography autosampler vials (Thermo Fisher). Store at 4 °C until further analysis

2.8 The supernatants are analyzed using RP-HPLC on a C18 analytical column. Inject 100 μL of each supernatant and run at a 0.3 mL/min flow rate with 2%/min linear gradient of solvent B against solvent A up to 60% of solvent B. Peptides are detected by monitoring UV absorbance at 214 nm. The original cyclotide stock is also analyzed using the same chromatography method as a reference for the peptide retention time (Fig. 5A).

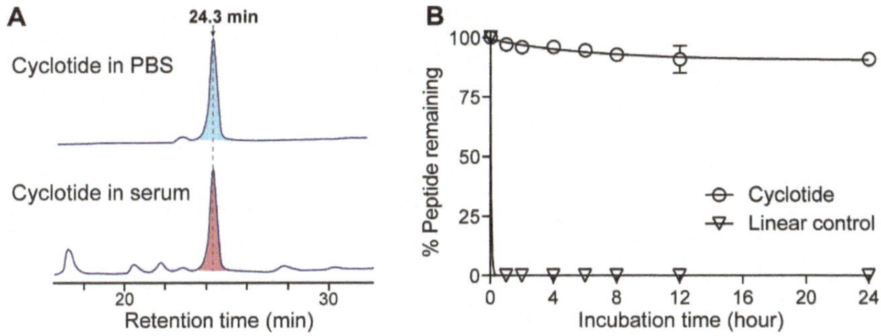

Fig. 5 (A) RP-HPLC traces of the original cyclotide stock (top) and the serum-treated cyclotide (bottom) monitored at 214nm. The retention time of the cyclotide in PBS is used a reference for chromatographic peak identification. (B) The percentage of peptide remaining for a given time interval is plotted against the corresponding time. Error bars represent the standard deviations of three replicates.

2.9 The percentage of remaining peptide in samples from each time interval is subsequently calculated using the following formula: $Area_{TN}/Area_{T0} \times 100$, where TN is the area of the integration peptide peak at a time point and T0 is the area of the integration peptide peak at Time 0. Plot the data against time as shown in Fig. 5B. Experiments are carried out in triplicate

4.3 Cytotoxicity of cyclotides

4.3.1 Measuring hemolytic activity (cytotoxic activity against red blood cells)

3.1.1 Equipment
Benchtop centrifuge
Plate centrifuge
Microplate absorbance reader
Multichannel pipette

3.1.2 Prepare the following solutions and items prior to the assay:
Whole blood
PBS
Cyclotide stock solutions prepared in PBS at 300 μM (6 × stock)
Melittin stock solution prepared in PBS at 20 μM (6 × stock)
1% Triton X-100
Round-bottom plate and flat-bottom plate (Thermo Fisher)

3.1.3 Aliquot 200 μL of whole blood into low binding microcentrifuge tubes containing 1 mL of PBS, invert the tubes up and down until fully mixed. Centrifuge the tubes at 4000 rpm for 1 min. Do not exceed the spin speed or time as prolonged centrifugation at higher speed may cause structural damage of red blood cells (RBCs). Carefully discard the supernatant without disturbing the cells at the bottom. Rinse RBCs with PBS as described above for 4–5 times

3.1.4 Remove most of the supernatant while leaving a sufficient amount to cover the cell pellet to facilitate the transfer of the cells. Carefully resuspend the RBCs and dilute in PBS to make a 0.25% RBC solution. For example, 25 μL of RBC is required to make 10 mL of 0.25% RBC. Dispense 100 μL of 0.25% RBCs per well into a 96-well round-bottom plate prior to the assay

3.1.5 Prepare 140 μL of cyclotide stock solutions and perform two-fold serial dilutions from 300 μM down to 2.34 μM (eight concentrations in total). Since 20 μL of peptide solution is required each well in triplicate, the minimum amount of peptide solutions per concentration is 60 μL (20 μL × 3). Pipette 20 μL of peptide solutions into wells containing 100 μL of 0.25% RBC in the 96-well round bottom plate, so the final concentrations of cyclotide are 50–0.39 μM. Melittin, a membrane lytic peptide used as a reference in hemolytic assays, is diluted in the same manner with 20 μM as the highest concentration instead. PBS and 1% Triton-X are included as negative and positive controls for this assay

3.1.6 Incubate the plate in a 37 °C incubator for an hour and then spin at 1000 rpm for 5 min on a microplate centrifuge. Use a multi-channel pipette to transfer 100 μL of supernatant from each well (without disturbing cell pellets at the bottom) into a fresh flat-bottomed 96-well plate

3.1.7 Measure the absorbance at 415 nm on a microplate reader and calculate the percentage of hemolysis using the following formula: % hemolysis $= (A_{sample} - A_{PBS})/(A_{Triton} - A_{PBS}) \times 100$, where A_{sample} is the absorbance of the sample, A_{PBS} and A_{Triton} are the readout at 415 nm for samples treated with PBS or 1% Triton-X, individually

4.3.2 Measuring toxicity against mammalian cells

3.2.1. Equipment

Hemocytometer

Microplate reader

Multichannel pipette

3.2.2. Prepare the following solutions prior to the assay

HeLa cells—derived from human cervical cancer cells, adherent

DMEM medium supplemented with 1% penicillin/streptomycin and 10% fetal bovine serum (FBS)—growth medium for HeLa cells
Trypsin-EDTA (0.25%), phenol red (Thermo Fisher)
K562 cells—derived from human erythromyeloblastoid leukemia, suspension
RPMI-1640 supplemented with 1% penicillin/streptomycin, 20 mM L-Glutamine and 10% FBS—growth medium for K562 cells
Trypan blue solution (0.4%)
Cyclotide stock solutions prepared in PBS at 640 μM (10 × stock)
Sterile PBS
1% Triton X-100
Resazurin solution (0.05%, *w/v*)
Sterile flat-bottom plate (Thermo Fisher)

3.2.3 This protocol is a high-throughput, quantitative resazurin-based microdilution assay to monitor cell viability or cytotoxicity of mammalian cell lines, including adherent and suspension cells. We use HeLa and K562 cell lines as examples for adherent and suspension cells individually. HeLa and K562 cells are cultured in complete DMEM or RPMI-1640 growth medium individually until 80% confluent

3.2.4 For cytotoxicity assay using HeLa cells, cells are removed from the culture flasks via trypsinization, rinsed, spun down and are resuspended in fresh complete DMEM medium. Take 10 μL of cell suspension, mix with equal amount of trypan blue solution and then apply 10 μL of trypan blue-treated cells onto a hemocytometer. Count live cells (unstained cells) under the microscope in four sets of 16 squares on the hemocytometer and calculate the number of viable cells per mL in the original suspension according to the manufacturer's instructions. HeLa cells are diluted to 2.5×10^4 cells/mL and seeded at 2500 cells/100 μL/well in a 96-well flat-bottomed plate 24 h prior to the experiment. Cyclotide stock solutions (10 ×) are diluted with PBS using two-fold serial dilutions ranging from 640 to 5 μM (eight concentrations in total). Before treating with cyclotide solutions, replace medium in each well with 90 μL of fresh serum-free medium and then add 10 μL of 10 × cyclotide solutions using a multichannel pipette. The final concentrations of cyclotide solution are ranging from 64 to 0.5 μM. Cells treated with PBS or 1% Triton X-100 are included as negative and positive controls, respectively. Assays are carried out in triplicate. Incubate the plate for 24 h at 37 °C

3.2.5 After 24-h incubation, add 10 μL of resazurin solution into each well and incubate for another 18 h. Measure the absorbance of the plate at 540 and 620 nm on a microplate reader. The percentage of cytotoxicity is then calculated using the absorbance ratio R, where R equals to Abs_{620}/Abs_{540}. Calculate the cytotoxicity using the following equation: % Cell death $= (R_{sample} - R_{PBS})/(R_{Triton} - R_{PBS}) \times 100$

3.2.6 For cytotoxicity assay using suspension cell lines, for example K562, cells can be plated right before the assay. Pipette cells from the culture flasks, rinse and spin cells down before seeding at 5000 cells/90 μL/well in a 96-well flat-bottomed plate on the day of experiment. Add 10 μL of 10 × cyclotide solutions into each well to reach final concentrations ranging from 64 to 0.5 μM. Cells treated with PBS or 1% Triton X-100 are included as negative and positive controls, respectively. Perform cytotoxicity assay in accordance with steps described in subsection 3.2.5. Assays are carried out in triplicate

4.3.3 Measuring toxicity against insect cells

The following protocol is convenient for measuring the toxicity of cyclotides against insect cells as described recently (Oguis et al., 2020).

3.3.1 Equipment
Microplate reader
Multichannel pipette
27 °C incubator

3.3.2 Prepare the following solutions prior to the assay
Sf9 cells—insect cell line derived from ovarian tissue of *Spodoptera frugiperda*
Grace's insect medium supplemented with 1% penicillin/streptomycin and 10% FBS
Cyclotide stock solutions prepared in PBS at 640 μM (10 × stock)
MTT solution at 5 mg/mL (3-(4,5-dimethylthiazol-2-yl)-2,5-diphenyltetrazolium bromide)
Dimethyl sulfoxide (DMSO)

3.3.3 Culture Sf9 cells until 80% confluent, rinse and seed at 10,000 cells/100 μL/well in a 96-well flat-bottomed plate 48 h prior to the experiment. The additional incubation time is to allow sf9 cells to adhere to the bottom of the wells. Dilute 10 × cyclotide stocks with PBS using a two-fold serial dilution as described above (eight concentrations in total), replace medium in the wells and treat sf9 cells with cyclotides of interest for 24 h

3.3.4 Add 10 μL of MTT solution into each well and incubate for 4 h before carefully removing supernatants using a multichannel pipette. Insoluble, purple formazan crystals can be observed under the inverted microscope. Add 100 μL of DMSO into each well and shake for 10 min or until all the formazan crystals are fully dissolved. Measure the absorbance at 570 nm using a microplate reader. Percentage of cytotoxicity is calculated using the following formula: % Inhibition = $(A_{sample} - A_{Triton})/(A_{PBS} - A_{Triton}) \times 100$

5. Analysis

For evaluating and comparing the biological activities of cyclotides it is important to determine the concentration accurately. In general, the most routinely used approach in the laboratory for cyclotide quantification is the UV-light absorption method, and it is applicable to pure peptide samples containing Trp, Tyr or a Cys-Cys disulfide bond that exhibit absorbance at 280 nm. Besides using a UV–Vis spectrophotometer, the UV absorbance of peptide samples can be measured directly using a Nanodrop instrument, which requires 2 μL of sample in pedestal mode and has an adjustable optical path length to accommodate concentrated cyclotide stocks. Even though the analytical RP-HPLC method was proved to be more precise in peptide quantification (Conibear et al., 2012), it requires a peptide stock (of similar amino acid content to the analyte peptide) with pre-determined concentration by amino acid analysis to build a standard curve, which is time-consuming and expensive, and therefore not practical for daily use. In addition, over 95% of cyclotides carry at least one or more Trp or Tyr residues. Therefore, direct spectrophotometric determination of cyclotide concentration at 280 nm is a straightforward and preferred method for quantifying pure cyclotides.

The exceptional stability of cyclotides against extreme thermal, chemical or proteolytic conditions has made them suitable candidates for peptide-based drug design. The serum stability of native cyclotides and cyclotides engineered with pharmacophores is routinely evaluated and used as a guide to provide important biological insights, e.g., their *in vivo* stability. Therefore, a standardized method for serum stability evaluation is important. Since the cyclic cystine knot confers cyclotides extra stability, their serum half-lives are usually longer than 24 h and hence a linear peptide control that is susceptible to serum proteases would be a good addition to demonstrate serum activity.

As aforementioned, cyclotides exhibit a range of biological activities, including insecticidal and anti-cancer properties (Huang, Du, & Craik, 2019). The cytotoxicity against human cancer and insect cell lines is an *in vitro* pilot testing as an indication for biological evaluation of cyclotides. Two types of assays are described here to monitor cell viability, i.e., the resazurin and MTT assays, and both are high-throughput, sensitive and inexpensive. The major advantage of the resazurin assay is that fewer wash steps are involved and therefore is more suitable for suspension cell lines like K562. Various amounts of cells are used per well for different cell lines depending on their cellular metabolic rate and cell proliferation. For instance, given that the proliferation of HeLa cells is abnormally rapid and significantly higher than that of K562, less cells are used per sample in the cytotoxicity assay (2500 vs. 5000 cells/well).

6. Summary

In summary, cyclotides have a wide range of applications in medicine and agriculture. Despite the fact that most of the molecular types applied in such applications previously have been of the small molecule, rather than peptide type, peptides are showing great promise. However, the bioanalytics for measuring cyclotides in products, biofluids or the environment are lagging behind those of small molecules for which there is vastly much more experience currently in the pharmaceutical or agrichemical industries. We hope the protocols described here will provide basic information for the quantification of cyclotides for assessing their stabilities and cytotoxicities.

Acknowledgments

Work in our laboratory on cyclotides is funded by a grant from the Australian Research Council (ARC) through the ARC Centre of Excellence for Innovations in Peptide and Protein Science (CE200100012).

References

Barry, D. G., Daly, N. L., Clark, R. J., Sando, L., & Craik, D. J. (2003). Linearization of a naturally occurring circular protein maintains structure but eliminates hemolytic activity. *Biochemistry*, *42*, 6688–6695.

Camarero, J. A., & Campbell, M. J. (2019). The potential of the cyclotide scaffold for drug development. *Biomedicine*, 7, 31.

Cheneval, O., Schroeder, C. I., Durek, T., Walsh, P., Huang, Y.-H., Liras, S., et al. (2014). Fmoc-based synthesis of disulfide-rich cyclic peptides. *The Journal of Organic Chemistry*, *79*, 5538–5544.

Colgrave, M. L., Huang, Y. H., Craik, D. J., & Kotze, A. C. (2010). Cyclotide interactions with the nematode external surface. *Antimicrobial Agents and Chemotherapy*, *54*, 2160–2166.

Colgrave, M. L., Kotze, A. C., Huang, Y. H., O'Grady, J., Simonsen, S. M., & Craik, D. J. (2008). Cyclotides: Natural, circular plant peptides that possess significant activity against gastrointestinal nematode parasites of sheep. *Biochemistry*, *47*, 5581–5589.

Colgrave, M. L., Kotze, A. C., Ireland, D. C., Wang, C. K., & Craik, D. J. (2008). The anthelmintic activity of the cyclotides: Natural variants with enhanced activity. *ChemBioChem*, *9*, 1939–1945.

Colgrave, M. L., Kotze, A. C., Kopp, S., McCarthy, J. S., Coleman, G. T., & Craik, D. J. (2009). Anthelmintic activity of cyclotides: *In vitro* studies with canine and human hookworms. *Acta Tropica*, *109*, 163–166.

Conibear, A. C., Daly, N. L., & Craik, D. J. (2012). Quantification of small cyclic disulfide-rich peptides. *Biopolymers: Peptide Science*, *98*, 518–524.

Craik, D. J. (2015). *Advances in botanical research: Plant Cyclotides. Vol. 76*. London: Academic Press.

Craik, D. J., Daly, N. L., Bond, T., & Waine, C. (1999). Plant cyclotides: A unique family of cyclic and knotted proteins that defines the cyclic cystine knot structural motif. *Journal of Molecular Biology*, *294*, 1327–1336.

Craik, D. J., & Du, J. (2017). Cyclotides as drug design scaffolds. *Current Opinion in Chemical Biology*, *38*, 8–16.

Daly, N. L., & Craik, D. J. (2000). Acyclic permutants of naturally occurring cyclic proteins. Characterization of cystine knot and β-sheet formation in the macrocyclic polypeptide kalata B1. *Journal of Biological Chemistry*, *275*, 19068–19075.

Daly, N. L., Koltay, A., Gustafson, K. R., Boyd, M. R., Casas-Finet, J. R., & Craik, D. J. (1999). Solution structure by NMR of circulin A: A macrocyclic knotted peptide having anti-HIV activity. *Journal of Molecular Biology*, *285*, 333–345.

Daly, N. L., Love, S., Alewood, P. F., & Craik, D. J. (1999). Chemical synthesis and folding pathways of large cyclic polypeptides: Studies of the cystine knot polypeptide kalata b1. *Biochemistry*, *38*, 10606–10614.

de Veer, S. J., Kan, M.-W., & Craik, D. J. (2019). Cyclotides: From structure to function. *Chemical Reviews*, *119*, 12375–12421.

Du, J., Yap, K., Chan, L. Y., Rehm, F. B. H., Looi, F. Y., Poth, A. G., et al. (2020). A bifunctional asparaginyl endopeptidase efficiently catalyzes both cleavage and cyclization of cyclic trypsin inhibitors. *Nature Communications*, *11*, 1575.

Gerlach, S. L., & Mondal, D. (2012). The bountiful biological activities of cyclotides. *Chronicles of Young Scientists*, *3*, 169–177.

Gilding, E. K., Jackson, M. A., Poth, A. G., Henriques, S. T., Prentis, P. J., Mahatmanto, T., et al. (2016). Gene coevolution and regulation lock cyclic plant defence peptides to their targets. *New Phytologist*, *210*, 717–730.

Göransson, U., & Craik, D. J. (2003). Disulfide mapping of the cyclotide kalata B1. Chemical proof of the cyclic cystine knot motif. *Journal of Biological Chemistry*, *278*, 48188–48196.

Göransson, U., Luijendijk, T., Johansson, S., Bohlin, L., & Claeson, P. (1999). Seven novel macrocyclic polypeptides from *Viola arvensis*. *Journal of Natural Products*, *62*, 283–286.

Göransson, U., Sjögren, M., Svangård, E., Claeson, P., & Bohlin, L. (2004). Reversible antifouling effect of the cyclotide cycloviolacin O2 against barnacles. *Journal of Natural Products*, *67*, 1287–1290.

Gran, L. (1970). An oxytocic principle found in *Oldenlandia affinis* DC. *Meddelelser fra Norsk Farmaceutisk Selskap*, *12*, 173–180.

Gran, L. (1973a). On the effect of a polypeptide isolated from "Kalata-Kalata" (*Oldenlandia affinis* DC) on the oestrogen dominated uterus. *Acta Pharmacologica et Toxicologica*, *33*, 400–408.

Gran, L. (1973b). Oxytocic principles of *Oldenlandia affinis*. *Lloydia*, *36*, 174–178.

Gruber, C. W., Elliott, A. G., Ireland, D. C., Delprete, P. G., Dessein, S., Göransson, U., et al. (2008). Distribution and evolution of circular miniproteins in flowering plants. *Plant Cell*, *20*, 2471–2483.

Gründemann, C., Stenberg, K. G., & Gruber, C. W. (2019). T20K: An immunomodulatory cyclotide on its way to the clinic. *International Journal of Peptide Research and Therapeutics*, *25*, 9–13.

Gustafson, K. R., McKee, T. C., & Bokesch, H. R. (2004). Anti-HIV cyclotides. *Current Protein and Peptide Science*, *5*, 331–340.

Gustafson, K. R., Sowder, R. C., Henderson, L. E., Parsons, I. C., Kashman, Y., Cardellina, J. H., et al. (1994). Circulins A and B: Novel HIV-inhibitory macrocyclic peptides from the tropical tree *Chassalia parvifolia*. *Journal of the American Chemical Society*, *116*, 9337–9338.

Handley, T. N. G., Wang, C. K., Harvey, P. J., Lawrence, N., & Craik, D. J. (2020). Cyclotide structures revealed by NMR, with a little help from X-ray crystallography. *ChemBioChem*, *21*, 3463–3475.

Harris, K. S., Durek, T., Kaas, Q., Poth, A. G., Gilding, E. K., Conlan, B. F., et al. (2015). Efficient backbone cyclization of linear peptides by a recombinant asparaginyl endopeptidase. *Nature Communications*, *6*, 10199.

Henriques, S. T., Huang, Y.-H., Chaousis, S., Wang, C. K., & Craik, D. J. (2014). Anticancer and toxic properties of cyclotides are dependent on phosphatidylethanolamine phospholipid targeting. *ChemBioChem*, *15*, 1956–1965.

Henriques, S. T., Peacock, H., Benfield, A. H., Wang, C. K., & Craik, D. J. (2019). Is the mirror image a true reflection? Intrinsic membrane chirality modulates peptide binding. *Journal of the American Chemical Society*, *141*, 20460–20469.

Huang, Y.-H., Du, Q., & Craik, D. J. (2019). Cyclotides: Disulfide-rich peptide toxins in plants. *Toxicon*, *172*, 33–44.

Huang, Y.-H., Du, Q., Jiang, Z., King, G., Collins, B. M., Wang, C. K., et al. (2021). Enabling efficient folding and high-resolution crystallographic analysis of bracelet cyclotides. *Molecules*, *26*, 5554.

Huang, Y.-H., Henriques, S. T., Wang, C. K., Thorstholm, L., Daly, N. L., Kaas, Q., et al. (2015). Design of substrate-based BCR-ABL kinase inhibitors using the cyclotide scaffold. *Scientific Reports*, *5*, 12974.

Ireland, D. C., Wang, C. K., Wilson, J. A., Gustafson, K. R., & Craik, D. J. (2008). Cyclotides as natural anti-HIV agents. *Biopolymers: Peptide Science*, *90*, 51–60.

Jackson, M. A., Nguyen, L. T. T., Gilding, E. K., Durek, T., & Craik, D. J. (2020). Make it or break it: Plant AEPs on stage in biotechnology. *Biotechnology Advances*, *45*, 107651.

Jennings, C. V., Rosengren, K. J., Daly, N. L., Plan, M., Stevens, J., Scanlon, M. J., et al. (2005). Isolation, solution structure, and insecticidal activity of kalata B2, a circular protein with a twist: Do möbius strips exist in nature? *Biochemistry*, *44*, 851–860.

Jennings, C., West, J., Waine, C., Craik, D., & Anderson, M. (2001). Biosynthesis and insecticidal properties of plant cyclotides: The cyclic knotted proteins from *Oldenlandia affinis*. *Proceedings of the National Academy of Sciences of the United States of America*, *98*, 10614–10619.

Ji, Y., Majumder, S., Millard, M., Borra, R., Bi, T., Elnagar, A. Y., et al. (2013). In vivo activation of the p53 tumor suppressor pathway by an engineered cyclotide. *Journal of the American Chemical Society*, *135*, 11623–11633.

Kan, M.-W., & Craik, D. J. (2018). Trends in cyclotide research. In J. Koehnke, J. Naismith, & W. A. van der Donk (Eds.), *Cyclic peptides: From bioorganic synthesis to applications (chapter 14)* (pp. 302–339). Cambridge: The Royal Society of Chemistry.

Kirkpatrick, C. L., Broberg, C. A., McCool, E. N., Lee, W. J., Chao, A., McConnell, E. W., et al. (2017). The "PepSAVI-MS" pipeline for natural product bioactive peptide discovery. *Analytical Chemistry, 89*, 1194–1201.
Koehbach, J., & Gruber, C. W. (2015). Chapter Three—Cyclotides in the Rubiaceae. In D. J. Craik (Ed.), *Vol. 76. Advances in botanical research: Plant Cyclotides* (pp. 51–78). London: Academic Press.
Lindholm, P., Göransson, U., Johansson, S., Claeson, P., Gulbo, J., Larsson, R., et al. (2002). Cyclotides: A novel type of cytotoxic agents. *Molecular Cancer Therapeutics, 1*, 365–369.
Mehta, L., Dhankhar, R., Gulati, P., Kapoor, R. K., Mohanty, A., & Kumar, S. (2020). Natural and grafted cyclotides in cancer therapy: An insight. *Journal of Peptide Science, 26*, e3246.
Narayani, M., Babu, R., Chadha, A., & Srivastava, S. (2020). Production of bioactive cyclotides: A comprehensive overview. *Phytochemistry Reviews, 19*, 787–825.
Oguis, G. K., Gilding, E. K., Huang, Y.-H., Poth, A. G., Jackson, M. A., & Craik, D. J. (2020). Insecticidal diversity of butterfly pea (*Clitoria ternatea*) accessions. *Industrial Crops and Products, 147*, 112214.
Parsley, N. C., Kirkpatrick, C. L., Crittenden, C. M., Rad, J. G., Hoskin, D. W., Brodbelt, J. S., et al. (2018). PepSAVI-MS reveals anticancer and antifungal cycloviolacins in *Viola odorata*. *Phytochemistry, 152*, 61–70.
Pinto, M. F. S., Fensterseifer, I. C. M., Migliolo, L., Sousa, D. A., de Capdville, G., Arboleda-Valencia, J. W., et al. (2012). Identification and structural characterization of novel cyclotide with activity against an insect pest of sugar cane. *Journal of Biological Chemistry, 287*, 134–147.
Plan, M. R. R., Saska, I., Cagauan, A. G., & Craik, D. J. (2008). Backbone cyclised peptides from plants show molluscicidal activity against the rice pest *Pomacea canaliculata* (golden apple snail). *Journal of Agricultural and Food Chemistry, 56*, 5237–5241.
Poth, A. G., Chan, L. Y., & Craik, D. J. (2013). Cyclotides as grafting frameworks for protein engineering and drug design applications. *Biopolymers: Peptide Science, 100*, 480–491.
Poth, A. G., Chiu, F. C. K., Stalmans, S., Hamilton, B. R., Huang, Y.-H., Shackleford, D. M., et al. (2021). Effects of backbone cyclization on the pharmacokinetics and drug efficiency of the orally active analgesic conotoxin cVc1.1. *Medicine in Drug Discovery, 10*, 100087.
Poth, A. G., Colgrave, M. L., Lyons, R. E., Daly, N. L., & Craik, D. J. (2011). Discovery of an unusual biosynthetic origin for circular proteins in legumes. *Proceedings of the National Academy of Sciences, 108*, 10127–10132.
Poth, A. G., Huang, Y.-H., Le, T. T., Kan, M.-W., & Craik, D. J. (2019). Pharmacokinetic characterization of kalata B1 and related therapeutics built on the cyclotide scaffold. *International Journal of Pharmaceutics, 565*, 437–446.
Saether, O., Craik, D. J., Campbell, I. D., Sletten, K., Juul, J., & Norman, D. G. (1995). Elucidation of the primary and three-dimensional structure of the uterotonic polypeptide kalata B1. *Biochemistry, 34*, 4147–4158.
Slazak, B., Kapusta, M., Strömstedt, A. A., Słomka, A., Krychowiak, M., Shariatgorji, M., et al. (2018). How does the sweet violet (*Viola odorata* L.) fight pathogens and pests—Cyclotides as a comprehensive plant host defense system. *Frontiers in Plant Science, 9*, 1296.
Tam, J. P., Lu, Y. A., Yang, J. L., & Chiu, K. W. (1999). An unusual structural motif of antimicrobial peptides containing end-to-end macrocycle and cystine-knot disulfides. *Proceedings of the National Academy of Sciences of the United States of America, 96*, 8913–8918.
Tam, J. P., Lu, Y.-A., & Yu, Q. (1999). Thia zip reaction for synthesis of large cyclic peptides: Mechanisms and applications. *Journal of the American Chemical Society, 121*, 4316–4324.

Trabi, M., & Craik, D. J. (2004). Tissue-specific expression of head-to-tail cyclized miniproteins in Violaceae and structure determination of the root cyclotide *Viola hederacea* root cyclotide1. *The Plant Cell*, *16*, 2204–2216.

Trabi, M., Svangård, E., Herrmann, A., Göransson, U., Claeson, P., Craik, D. J., et al. (2004). Variations in cyclotide expression in *Viola* species. *Journal of Natural Products*, *67*, 806–810.

Wang, C. K., & Craik, D. J. (2018). Designing macrocyclic disulfide-rich peptides for biotechnological applications. *Nature Chemical Biology*, *14*, 417–427.

Wang, C. K., Gruber, C. W., Čemažar, M., Siatskas, C., Tagore, P., Payne, N., et al. (2014). Molecular grafting onto a stable framework yields novel cyclic peptides for the treatment of multiple sclerosis. *ACS Chemical Biology*, *9*, 156–163.

Wang, C. K., Kaas, Q., Chiche, L., & Craik, D. J. (2008). CyBase: A database of cyclic protein sequences and structures, with applications in protein discovery and engineering. *Nucleic Acids Research*, *36*, D206–D210.

Wang, C. K., King, G. J., Northfield, S. E., Ojeda, P. G., & Craik, D. J. (2014). Racemic and quasi-racemic X-ray structures of cyclic disulfide-rich peptide drug scaffolds. *Angewandte Chemie International Edition*, *53*, 11236–11241.

Weidmann, J., & Craik, D. J. (2016). Discovery, structure, function, and applications of cyclotides: Circular proteins from plants. *Journal of Experimental Botany*, *67*, 4801–4812.

CHAPTER THREE

Creating optimized peptide libraries for AMP discovery via PepSAVI-MS

Amanda M. Brechbill, Tessa B. Moyer, Nicole C. Parsley, and Leslie M. Hicks*

Department of Chemistry, University of North Carolina at Chapel Hill, Chapel Hill, NC, United States
*Corresponding author: e-mail address: lmhicks@unc.edu

Contents

Methods in Enzymology, Volume 663
ISSN 0076-6879
https://doi.org/10.1016/bs.mie.2021.10.024

Abstract

Antimicrobial peptides (AMPs) are host defense peptides with a range of functions/activities/modes of action that are ubiquitously expressed across all forms of life. Continued discovery of novel AMPs presents exciting opportunities to address evolving resistance to existing treatments in multiple fields, including agricultural pathogens/pests as well as antimicrobial and chemotherapeutics for human health. However, typical discovery methods including bioassay-guided fractionation and genome mining generally lack the capacity for robust AMP discovery in non-model/unsequenced organisms. PepSAVI-MS (Statistically guided bioactive *pep*tides prioritized via mass spectrometry) was developed as an AMP discovery approach that addresses some of the limitations associated with these standard methods. PepSAVI-MS is a multi-pronged pipeline that includes peptide library creation, bioactivity screening, LC-MS analysis, and statistical modeling for putative AMP identification. The original implementation of PepSAVI-MS outlined strategies for the fractionation of plant extracts with strong cation exchange chromatography (SCX). Herein, we elaborate on recent improvements to peptide library creation through the use of orthogonal fractionation methods, specifically crude SCX chromatography and reversed-phase liquid chromatography (RPLC). This optimization of the "peptide library creation" step has demonstrated improvements for processing and AMP identifications via PepSAVI-MS.

1. Introduction

Antimicrobial peptides (AMPs) are host defense peptides that are expressed by all taxa and possess a broad range of activity profiles (Mahlapuu, Håkansson, Ringstad, & Björn, 2016; Wang, Li, & Wang, 2015). Endogenous AMPs are typically small (<10 kDa) and cationic with a hydrophobic region (Tam, Wang, Wong, & Tan, 2015). Additionally, AMPs are generally cysteine-rich with multiple disulfide bonds for enhanced stability against proteases, temperature, and pH. Overall these characteristics aid in the broad bioactivity profile of AMPs to position their emerging use as therapeutic agents (Harvey, 2008; Mahlapuu et al., 2016; Tam et al., 2015). Compilations of AMPs from bacteria, archaea, protists, fungi, plants, and animals can be found in ever-growing publicly available databases including the Antimicrobial Peptide Database (APD, https://aps.unmc.edu/AP/) (Wang, 2021; Wang, Li, & Wang, 2016). The APD includes bioactive natural AMPs with known amino acid sequences less than 100 residues in length and contains 3273 AMPs from six kingdoms (Wang et al., 2016). Other databases include the Data Repository of Antimicrobial Peptides (DRAMP, http://dramp.cpu-bioinfor.org/) (Kang et al., 2019), the Collection of Anti-Microbial Peptides ($CAMP_{R3}$, http://www.

camp3.bicnirrh.res.in/) (Waghu, Barai, Gurung, & Idicula-Thomas, 2016), and the Database Linking Antimicrobial Peptides (LAMP, http://biotechlab.fudan.edu.cn/database/lamp/) (Zhao, Wu, Lu, Li, & Huang, 2013).

Traditionally, AMP discovery is facilitated through bioassay-guided fractionation and/or genome mining approaches. Bioassay-guided fractionation requires iterative rounds of fractionation and bioactivity screening to isolate the active compound (Sharma & Gupta, 2015; Strömstedt, Felth, & Bohlin, 2013). Significant material and activity losses can occur with each round of fractionation resulting in depletion of the bioactive components within the extract (Weller, 2012). Bioactivity screening is both concentration and pathogen-dependent, so some present AMPs may remain undetected via bioassay-guided fractionation and often results in the identification of previously characterized and highly abundant AMPs (Weller, 2012). Overall, this method is laborious and requires large amounts of material without the guarantee of identifying novel bioactive AMPs (Sharma & Gupta, 2015; Strömstedt et al., 2013).

AMP discovery via genome mining relies on DNA/RNA sequencing of the organism, knowledge regarding biosynthetic pathways, and computational resources to aid in the identification of cryptic biosynthetic gene clusters producing bioactive peptides (Skinnider et al., 2015, 2016). Though this approach has been readily applied in bacterial and fungal species with well-annotated and predictable bioactive natural product biosynthetic gene clusters, various challenges remain when implementing this process in higher-order and non-model organisms, including plants. For example, there is limited access to high-quality genomic and/or transcriptomic resources for non-model organisms which limits genome mining capabilities. Additionally, higher-order organisms generally lack the typical tight clustering of AMP-producing gene clusters that is common in bacteria (Russell & Truman, 2020). Additionally, genome mining approaches typically do not employ any means to detect bioactivity.

While both bioassay-guided fractionation and genome mining are powerful approaches for discovery, methods that address their limitations help facilitate novel AMP discovery, especially in non-model organisms. PepSAVI-MS (Statistically guided bioactive *pep*tides prioritized via mass spectrometry) screens and identifies bioactive peptides from natural sources while addressing the aforementioned limitations (Kirkpatrick et al., 2017). This pipeline includes the creation of a peptide library via the extraction of peptides from source material followed by crude fractionation, bioactivity screening of these fractions, LC-MS-based peptidomics, and statistical analysis for higher-throughput discovery of putative bioactive AMPs. PepSAVI-MS has enabled the discovery of several novel bioactive AMPs with antibacterial, antifungal, and/or anticancer properties from various plants and microbial secretomes, and has revealed new bioactivities for known AMPs (Kirkpatrick et al., 2017; Kirkpatrick, Parsley, Bartges, Cooke, et al., 2018; Kirkpatrick, Parsley, Bartges, Wing, et al., 2018; Moyer et al., 2019; Parsley et al., 2018). The utilization of mass spectrometry is very beneficial for high-throughput AMP identification and quantification as it can determine peptidyl abundance which is then correlated with the bioactive regions found from the bioassay. Additionally, PepSAVI-MS encompasses an AMP discovery workflow that utilizes direct bioactivity measurements with minimal fractions and avoids reliance on genomic information which is not always available or well-annotated for non-model organisms.

Initially, PepSAVI-MS employed a single iteration of crude fractionation via strong cation exchange chromatography (SCX) for peptide library creation (Kirkpatrick et al., 2017). This minimizes material requirements, decreases the probability of peptide degradation via denaturation, and increases bioassay throughput in comparison to the bioassay-guided fractionation method. However, we have vastly improved downstream processing by instead implementing a two-step orthogonal SCX and reversed-phase liquid chromatography (RPLC) strategy for peptide library creation. SCX is used to remove the bulk of unretained small molecules (with the exception of N-containing small molecules) and create a bulk "superfraction" of cationic peptides. This SCX "superfraction" is then fractionated via RPLC generating the peptide library for bioactivity screening and mass spectrometry (Fig. 1). RPLC has multi-faceted advantages including (1) reducing labor-intensive sample handling steps to sublimate salts from SCX buffers, (2) decreasing false positives in bioassays resulting from residual salts in SCX peptide fractions, and (3) providing higher resolution separation compatible with downstream isolation efforts for characterization. Specifically, this improved library creation allows for easier identification of lower abundance peptides and/or multiple contributors to bioactivity. The orthogonal fractionation utilizing SCX and RPLC can be adjusted to employ different chromatography that would best suit the experiments of interest. Though there is a small risk for bioactivity loss following exposure to organic solvent in RPLC, AMPs that denature during RPLC would be difficult to isolate and characterize in requisite downstream experiments. Altogether, this additional fractionation is advantageous for the aforementioned PepSAVI-MS

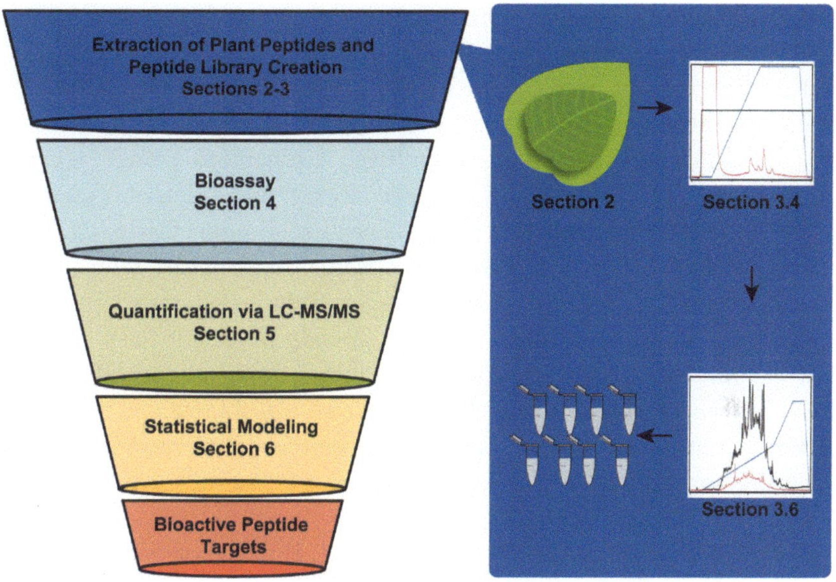

Fig. 1 PepSAVI-MS pipeline highlighting the optimization of peptide library creation using SCX for "superfraction" collection and RPLC for fractionation.

pipeline. This optimized peptide library creation for implementation in the PepSAVI-MS pipeline can be combined with *in silico* AMP predictions and downstream MS analyses as described in Culver and Hicks (2021). Herein, we detail the protocols for optimal peptide library creation from plants to be implemented within PepSAVI-MS or for other similar bioactivity-guided approaches aimed at novel AMP discovery.

2. Extraction of plant peptides

The first step of the PepSAVI-MS pipeline is to extract peptides from plant material. Throughout the extraction, it is important to prevent peptide denaturation/degradation from warm temperatures and protease activity. It is necessary to prepare the extract for fractionation with dialysis as a preliminary means to remove small molecules. This process yields a crude extract for further fractionation to create the peptide library (Section 3).

2.1 Equipment

2.1.1. Porcelain mortar and pestle or spice grinder (Hamilton Beach Fresh Grind Electric Coffee Grinder, ASIN B005EPRFKO).

2.1.2. Analytical balance (Mettler Toledo, New Classic MF, Model #MS104S, Mass range 0.1 mg–120 g).

2.1.3. Stirring hot plate (Corning, 6795-420D, Item # UX-84303-40).

2.1.4. Centrifuges

2.1.4.1. Superspeed centrifuge; temperature range: –20 °C–+40 °C; max speed 22,000 rpm (Thermo Scientific Sorvall RC 6 Plus Centrifuge, Product Code 12121680).

2.1.4.2. Large benchtop centrifuge; temperature range: –9 °C–+40 °C; speed range: 200–14,000 rpm (Eppendorf Centrifuge 5810 R, Catalog # 022625101).

2.1.4.3. Microcentrifuge; max speed 15,060 rpm (Eppendorf Centrifuge 5425 R, Catalog # 2231000815).

2.1.5. Centrifuge bottles (Thermo Scientific Nalgene Oak Ridge High-Speed Polycarbonate Centrifuge Tubes, Catalog # 3118-0085PK).

2.1.6. Stericup filtration (Corning, 150 mL Bottle Top Vacuum Filter, 0.22 μm Pore, PES Membrane, Low Binding Protein, Product # 431160).

2.1.7. Vacuum pump (BrandTech Scientific, Vacuubrand ME1C, MFG # 2071103).

2.1.8. Centrifugal filter devices (Millipore, Amicon Ultra-15, 30 kDa nominal MWCO, Regenerated Cellulose Membrane, Catalog # UFC203024).

2.1.9. Plastic test tubes (Globe Scientific, 13 × 100 mm Plastic Test Tubes, 8 mL, Polypropylene, Item # 110445).

2.1.10. Vacuum centrifuge (Labconco Acid-Resistant CentriVap Centrifugal Concentrator Catalog # 7810016 with CentriVap Cold Trap –50 °C Catalog #7811021).

2.1.11. pH meter (Thermo Scientific, Orion Combination pH Electrode, Refillable Ag/AgCl, Catalog # 9156BNWP).

2.1.12. Dialysis tubing (Spectrum Labs, SpectraPor Float-A-Lyzer G2 500–1000 MWCO 10 mL devices, Catalog # 888-97239).

2.1.13. Extra-long transfer pipettes (Thermo Fisher Scientific, Samco, 9", Catalog # 262PK).

2.1.14. Safe-Lock Microcentrifuge Tubes, Polypropylene (Eppendorf, 2 mL, Natural color, Catalog # 20901-540).

2.2 Chemicals

2.2.1. Dry ice

2.2.2. Liquid nitrogen

2.2.3. Milli-Q water
2.2.4. Acetic acid (Millipore Sigma; CAS # 64-19-7)
2.2.5. Roche EDTA-free protease inhibitor cocktail (Millipore Sigma; SKU 11873580001)
2.2.6. Pepstatin (Sigma-Aldrich; CAS # 26305-03-3)
2.2.7. Polyvinylpolypyrrolidone (Millipore Sigma; CAS # 9003-39-8)
2.2.8. Ammonium formate (Fisher Scientific; CAS # 540-69-2)
2.2.9. Formic acid (Millipore Sigma; CAS # 64-18-6)
2.2.10. Ethanol (Fisher Scientific; CAS # 64-17-5)

2.3 Buffers/solutions to be prepared

2.3.1. 0.5 mg/mL pepstatin stock
  2.3.1.1. Resuspend pepstatin in 80% ethanol and 20% Milli-Q water with final concentration of 0.5 mg/mL.
2.3.2. Extraction buffer
  2.3.2.1. 3 mL of 10% acetic acid solution per gram of pulverized material (i.e., 300 mL for 100 g of tissue)
  2.3.2.2. 1 tablet/50 mL buffer Roche EDTA-free protease inhibitor cocktail
  2.3.2.3. 1 mg/L pepstatin
  2.3.2.4. 3% polyvinylpolypyrrolidone (g/mL)

Tip: If using plant seeds, you should prepare the buffer at 30 mL/g of pulverized seeds (i.e., 150 mL for 5 g of seeds) due to the lower water content in seeds vs plant tissue.

2.3.3. Dialysis buffer
  2.3.3.1. 5 mM ammonium formate, pH 2.7 using formic acid (tested with pH meter), 1 L for every 10 mL of the concentrated extract (i.e., 3 L for 30 mL concentrated extract).

Tip: Need to prepare two volumes of dialysis buffer for the two dialysis steps.

2.4 Protocol: Extraction

2.4.1. Grind plant tissue (aerial or root) material with a pre-chilled mortar and pestle that are nestled in dry ice, adding liquid nitrogen as needed such that the plant material remains frozen throughout the grinding process.

Tip: If using plant seeds instead, use a room temperature spice grinder to pulverize the seeds for extraction.

2.4.2. Transfer pulverized plant material to a tared and chilled weigh boat to measure and record the mass of the pulverized plant material.

Tip: The target total mass for pulverized aerial/root tissue and seeds is approximately 150 and 5 g, respectively, which could take iterative rounds of grinding to this pulverize total mass. This should be a good estimate of tissue or seed mass for a preliminary exploration of AMPs in plant tissue, but additional material is likely needed for additional isolation and characterization if intriguing AMPs are found from the first iteration with the PepSAVI-MS pipeline.

2.4.3. Store pulverized plant material at –80 °C until next steps.

2.4.4. Add plant material to extraction buffer (3 mL/g for aerial/root tissue or 30 mL/g for seeds) and stir vigorously via stir plate for 15 min at room temperature.

Tip: Use an oversized glass beaker to accommodate the volume of plant material and extraction buffer.

2.4.5. Transfer to a cold room at 4 °C and continue stirring vigorously via stir plate for 4 h.

2.4.6. Transfer the plant extract into centrifuge bottles and centrifuge for 45 min, 13,200 RCF, and 4 °C to pellet insoluble material.

Tip: If a high-speed centrifuge is not available, it is possible to have a longer centrifugation step at slower speeds. This will generally increase the time required for the Stericup filtration (Step 2.4.7).

2.4.7. Carefully decant supernatant into a Stericup filtration unit and vacuum filter liquid to remove residual insoluble material. Discard the remaining pellet of insoluble material.

Tip: Depending on the viscosity of plant material, this filtration step can range from 5 min to several days for all material to be filtered. It is typically on the time scale of minutes, however, keep extract cold in an ice bath for a multi-day filtration step.

2.4.8. Precondition 30 kDa MWCO centrifuge filter devices with Milli-Q water per manufacturer's protocol. Discard the Milli-Q water flow-through before adding the sample (Stericup filtrate).

Tip: The size of the MWCO centrifuge filter devices can change depending on the size of the peptide of interest.

2.4.9. Centrifuge for 1 h at 3200 RCF at 4 °C. Repeat until all input is filtered removing large proteins from the extract.

2.4.10. Combine all MWCO sample filtrate and record the volume. Distribute the extract equally into polypropylene test tubes and concentrate in a vacuum centrifuge to a 10-fold decrease in volume (i.e., 30 mL remaining from 300 mL starting volume).

Tip: Monitor the volume of liquid remaining every few hours and pool material to ensure that it does not concentrate to dryness as this may lead to precipitation of peptides.

2.4.11. Prepare the dialysis tubing according to the manufacturer's instructions by filling the dialysis tubing with 10% ethanol and slowly stirring on a stir plate for 10 min. Repeat by filling the tubing with Milli-Q water and slowly stirring on a stir plate for 20 min.

2.4.12. Fill dialysis tubing with the concentrated extract.

Tip: If any precipitation is visible in concentrated extract, combine in a conical tube. Try to solubilize any precipitate via vortexing and then centrifuge to pellet any insolubilized precipitation. Add supernatant to the dialysis tubing and discard the pellet.

2.4.13. Place filled dialysis tubing in fresh dialysis buffer (i.e., 3 L beaker) and dialyze with gentle stirring via stir plate in a cold room at 4 °C for 4–6 h. Transfer to fresh dialysis buffer and continue to dialyze overnight (~16 h).

2.4.14. Remove dialysis tubing from the buffer and remove dialyzed samples from the tubing using long transfer pipettes. Distribute the dialyzed sample equally into polypropylene test tubes and concentrate in a vacuum centrifuge. Pool material into 2 mL microcentrifuge tubes as the volume decreases to produce a concentrated crude extract (i.e., concentrate the 300 mL extraction buffer to 4 mL crude plate extract) that can be considered the "1X" concentration.

Tip: Monitor the volume of liquid remaining every few hours and pool material to ensure that it does not concentrate to dryness as this may lead to precipitation of peptides.

2.4.15. Store "1X" aliquot at –80 °C until further use.

3. Peptide library creation

Following the extraction of peptides from plant material into the "1X" crude extract aliquot, two fractionation steps are needed to yield the peptide library for bioactivity screening via bioassay (Section 4) and peptidomic analysis (Section 5). Herein, we discuss the use of strong cation

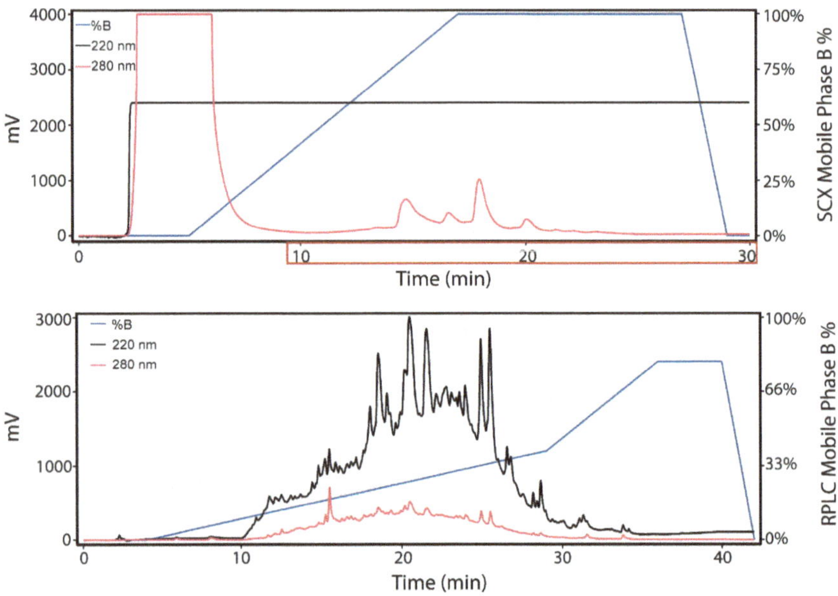

Fig. 2 Sample gradients and UV traces (A) for SCX injection with 10-min fractions collected. The first 10 min are unretained small molecules, but the rest (10–30) are combined to create SCX "superfraction" as highlighted with the red box. (B) for RPLC injection from SCX "superfraction" with 1-min fractions collected to create peptide library. UV detector 220 nm (black) and 280 (pink) and Mobile Phase B gradient (blue).

exchange chromatography (SCX) and reversed-phase liquid chromatography (RPLC) as orthogonal fractionations for peptide library creation, but recall that this can be adjusted to employ different chromatography as needed. The first step for peptide library creation is SCX fractionation (Fig. 2A), which yields a "superfraction" of the late-eluting fractions after the initial removal of unretained small molecules. This "superfraction" is then desalted via solid-phase extraction (SPE) in preparation for further fractionation via RPLC (Fig. 2B) to yield the small-increment fractions which compose the peptide library. The key advantages for the additional fractionation via RPLC are for removing residual salts from the SCX eluate and having higher resolution isolation.

3.1 Equipment

3.1.1 Large benchtop centrifuge; temperature range: -9 °C–+40 °C; speed range: 200–14,000 rpm (Eppendorf Centrifuge 5810 R, Catalog # 022625101).

3.1.2 Analytical balance (Mettler Toledo, New Classic MF, Mass range 0.1 mg–120 g, Model #MS104S).

3.1.3 Stirring hot plate (Corning, 6795-420D, Item # UX-84303-40).

3.1.4 pH meter (Thermo Scientific, Orion Combination pH Electrode, Refillable Ag/AgCl, Catalog # 9156BNWP).

3.1.5 Ultrasonic Cleaner (VWR, Symphony, Operating Frequency: 35 kHz, Catalog Code 97043-944-EA).

3.1.6 High-Performance Liquid Chromatograph (Shimadzu, LC-20AT, with UV-Vis Detector SPD-20A).

3.1.7 Fraction collector (Bio-Rad, Model 2110 Fraction Collector, Catalog # 7318122).

3.1.8 Plastic test tubes (Globe Scientific, 13 x 100 mm Plastic Test Tubes, 8 mL, Polypropylene, Item # 110445).

3.1.9 PolySULFOETHYL A column (PolyLC, 100 mm x 4.6 mm, 3 μm particles, Item # 104SE0303).

3.1.10 pH Test Strips (VWR Chemicals BDH, pH range 1.7–3.8, Catalog # BDH35315.607).

3.1.11 Vacuum pump (BrandTech Scientific, Vacuubrand ME1C, MFG # 2071103).

3.1.12 Vacuum manifold (Phenomenex, 24-port Vacuum SPE Manifold, Part # VM24).

3.1.13 Sep-Pak C_{18} Vac Cartridge (Waters, 500 mg Sorbent per Cartridge, 6cc, SKU WAT043395).

3.1.14 C_{18} column (Phenomenex Jupiter, 5 μm, 300 Å, LC Column 150 x 4.6 mm, Fully Porous Silica, Part # 00F-4053-E0).

3.1.15 Safe-Lock Microcentrifuge Tubes, Polypropylene (Eppendorf, 2 mL, Natural color, Catalog # 20901-540).

3.1.16 LC-MS Total Recovery Vials (Waters, 12 x 32 mm glass screw neck vial, PTFE/silicone septa, Catalog # 186000385c).

3.2 Chemicals

3.2.1 Ammonium formate (Fisher Scientific; CAS # 540-69-2)

3.2.2 Formic acid (Millipore Sigma; CAS # 64-18-6)

3.2.3 Acetonitrile (HPLC Grade; Fisher Scientific; CAS # 75-05-8)

3.2.4 Water (HPLC Grade and LC-MS; Fisher Scientific; CAS # 7732-18-5)

3.2.5 Trifluoroacetic acid (LC-MS Grade; Fisher Scientific; CAS # 76-05-1)

3.3 Buffers/solutions to be prepared

3.3.1. 5% formic acid: 5% formic acid in LC-MS Grade water.

3.3.2. SCX Mobile Phase A: 5 mM ammonium formate, 20% acetonitrile, pH 2.7 using formic acid (tested with pH meter).

3.3.3. SCX Mobile Phase B: 500 mM ammonium formate, 20% acetonitrile, pH 3.0 using formic acid (tested with pH meter).

3.3.4. SPE Wash: 0.1% formic acid.

3.3.5. SPE Elution: 80% acetonitrile with 0.1% formic acid.

3.3.6. RPLC Mobile Phase A: 5% acetonitrile with 0.1% TFA.

3.3.7. RPLC Mobile Phase B: Acetonitrile with 0.1% TFA.

Tip: Sonicate all mobile phases for 30 min to de-gas before use.

3.4 Protocol: Generate "Superfraction" via strong cation exchange (SCX)

3.4.1. Thaw "1X" aliquot of extract and centrifuge in a large benchtop centrifuge for 5 min at 3,220 RCFto pellet any protein precipitate.

3.4.2. Perform SCX to remove neutral and negatively charged small molecules from the "1X" crude extract. Extract (450 μL) is injected on PolySULFOETHYL A column with a flow rate of 0.5 mL/min with UV detector set at 220 nm and 280 nm. The 30-min gradient has the following steps: 0–5 min (0% SCX Mobile Phase B), 5–17 min (linear ramp from 0% to 100% B), 17–27 min (100% B), 27–29 min (linear ramp from 100 to 0% B), and 29–30 min (0% B) (Fig. 2A). Between injections, allow column re-equilibration. UV absorbance is monitored at 220 nm and 280 nm to monitor the elution of peptides containing aromatic residues (280 nm) and confirm the increasing concentration of ammonium formate throughout the gradient (220 nm).

Tip: Peptides that do not contain aromatic residues will be masked by ammonium formate.

3.4.3. Using a fraction collector, collect 10-min fractions in polypropylene test tubes. The first 10-min fraction is unretained small molecules and is discarded. The three fractions collected after 10 minutes are combined as a "superfraction" of approximately 15 mL (Fig. 2A).

Tip: The timing of fractions collected is adaptable depending on what the UV traces look like since there may be a more convenient place to switch tubes.

3.4.4. Dry the "superfraction" in a vacuum centrifuge to remove all of the acetonitrile.

3.4.5. Resuspend dried "superfraction" in 5 mL of 0.1% formic acid.

3.5 Protocol: Desalting via solid phase extraction (SPE)

3.5.1. Acidify "superfraction" by adding small aliquots of 5% formic acid until pH is less than 3 when measured with a pH test strip.

Tip: The volume needed will depend on the acidity of the sample, but this can be upwards of 1 mL to acidify to pH 3.

3.5.2. Desalt the extract with solid-phase extraction on C_{18} Sep-Pak (500 mg) with vacuum manifold.

- **3.5.2.1.** Resin is conditioned with 6 mL of SPE Elution.
- **3.5.2.2.** Equilibrate with 6 mL of SPE Wash.
- **3.5.2.3.** Load acidified sample onto the cartridges and collect the flowthrough.
- **3.5.2.4.** Repeat 3.5.2.3 twice to ensure all material is bound to the resin.
- **3.5.2.5.** Wash resin with 18 mL of SPE Wash.
- **3.5.2.6.** Elute peptides with 6 mL of SPE Elution.

3.5.3. Dry eluate in a vacuum centrifuge.

3.6 Protocol: Reversed-phase liquid chromatography (RPLC) library creation

3.6.1. Resuspend desalted SCX eluate in 500 μL of RPLC Mobile Phase A.

3.6.2. Perform RPLC on the entire desalted SCX eluate. Add the 500 μL of resuspended SCX eluate to an LC-MS vial and inject it on a Phenomenex Jupiter C_{18} column at a flow rate of 1.0 mL/min with UV detector monitoring 220 nm and 280 nm and column oven temperature at 40 °C. The 42-min gradient has the following steps: 0–4 min (0% RPLC Mobile Phase B), 4–29 min (linear ramp from 0 to 40% B), 29–36 min (linear ramp from 40 to 80% B), 36–40 min (80% B), and 40–42 min (linear ramp from 80 to 0% B) (Fig. 2B). Between injections, allow column re-equilibration.

3.6.3. Using a fraction collector, collect 1-min fractions across full injection time into 2 mL microcentrifuge tubes and dry in a vacuum centrifuge.

3.6.4. Reconstitute dried fractions in LC-MS grade water for the formation of the peptide library. Fractions 11–40 will be needed for the bioassay screening (Section 4) and LC-MS analysis (Section 5) to obtain peptide abundance data for statistical modeling.

4. Bioassay

Following the creation of the peptide library, all fractions should be screened for bioactivity. A variety of pathogens can be cultivated in the lab and assayed in a 96 well plate format, including antibacterial (Kirkpatrick, Parsley, Bartges, Wing, et al., 2018; Moyer et al., 2019; Moyer, Allen, Shaw, & Hicks, 2021; Moyer, Purvis, Wommack, & Hicks, 2021; Parsley, Sadecki, Hartmann, & Hicks, 2019), antifungal (Kirkpatrick, Parsley, Bartges, Cooke, et al., 2018; Parsley et al., 2018), and anticancer (Parsley et al., 2018) activities. Herein, we detail the use of a 96-well plate bioassay against *E. coli* utilizing optical density to determine cell growth and resazurin to estimate cell death. The relative bioactivity in library fractions will identify regions containing putative AMPs and guide statistical modeling efforts (Section 6).

4.1 Equipment

4.1.1. Incubators

4.1.1.1. Isotemp Incubator (Fischer Scientific, Model 650F, Catalog # 11-690-650F).

4.1.1.2. Incubatory Shaker (New Brunswick Scientific, Classic C25KC, MFG # M1248-0010).

4.1.2. Biosafety Cabinet (The Baker Company, SterilGARD Class II Type A2, Model VBM-400).

4.1.3. Petri dish (Fisher Scientific, 100 x 15 mm, Sterile, Polystyrene, Stackable Lid Catalog # FB0875712).

4.1.4. Inoculating Combi Loop (Fisher Scientific, HIPS, White, Flexible 1 µL/10 µL IWP, Catalog #22-363-601).

4.1.5. Culture Test Tubes (Fisher Scientific, Sterile, Translucent Polypropylene with blue snap cap, 17 x 100 mm, Catalog # 14-956-1J).

4.1.6. Spectrophotometer Cuvette (Globe Scientific, Micro, 1.5 mL, Polystyrene, 2 clear sides, Item # 112137).

4.1.7. Plate and cuvette reader (Cambridge Scientific, Molecular Devices SpectraMax M5 Multimode Microplate Reader, ID # 17000).

4.1.8. 96 Well Plates (Greiner, Microplate, 96 well, Polypropylene, Flat Bottom, Natural, Item # 655261).

4.1.9. Microplate Lids (Corning, General Assay Microplate Low Evaporation Lids, with corner notch, Catalog # 07-200-598).

4.1.10. 15 mL Falcon tubes (Corning, 15 mL High-Clarity Polypropylene Conical Tube, Produce # 352096).

4.1.11. Pipetting Reservoirs (Argos Technologies, 25 mL, Sterile, Individually Wrapped, Polystyrene, Catalog # EW-04395-23).

4.1.12. Multi Channel Pipettes

4.1.12.1. 12-Channel 30–300 μL (Eppendorf Research Plus MFG # 3125000060).

4.1.12.2. 12-Channel 0.5–10 μL (Eppendorf Research Plus MFG # 3125000028).

4.1.13. Large benchtop centrifuge; temperature range: –9 °C – +40 °C; speed range: 200–14,000 rpm (Eppendorf Centrifuge 5810 R, Catalog # 022625101).

4.1.14. Microplate Rack (Fisher Scientific, Eppendorf, Catalog # 14-278-200).

4.1.15. Parafilm (Fisher Scientific, Bemis Parafilm M Laboratory Wrapping Film, Catalog # 13-374-16).

4.2 Chemicals

4.2.1. Müeller Hinton Agar (Fisher Scientific, BD Difco, Dehydrated Culture Media, Catalog # DF0252-17-6)

4.2.2. Müeller Hinton Broth (Fisher Scientific, BD Difco, Dehydrated Powder, Catalog # DF0757-17-6)

4.2.3. Milli-Q water

4.2.4. *E. coli* freezer stock (Strain ATCC 29522)

4.2.5. Ampicillin, sodium salt (VWR, CAS # 69-52-3)

4.2.6. Water (LC-MS Grade; Fisher Scientific; CAS # 7732-18-5)

4.2.7. Resazurin sodium salt (Millipore Sigma, BioReagent, CAS # 6275-13-8)

4.3 Buffers/solutions to be prepared

4.3.1. Müeller Hinton Agar Plates

4.3.1.1. Suspend 38 g Müeller Hinton Agar powder in 1 L Milli-Q water.

4.3.1.2. Heat to boiling with gentle stirring to dissolve powder completely.

4.3.1.3. Sterilize media in autoclave at 121 °C for 15 min.

4.3.1.4. Pour a thin layer (~20 mL) of media into Petri dishes such that the entire plate is covered. Allow cooling in the biosafety cabinet until solidified.

4.3.1.5. Store at 4 °C.

4.3.2. 1x Müeller Hinton Broth

4.3.2.1. Suspend 21 g Müeller Hinton Broth powder in 1 L Milli-Q water.

4.3.2.2. Warm with gentle stirring to dissolve powder completely.

4.3.2.3. Sterilize media in autoclave at 121 °C for 15 min.

4.3.2.4. Store at room temperature.

4.3.3. 2x Müeller Hinton Broth

4.3.3.1. Suspend 42 g Müeller Hinton Broth powder in 1 L Milli-Q water.

4.3.3.2. Warm with gentle stirring to dissolve powder completely.

4.3.3.3. Sterilize media in autoclave at 121 °C for 15 min.

4.3.3.4. Store at room temperature.

4.3.4. Amp_{100} stock

4.3.4.1. Dissolve ampicillin sodium salt in LC-MS grade water to yield a final concentration of 100 mg/mL as the Amp_{100} stock.

4.3.4.2. Dilute Amp_{100} stock 1:100 with water for a working concentration of 1 mg/mL ampicillin.

4.3.5. 50 mM resazurin

4.3.5.1. Dissolve resazurin sodium salt with water to yield a final concentration of 50 mM.

4.3.5.2. Filter sterilize 50 mM resazurin with 0.22-micron filter.

4.4 Protocol: Bioassay

Tip: It is important to use sterile technique for all aspects of the bioassay protocol.

4.4.1. Streak Müeller Hinton agar plate with *E. coli* from a freezer glycerol stock. Use the quadrant streaking method to yield isolated single colonies. Incubate overnight at 37 °C.

4.4.2. Store the streaked plate at 4 °C for up to 2 weeks.

4.4.3. Transfer a single colony from the streaked plate into a culture tube with 5 mL of Müeller Hinton Broth (MHB) using an inoculating stick. Incubate overnight at 37 °C with shaking (250 rpm).

4.4.4. Perform 10x dilution of overnight culture in a cuvette (1 cm path length) and measure optical density (OD) at 600 nm. Back calculate to determine the undiluted OD600.

4.4.5. Subculture from the overnight culture into a new culture tube with 5 mL total and a final OD600 of 0.25.

Tip: Use $M_1V_1 = M_2V_2$ to determine the volume of culture needed.

4.4.6. Incubate subculture for 1 h at 37 °C with shaking (250 rpm). At 1 h, the OD_{600} should be approximately 0.5.

4.4.7. During the incubation period, prepare sterile 96-well plates with the peptide library fractions, controls, and blanks. Add 10 μL of fraction, 1:100 dilution of Amp_{100} stock (positive control), or water (negative control) to test wells. 50 μL of 1x MHB is used as a blank. Each plate should have four wells of positive control, negative control, and blanks.

Tip: To obtain results in triplicate, three plates are needed.

4.4.8. Following the 1-h incubation, confirm culture density is approximately 0.5 OD_{600} in a cuvette (1 cm path length).

4.4.9. Using this starter culture, 2x MHB, and 1x MHB, make a "master mix" of OD_{600} 0.125 culture. For three replicates, prepare 8 mL of culture as follows:

4.4.9.1. Calculate the volume of culture required to prepare 8 mL of OD_{600} 0.125 "master mix" and add to a 15 mL Falcon tube.

Tip: Use $M_1V_1 = M_2V_2$ to determine the volume of culture needed.

4.4.9.2. Add 2 mL of 2x MHB to Falcon Tube.

4.4.9.3. Bring the "master mix" to 8 mL with 1x MHB.

4.4.10. Using the sterile pipetting reservoir and multichannel pipette, add 40 μL of "master mix" to all wells with peptide fractions, positive control, and negative control, discarding tips after each transfer.

Tip: The "master mix" is not needed in the blank controls which confirm that the media is not contaminated and be used as a measure of background absorbance.

4.4.11. Spin plates briefly (15 s) in benchtop centrifuge to remove bubbles and to ensure the wells are coated with the sample.

4.4.12. Check and record the initial optical density of the 96-well plates.

4.4.13. Incubate the plates at 37 °C and with shaking (250 rpm) for 4 h to grow cells to mid-log phase ($OD_{600} = 0.5$). Wrap plate in parafilm to prevent wells from drying out and avoid splashing.

4.4.14. Check and record the optical density following the 4-h incubation. The OD_{600} should be approximately 0.5.

4.4.15. Using the sterile pipetting reservoir and multichannel pipette, add 1 μL of 50 mM resazurin to every well.

Tip: Examine the level of resazurin in pipette tips before transfer to ensure equal amounts are added to each well.

4.4.16. Spin plates briefly (15 s) in benchtop centrifuge to remove bubbles and to ensure the wells are coated with the sample.

4.4.17. Incubate the plates at 37 °C with shaking (250 rpm) for 1 h. Wrap plate in parafilm to prevent wells from drying out and avoid splashing.

4.4.18. Using plate reader, read fluorescence with λ_{ex} at 544 nm and λ_{em} at 590 nm. Save readings as *.csv.

4.4.19. Determine the percent activity of each fraction using the below equation.

$$\text{Percent Activity} = 100 \times \left[1 - \left(\frac{\text{Fraction}_{raw} - \text{Positive Control}_{avg}}{\text{Negative Control}_{avg} - \text{Positive Control}_{avg}}\right)\right]$$

Tip: It is beneficial to use Excel to template these calculations.

4.4.20. Determine the average percent activity of each fraction. Format this data in a new *.csv file with row 1 with the fraction numbers (11–40) and row 2 with average percent activity from each fraction. This format will be needed for the PepSAVI-MS modeling.

5. Quantitation via LC-MS

Following the creation of the peptide library, fractions 11–40 are analyzed using LC-MS analysis to measure the relative abundance of peptidyl species. The LC-MS data is compiled using Progenesis QI for Proteomics software to provide chromatographic peak alignment, deisotoped peak lists, and quantification of peptides in each fraction. This data will be compiled with the bioassay results (Section 4) for statistical modeling (Section 6). As listed in a review of computational approaches for proteomic analysis, other publicly available software packages can output peptide lists equivalent to Progenesis QI (Gracia & Husi, 2019). A key part of Progenesis that is needed for the PepSAVI-MS statistical modeling is that the ions are quantified before identification. Therefore, if using another integrated pipeline, this same approach to quantification should be implemented.

5.1 Equipment

5.1.1. LC-MS/MS platform

- **5.1.1.1.** nanoAcquity UPLC (Waters, Milford, MA).
- **5.1.1.2.** TripleTOF 5600 mass spectrometer (AB Sciex, Framingham, MA).

5.1.2. Symmetry C_{18} trap column (Waters, Acquity UPLC M-Class, 100 Å, 5 µm, 180 µm x 20 mm, 2G, V/M, SKU 186007496).

5.1.3. HSS T3 C_{18} analytical column (Waters, Acquity UPLC M-Class, 100 Å, 1.8 µm, 75 µm x 250 mm, SKU 186007474).

5.1.4. pH Test Strips (VWR Chemicals BDH, pH range 1.7-3.8, Catalog # BDH35315.607).

5.1.5. LC-MS Total Recovery Vials (Waters, 12 x 32 mm glass screw neck vial, PTFE/silicone septa, Catalog # 186000385c).

5.1.6. Progenesis QI for Proteomics software (Nonlinear Dynamics, v.2.0).

5.2 Chemicals

5.2.1. Formic acid (LC-MS Grade; Fischer Scientific; CAS # 64-18-6)

5.2.2. Acetonitrile (LC-MS Grade; Fisher Scientific; CAS # 75-05-8)

5.2.3. Water (LC-MS Grade; Fisher Scientific; CAS # 7732-18-5)

5.3 Buffers/solutions to be prepared

5.3.1. LC-MS/MS Mobile Phase A: Water with 0.1% formic acid

5.3.2. LC-MS/MS Mobile Phase B: Acetonitrile with 0.1% formic acid

Tip: Sonicate all mobile phases to de-gas before use.

5.4 Protocol: LC-MS analysis

5.4.1. Determine the most concentrated fraction based on the intensity of the 220 nm UV trace from the RPLC chromatogram. Acidify this peptide fraction and perform a dilution series with LC-MS grade

95% water/5% acetonitrile/0.1% formic acid, starting with the lowest concentration (i.e., 1:100) and increasing the concentration as needed until appropriate loading level is achieved. Confirm peptide fraction is acidified to pH less than 3 when measured with a pH test strip before injection.

5.4.2. Inject acidified dilution series (1–10 μL injections depending on sample loop size) onto the nano-LC-ESI-MS/MS platform monitoring signal intensity to identify an appropriate loading level (i.e., low e7 TIC for AB Sciex TripleTOF 5600). Prior to the analytical column, injections are made to the Symmetry C_{18} trap column for 3 min at a flow rate of 5 μL/min with 99% LC-MS/MS Mobile Phase A and 1% LC-MS/MS Mobile Phase B. Samples are then separated on the HSS T3 C_{18} analytical column using a 60-min gradient at a flow rate of 0.3 μL/min. The gradient has the following steps: 0–1 min (5% LC-MS/MS Mobile Phase B), 1–40 min (linear ramp from 5 to 50% B), 40–41 min (linear ramp from 50 to 85% B), 41–45 min (85% B), 45–47 min (linear ramp from 85 to 5% B), and 47–60 min (5% B). The mass spectrometer is operated in positive polarity and high sensitivity mode. MS survey scans have a mass range of 350–1600 kDa, a survey spectrum acquisition time of 250 ms, and information-dependent acquisition (IDA) of MS/MS data using an 8-second dynamic exclusion window and a cycle time of 2 seconds. Features above the intensity threshold of 150 counts/second, with a charge state of +2 to +5, were fragmented using rolling collision energy (±5%).

Tip: MS/MS data is supplemental for PepSAVI-MS statistical modeling but is not essential.

5.4.3. Dilute the fractions 11–40 of the peptide library the same as the optimized, representative fraction. Inject diluted samples following the LC-MS/MS parameters listed above and rerun the representative fraction.

Tip: It is essential for downstream data processing steps that all fractions are analyzed using consistent dilution ratios and injection volumes.

Tip: If there is limited access to LC-MS, but a clear bioactivity region is observed in the bioassay, it is possible to limit the LC-MS data to the fractions of the bioactive region with a few surrounding fractions instead of all 30 fractions.

5.5 Protocol: Progenesis

5.5.1. Open new Progenesis experiment and import the LC-MS/MS data from all fractions as *.wiff format for AB Sciex.

Tip: Other data files can be processed via Progenesis, including data from Thermo instruments. For Thermo data files, they will be imported in *. raw format.

5.5.2. Click "Start automatic processing" when the data is imported into Progenesis. Within this step, Progenesis will select an alignment reference, align the runs, and perform peak picking with maximum sensitivity. During this step, do not set up an experimental design as this will be completed in another section. After the automatic processing is complete, review the information and click "Section Complete" to progress to the next step.

5.5.3. Review the automatic peak alignment and choose a reference run that contains a region of activity from the bioassay screening.

5.5.4. Click "Section Complete" through the "Filtering" stage.

5.5.5. Complete the "Experiment Design Setup" by manually grouping all fractions to the same condition.

5.5.6. Click "Section Complete" through the "Review Peak Picking" stage.

5.5.7. Export data at the "Peptide Ion Statistics" stage and save as *.csv file. This file should be formatted to have columns for m/z, retention time, mass, charge, and intensity from each fraction (11–40).

6. Statistical modeling and analysis

Data from the bioassay results (Section 4) and LC-MS analysis (Section 5) are integrated utilizing the R programming language with the PepSAVIms CRAN package (https://cran.r-project.org/web/packages/PepSAVIms), to compile a list of the Top 20 ranked candidates that are most likely to contribute to the bioactive properties as seen from the bioassay results (Kirkpatrick et al., 2017). This data can be used to prioritize further investigation, including isolating or synthesizing the peptide of interest for determining minimum inhibitory concentration, probing biological activity, or other downstream characterization approaches (Balboa & Hicks, 2021).

6.1 Equipment

6.1.1. RStudio (v 3.5).

6.1.2. PepSAVIms CRAN package (https://cran.r-project.org/web/packages/PepSAVIms).

6.2 Protocol: PepSAVI-MS statistical analysis

6.2.1. Open PepSAVI-MS statistics CRAN package, publicly available at https://cran.r-project.org/web/packages/PepSAVIms (Kirkpatrick et al., 2017).

6.2.2. Load the *.csv files from Progenesis and formatted bioassay results into R.

6.2.3. Use the *binMS* function to combine peptide ion dataset entries that are the same. The R script notation for this function is as follows:

binMS (mass_spec, mtoz, charge, mass, time_peak_reten, ms_inten, time_range, mass_range, charge_range, mtoz_diff, time_diff)

In the above notation, "mass_spec" represents the mass spectrometry data matrix from the *.csv file. The next five parameters ("mtoz," "charge," "mass," "time_peak_reten," and "ms_inten") correspond to the columns from the data matrix (m/z, charge, mass, retention time, and intensity from each fraction). Filters for retention time, mass, and charge are employed with the "time_range," "mass_range," and "charge_range" parameters, respectively. The last two parameters ("mtoz_diff" and "time_diff") are used to specify the largest difference in m/z rations and retention time shifts that is allowed for binning rows together.

6.2.4. Typical AMPs have masses, charge states, and retention times of 1,000–15,000 Da, +2-9, and 15–45 min, respectively. These values should be used in the above notation for binning.

6.2.5. The *summary(binMS_out)* function will summarize what was consolidated and report the number of features remaining after each data reduction criteria.

6.2.6. Use the *filterMS* function to refine the binned mass spectrometry data for the compounds that are most likely to contribute to the bioactive regions. The R script notation for this function is as follows:

filterMS (msObj, region, border, bord_ratio, min_inten, max_chg)

In the above notation, the output *.csv from the *binMS* function is the "msObj" parameter. The "region" parameter is used to specify the filtering region based on the observed bioactivity results and all other fractions are considered in the "border" parameter. The "bord_ratio" parameter is used to specify the maximum ratio of a compound's abundance outside of the bioactivity region compared to the observed abundance of that feature. This step is critical for filtering out compounds that are unlikely to be the cause of the observed bioactivity. The last two parameters, "min_inten" and "max_chg" are used to eliminate noise by filtering a minimum intensity and maximum charge value, respectively.

6.2.7. The *summary(filterMS_out)* function will summarize what was consolidated and report the number of features remaining after each data reduction criteria.

6.2.8. Use the *rankEN* function for penalized linear regression to determine which of the candidate compounds are best matched to the observed bioactivity profile. The R script notation for this function is as follows:

rankEN (msObj, bioact, region_ms, region_bio, lambda, pos_only, ncomp)

In the above notation, the output *.csv from the *filterMS* function is the "msObj" parameter. The "bioact" parameter is the *.csv formatted bioassay results. The next two parameters ("region_ms" and "region_bio") are used to specify that the mass spectrometry data and bioassay results need to be compared. The "lambda" parameter is used to set the penalty parameter used for estimation; it is typically 0.01 for our data sets. To ensure only compounds that have a positive correlation are considered for modeling, the "pos_only" parameter is needed. To specify the maximum number of compounds that will be reported, use the "ncomp" parameter.

6.2.9. This yields a list of ranked candidate compounds and the Top 20 are reported in the script. Candidates with multiple charge states ranked in the Top 20 should be prioritized for further investigation.

Tip: Possible further investigation includes peptide isolation or synthesis for the determination of minimum inhibitory concentration (MIC) studies. AMP identification can be confirmed via downstream analyse (Culver & Hicks, 2021).

7. Summary

AMPs compose a diverse class of compounds with a wide range of potential therapeutic applications. Unfortunately, common discovery workflows, including bioassay-guided fractionation and genome mining, can be hampered by low-throughput processes, compound rediscovery, lack of requisite genomic/transcriptomic databases, or predictive approaches lacking a direct bioactivity measurement. These challenges are often exacerbated when working with non-model organisms, including plants, which lack high-quality resources and knowledge of biosynthetic pathways. This chapter highlights the improvements to the PepSAVI-MS pipeline, an alternative AMP discovery method, which seeks to address the abovementioned challenges in AMP discovery. PepSAVI-MS optimization includes the adaption of sample preparation to utilize RPLC fractionation to reduce bioassay interference from SCX buffers and improve separation quality. This will allow for more confident identification of bioactivity, including the peptide features that are most likely contributing to this activity. Putative AMPs prioritized via the PepSAVI-MS pipeline are then subjected to further characterization as desired. Overall, the flexibility of the PepSAVI-MS pipeline is key for its widespread use. This flexibility has been noted with its adaptable use of orthogonal fractionation with various chromatography as needed. Additionally, though we detail the use of the PepSAVI-MS pipeline with Sciex TripleTOF 5600, other available MS platforms with sufficient resolving power can be utilized for comparable data. This optimized workflow of PepSAVI-MS can also be implemented in combination with different processes, including *in silico* AMP predictions and downstream mass spectrometry analysis (Culver & Hicks, 2021).

Acknowledgments

This work was supported by NIH-NIGMS under award R01 GM125814 to L.M.H. We thank Dr. Christine Kirkpatrick and Lilian Heil for their work developing and optimizing this method.

References

Balboa, S. J., & Hicks, L. M. (2021). Revealing AMP mechanisms of action through resistance evolution and quantitative proteomics. *Methods in Enzymology, 663*.

Culver, K. D., & Hicks, L. M. (2021). Identification of putative antimicrobial peptides from botanical extracts. *Methods in Enzymology, 663*.

Gracia, K. C., & Husi, H. (2019). Computational approaches in proteomics. In H. Husi (Ed.), *Computational Biology* (pp. 119–142). Brisbane, Australia: Codon Publications.

Harvey, A. L. (2008). Natural products in drug discovery. *Drug Discovery Today, 13*(19–20), 894–901. https://doi.org/10.1016/j.drudis.2008.07.004.

Kang, X., Dong, F., Shi, C., Liu, S., Sun, J., Chen, J., et al. (2019). DRAMP 2.0, an updated data repository of antimicrobial peptides. *Scientific Data, 6*(1), 1–10. https://doi.org/10.1038/s41597-019-0154-y.

Kirkpatrick, C. L., Broberg, C. A., McCool, E. N., Lee, W. J., Chao, A., McConnell, E. W., et al. (2017). The "PepSAVI-MS" pipeline for natural product bioactive peptide discovery. *Analytical Chemistry, 89*, 1194–1201. https://doi.org/10.1021/acs.analchem.6b03625.

Kirkpatrick, C. L., Parsley, N. C., Bartges, T. E., Cooke, M. E., Evans, W. S., Heil, L. R., et al. (2018). Fungal secretome analysis via PepSAVI-MS: Identification of the bioactive peptide KP4 from Ustilago maydis. *Journal of the American Society for Mass Spectrometry, 29*(5), 859–865. https://doi.org/10.1007/s13361-017-1880-z.

Kirkpatrick, C. L., Parsley, N. C., Bartges, T. E., Wing, C. E., Kommineni, S., Kristich, C. J., et al. (2018). Exploring bioactive peptides from bacterial secretomes using PepSAVI-MS: Identification and characterization of Bac-21 from Enterococcus faecalis pPD1. *Microbial Biotechnology, 11*(5), 943–951. https://doi.org/10.1111/1751-7915.13299.

Mahlapuu, M., Håkansson, J., Ringstad, L., & Björn, C. (2016). Antimicrobial peptides: An emerging category of therapeutic agents. *Frontiers in Cellular and Infection Microbiology, 6*(Dec), 1–12. https://doi.org/10.3389/fcimb.2016.00194.

Moyer, T. B., Allen, J. L., Shaw, L. N., & Hicks, L. M. (2021). Multiple classes of antimicrobial peptides in Amaranthus tricolor revealed by prediction, proteomics and mass spectrometric characterization. *Journal of Natural Products, 84*(2), 444–452.

Moyer, T. B., Heil, L. R., Kirkpatrick, C. L., Goldfarb, D., Lefever, W. A., Parsley, N. C., et al. (2019). PepSAVI-MS reveals a proline-rich antimicrobial peptide in Amaranthus tricolor. *Journal of Natural Products, 82*(10), 2744–2753. https://doi.org/10.1021/acs.jnatprod.9b00352.

Moyer, T. B., Purvis, A. L., Wommack, A. J., & Hicks, L. M. (2021). Proteomic response of Escherichia coli to a membrane lytic and iron chelating truncated Amaranthus tricolor defensin. *BMC Microbiology, 21*(1). https://doi.org/10.1186/s12866-021-02176-4.

Parsley, N. C., Kirkpatrick, C. L., Crittenden, C. M., Rad, J. G., Hoskin, D. W., Brodbelt, J. S., et al. (2018). PepSAVI-MS reveals anticancer and antifungal cycloviolacins in viola odorata. *Phytochemistry, 152*, 61–70. https://doi.org/10.1016/j.phytochem.2018.04.014.

Parsley, N. C., Sadecki, P. W., Hartmann, C. J., & Hicks, L. M. (2019). Viola "inconspicua" no more: An analysis of antibacterial cyclotides. *Journal of Natural Products, 82*(9), 2537–2543. https://doi.org/10.1021/acs.jnatprod.9b00359.

Russell, A. H., & Truman, A. W. (2020). Genome mining strategies for ribosomally synthesised and post-translationally modified peptides. *Computational and Structural Biotechnology Journal, 18*, 1838–1851. https://doi.org/10.1016/j.csbj.2020.06.032.

Sharma, S. B., & Gupta, R. (2015). Drug development from natural resource: A systematic approach. *Mini-Reviews in Medicinal Chemistry, 15*(1), 52–57. https://doi.org/10.2174/138955751501150224160518.

Skinnider, M. A., Dejong, C. A., Rees, P. N., Johnston, C. W., Li, H., Webster, A. L. H., et al. (2015). Genomes to natural products PRediction Informatics for Secondary Metabolomes (PRISM). *Nucleic Acids Research, 43*, 9645–9662. https://doi.org/10.1093/nar/gkv1012.

Skinnider, M. A., Johnston, C. W., Edgar, R. E., Dejong, C. A., Merwin, N. J., Rees, P. N., et al. (2016). Genomic charting of ribosomally synthesized natural product chemical space facilitates targeted mining. *Proceedings of the National Academy of Sciences of the United States of America, 113*(42), E6343–E6351. https://doi.org/10.1073/pnas.1609014113.

Strömstedt, A. A., Felth, J., & Bohlin, L. (2013). Bioassays in natural product research-strategies and methods in the search for anti-inflammatory and antimicrobial activity. *Phytochemical Analysis*. https://doi.org/10.1002/pca.2468.

Tam, J., Wang, S., Wong, K., & Tan, W. (2015). Antimicrobial peptides from plants. *Pharmaceuticals*, *8*(4), 711–757. https://doi.org/10.3390/ph8040711.

Waghu, F. H., Barai, R. S., Gurung, P., & Idicula-Thomas, S. (2016). CAMPR3: A database on sequences, structures and signatures of antimicrobial peptides. *Nucleic Acids Research*, *44*(D1), D1094–D1097. https://doi.org/10.1093/nar/gkv1051.

Wang, G. (2021). Unifying the classification of antimicrobial peptides in the antimicrobial peptide database. *Methods in Enzymology, 663*.

Wang, G., Li, X., & Wang, Z. (2015). APD3: The antimicrobial peptide database as a tool for research and education. *Nucleic Acids Research*, *44*, 1087–1093. https://doi.org/10.1093/nar/gkv1278.

Wang, G., Li, X., & Wang, Z. (2016). APD3: The antimicrobial peptide database as a tool for research and education. *Nucleic Acids Research*, *44*(D1), D1087–D1093. https://doi.org/10.1093/nar/gkv1278.

Weller, M. G. (2012). A unifying review of bioassay-guided fractionation, effect-directed analysis and related techniques. *Sensors*, *12*, 9181–9209. https://doi.org/10.3390/s120709181.

Zhao, X., Wu, H., Lu, H., Li, G., & Huang, Q. (2013). LAMP: A database linking antimicrobial peptides. *PLoS One*, *8*(6), 6–11. https://doi.org/10.1371/journal.pone.0066557.

CHAPTER FOUR

Screening for cysteine-stabilized scaffolds for developing proteolytic-resistant AMPs

Mariana Rocha Maximiano[a,b,†], Samilla Beatriz Rezende[a,†], Thuanny Borba Rios[a,b], Michel Lopes Leite[c], Liana Costa Pereira Vilas Boas[b,d], Nicolau Brito da Cunha[b], Állan da Silva Pires[b], Marlon Henrique Cardoso[a,b], and Octávio Luiz Franco[a,b,d,*]

[a]S-Inova Biotech, Pós-Graduação em Biotecnologia, Universidade Católica Dom Bosco, Campo Grande, Brazil
[b]Centro de Análises Proteômicas e Bioquímicas, Programa de Pós-Graduação em Ciências Genômicas e Biotecnologia, Universidade Católica de Brasília, Brasília, Brazil
[c]Departamento de Biologia Molecular, Instituto de Ciências Biológicas, Universidade de Brasília, Brasília, Brazil
[d]Programa de Pós-Graduação em Patologia Molecular, Universidade de Brasília, Brasília, Brazil
*Corresponding author: e-mail address: ocfranco@gmail.com

Contents

Abstract

Antimicrobial peptides (AMP) are present in all organisms and can present several activities and potential applications in human and animal health. Screening these molecules scaffolds represents a key point for discovering and developing novel biotechnological products, including antimicrobial, antiviral and anticancer drugs candidates and

† These authors contributed equally.

Methods in Enzymology, Volume 663
ISSN 0076-6879
https://doi.org/10.1016/bs.mie.2021.11.001

insecticidal molecules with potential applications in agriculture. Therefore, considering the amount of biological data currently deposited on public databases, computational approaches have been commonly used to predicted and identify novel cysteine-rich peptides scaffolds with known or unknown biological properties. Here, we describe a step-by-step *in silico* screening for cysteine-rich peptides employing molecular modeling (with a core focus on comparative modeling) and atomistic molecular dynamics simulations. Moreover, we also present the concept of additional tools aiming at the computer-aided screening of new Cs-AMPs based drug candidates. After the computational screening and peptide chemical synthesis, we also provide the reader with a step-by-step *in vitro* activity evaluation of these candidates, including antibacterial, antifungal, and antiviral assays.

1. Introduction

Antimicrobial peptide (AMP) are molecules that present up to 10 kDa and has been found in all organisms (Huan, Kong, Mou, & Yi, 2020; Nijnik & Hancock, 2009). Cysteine-stabilized peptides (Cs-AMPs) correspond to a group that presents conserved cysteine motifs (Table 1) and a broad variation in sequence and numerous structural profiles (β-sheets, α-helical, or a mixture of these two secondary structures) (Porto, Pires, & Franco, 2017). These peptides can present various activities (de Oliveira Dias & Franco, 2015; Srivastava et al., 2021), acting as promising model molecules for the generation of future peptide based-drug candidates with and potential applications in medicine, veterinary medicine, and agriculture, including the development of novel antimicrobial and antiviral properties

Table 1 Conserved cysteine motifs present in antimicrobial peptides families.

Peptide family	Conserved cysteine motifs	Reference
Lipid transporter proteins (LTPs)	$Cx_{(7-9)}Cx_{(12-14)}Cx_{(8-19)}Cx_{(1)}Cx_{(19-23)}Cx_{(13-15)}C$	Shelenkov, Slavokhotova, & Odintsova, 2020
Snakins	$CX_{(3)}CX_{(3)}CX_{(8)}CX_{(3)}CX_{(2)}CCX_{(2)}CX_{(1)}CX_{(11)}CX_{(1)}CX_{(12)}C$	
Heveins	$CX_{(3,8)}CX_{(4)}CCX_{(5)}\ CX_{(6)}CX_{(3,5)}CX_{(1,3)}C$	
Thionin	$CCX_{(11)}CX_{(9,15)}CX_{(5)}CX_{(6,11)}C$	
Cyclotides	$CX_{(3)}CX_{(4,5)}CX_{(4,6)}CX_{(1)}CX_{(4,5)}C$	
Cis-defensins	$CX_{(10)}CX_{(5)}CX_{(3)}CX_{(9)}CX_{(6)}CX_{(1)}\ CX_{(3)}C$	

(Boas, Campos, Berlanda, de Carvalho Neves, & Franco, 2019; Fleitas Martínez, Cardoso, Ribeiro, & Franco, 2019; Maximiano & Franco, 2021).

In this context, computational approaches are increasingly used to study peptide/protein functions from their primary sequence to their tridimensional (3D) structural arrangement, which can be investigated through *in vitro* and *in silico* techniques. *In silico* methodologies have assisted researchers to predict peptide/protein atomic coordinates in the absence of experimentally elucidated structures, which are commonly characterized by X-ray crystallography, solution nuclear magnetic resonance (NMR) or solid-state NMR, and cryo-electron microscopy (Cardoso, Oshiro, Rezende, Candido, & Franco, 2018).

Thus, comparative molecular modeling comprises a useful computational tool for 3D peptides' structure prediction (Fiser & Šali, 2003). Specifically, comparative molecular modeling uses one or more peptide/proteins of known structure (templates) to predict the 3D structure of a given peptide/protein sequence (target) based on sequence alignments (Webb & Sali, 2016). Currently, there are numerous web servers for automating comparative peptide/protein modeling (Table 2).

Table 2 Web servers are used to predict the 3D structures for peptides and proteins. **Web servers—comparative modeling**

Name	Link	References
3D-JIGSAW server	https://bmm.crick.ac.uk/~svc-bmm-djigsaw/help_crick.htmL	Bates, Kelley, MacCallum, and Sternberg (2001)
HHpred	https://toolkit.tuebingen.mpg.de/tools/hhpred	Söding, Biegert, and Lupas (2005)
IntFOLD	http://www.reading.ac.uk/bioinf/IntFOLD/IntFOLD6_form.html	McGuffin et al. (2019)
I-TASSER	https://zhanggroup.org/I-TASSER/	Roy, Kucukural, and Zhang (2010)
RaptorX	http://raptorx.uchicago.edu/	Xu, Mcpartlon, and Li (2021)
Robetta	http://robetta.bakerlab.org/	Song et al. (2013)
SWISS-MODEL	https://www.expasy.org/resources/swiss-model	Schwede, Kopp, Guex, and Peitsch (2003)

In this scenario, this work highlights the strategies to screen for Cs-AMPs structure, aiming at developing peptide-based drug candidates, with potential biotechnological applications in several areas, as medicine, veterinary medicine, and agriculture.

2. Materials and equipment

2.1 *In silico*

Minimal recommended configuration:

- Core I5 4GB RAM 500GB HD. Linux Ubuntu 20.04 LTS
- GROMACS 5.1.4
- MODELLER 10.1

2.2 *In vitro*

2.2.1 *Antimicrobial (antibacterial and antifungal) assay*

- Mueller–Hinton broth;
- Mueller–Hinton agar;
- Potato dextrose agar;
- RPMI 1640 medium;
- Spectrophotometer/microplate reader;
- Sterile 96-well microplates;
- Microcentrifuge tubes, 1.5 mL;
- Falcon tubes;
- Shaker;
- Sterile Petri plates.

2.2.2 *Antiviral assay*

- CO_2 incubator at 37 °C for cell growth
- Flask for culturing cells (e.g., glass flat bottom 60 mL);
- Inverted microscope
- 6, 24 and 96 wells plates
- DMEM—Dulbecco's Modified Eagle Medium
- Penicillin/Streptomycin solution for cell culture
- Fetal Bovine Serum for cell culture
- Sodium bicarbonate powder
- Carboxymethyl cellulose (CMC);
- Formaldehyde
- Crystal violet powder

3. Step-by-step method details

3.1 *In silico*

This topic presents how regular expressions (RegEx) can be applied for Cs-AMPs screening in public databases and repositories, including proteomes, genomes, and transcriptomes. Additionally, we present the concept of molecular modeling and how comparative modeling, with a core focus on MODELLER, has been used to predict the 3D structure of known and unknown Cs-AMPs sequences. Finally, we highlight how atomistic molecular dynamics simulations have profoundly contributed to calculating the trajectory of Cs-AMPs in molecular systems mimicking a desired biological condition.

3.1.1 Searching for Cs-AMPs in public databases

3.1.1.1 Preparing the computational environment and databases

(a) Perl or Python is necessary to perform the searches proposed here. An updated repository for installation on the Linux environment is available in the link: (https://www.cpan.org/src/5.0/perl-5.28.1.tar.gz, for Perl; and https://www.python.org/downloads/ for Python)

(b) A target database can be chosen based on specific preferences. For demonstrative purposes, here we will use the *Arabidopsis thaliana* proteome. The database is available on the Uniprot Consortium (https://www.uniprot.org/proteomes/UP000006548)

(c) There are several Cs-AMPs from plants (e.g., defensins, lipid transfer proteins (LPTs) and snakins), and the pipeline could be performed for most of them. However, this methodology can not be performed for families with a poor amount of sequences described such as β-barrelin. A RegEx that represents the peptide family is required to screen for Cs-AMP sequences within a target proteome/genome/transcriptome. Many of these families already have one or more RegEx described in the literature (Costa et al., 2020; Silverstein et al., 2007; Tomczak et al., 2012; Zhu, 2008). For this pipeline, the CSαβ-defensin RegEx: $CX_{2\text{-}18}CX_3CX_{2\text{-}10}$[GAPSIDERYW]$XCX_{4\text{-}17}$ CXC (each amino acid is represented by one-letter code; "X" represents any proteinogenic amino acid, and brackets indicate positions that allows one of those amino acids between brackets), described by Zhu (2008), will be used as an example. A commented Python script is available for data mining (Supplementary File 1 in the online version at https://doi.org/10.1016/bs.mie.2021.11.001). The packages *re* (Van Rossum, 2020) and Bio (Cock et al., 2009) are necessary for the proper performance of the script

3.1.2 Comparative modeling of Cs-AMPs by using MODELLER

3.1.2.1 Installing MODELLER

(a) MODELLER should be installed to proceed with the analyses. The software is available for Mac operating system (OS), Windows and Unix/Linux (https://salilab.org/modeller/download_installation.html).

(b) The files required to complete the basic protocol can be found on http://salilab.org/modeller/tutorial/basic-example.tar.gz (Unix/Linux) or http://salilab.org/modeller/tutorial/basic-example.zip (Windows).

(c) Here, the most recent version of MODELLER (v10.1) was used. The algorithms applied were tested for MODELLER v9.17 and superior.

3.1.2.2 Searching for suitable template structures

(a) Here, we will use as an example the following target Cs-AMP sequence:

>PDB ID: 1ti5

RT**C**MIRREGWGR**C**LIDTT**C**AHS**C**KNKGYIGGN**C**KGMTKT
CY**C**LVN**C**

Where cysteine residues are in bold type

(b) The template structures can be searched using different algorithms, including BLASTp (https://blast.ncbi.nlm.nih.gov/Blast.cgi?PROGRAM= blastp) and HHpred (https://toolkit.tuebingen.mpg.de/tools/hhpred). BLAST (Basic Local Alignment Tool) uses local sequence alignments to find proteins with high sequence similarity. By contrast, HHpred applies a secondary structure alignment to find similar proteins. In both cases, a target sequence is necessary as input

(c) In this process, alignment statistics such as sequences' identity, e-value, and sequence coverage are essential to select a template structure. Therefore, the identity values should be ~30% or higher for closely related Cs-AMP sequences. Moreover, the e-value defines the statistical measure of the alignment, representing the chances of random alignments

(d) All these parameters detect similarities between the target sequence and database template structures

(e) Once the template is selected, the .pdb file (the atomic coordinates of a structurally determined peptide) can be obtained from Protein Data Bank (https://www.rcsb.org/) (Fig. 1).

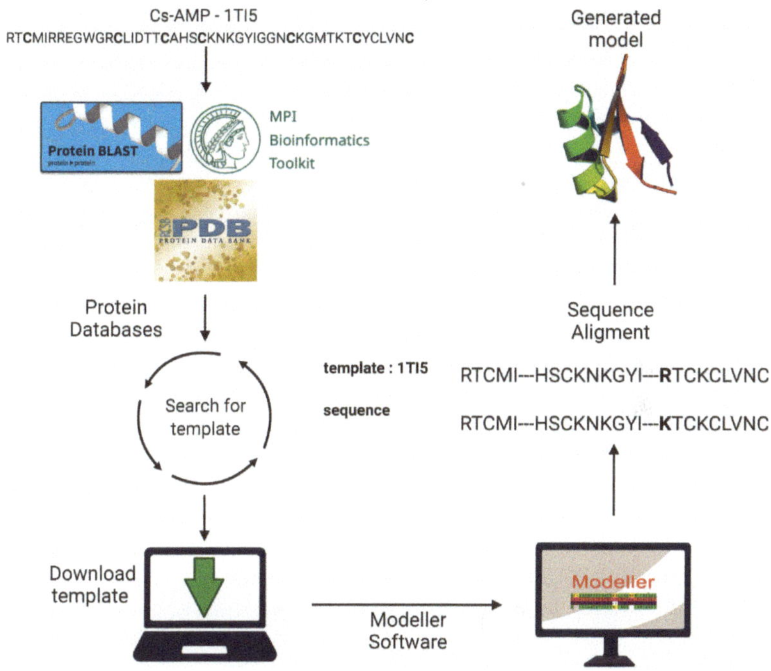

Fig. 1 Scheme demonstrating the steps to download .pdb files for a given peptide/protein structure from databases (Protein Data Bank Server, Blastp and HHpred). These experimentally determined structures are then used for comparative modeling. After downloading the template .pdb structure using a four-letter code (e.g., 1ti5), it is submitted to software MODELLER, based on sequence alignment simulations the model would be generated. Created with the support of Biorender (BioRender.com).

(f) In addition, MODELLER also provides a script for template searching. The script is named build_profile.py (https://salilab.org/modeller/tutorial/basic.html).

3.1.2.3 Checking necessary scripts to perform sequence alignment and generating 3D theoretical models

(a) TvLDH.ali—this PIR format file is necessary for MODELLER to proceed and write alignments. In this script, the target sequence is inserted in FASTA format. The first line consists of (>P1;code), followed by the sequence's identifier. After populating the sequence target, insert (*), which marks its end (Fig. 2)

(b) Align2d.py—this file is a script used to align the template (.pdb) sequence with the target sequence described in the TvLDH.ali script

```
>P1;TvLDH
sequence:target sequence:::::::0.00: 0.00
RTCMIRREGWGRCLIDTTCAHSCKNKGYIGGNCKGMTKTCYCLVNC*
```

Fig. 2 Representation of the TvLDH.ali script, which contains the target sequence in PIR format.

(c) model-single.py—this script generates the theoretical models based on the template atomic coordinates and spatial orientation. However, for Cs-AMPs, if the cysteines are not aligned in align2d.py output, another script should be used to specify each disulfide bond, as explained below

(d) model-disulfide.py—this script is a punctual script to generate models that contain disulfide bonds not covered by the template. The appropriate disulfide bond restraints are generated for the output model/ models

3.1.2.4 Aligning TvLDH.ali with the selected template

(a) Insert the target sequence according to the instructions above in the TvLDH.ali file

(b) With the template (e.g., 1ti5.pdb) structure in the same folder, open the file align2d.py. This script will align the target sequence in the TvLDH. ali file with the template structure described in the .pdb file

(c) Once the align2d.py script is opened, the first line will indicate the template structure (FILE) 1ti5.pdb. It is necessary to change the parameters in (ALIGN_CODES) to "1ti5", followed by (ALN.APPEND), where the variable (FILE) receives "TvLDH.ali", which contains the target sequence. Next, the sixth line will execute the command ALIGN2D to perform the alignment (Fig. 3).

```
from modeller import *

env = environ()
aln = alignment(env)
mdl = model(env, file='1ti5', model_segment=('FIRST:A','LAST:A'))
aln.append_model(mdl, align_codes='1ti5', atom_files='1ti5.pdb')
aln.append(file='TvLDH.ali', align_codes='TvLDH')
aln.align2d()
aln.write(file='TvLDH-1ti5.ali', alignment_format='PIR')
aln.write(file='TvLDH-1ti5.pap', alignment_format='PAP')
```

Fig. 3 Script align2d.py is used to align the target sequence with its selected template.

3.1.2.5 Command lines to execute the alignment

(a) Execute through the command line in Linux (Ubuntu), using the following command: mod10.1 (the number 10.1 depends on the installed version).

(b) To do the alignment, use the command mod10.1 align2d.py

Two output files will be produced: a PIR (TvLDH-1ti5.ali) and a PAP (TvLDH-1ti5.pap) format file. MODELLER uses the PIR format for the model-building stage, whereas the PAP alignment format is more intuitive to check the alignment results. Identical amino acid residues are underscored with an asterisk (Fig. 4A and B).

A **Script TvLDHi.ali**

```
>P1;1ti5
structureX:1ti5.pdb:    1 :A:+46   :A:MOL_ID  1; MOLECULE  PLANT DEFENSIN
RTCMIKKEGWGKCLIDTTCAHSCKNRGYIGGNCKGMTRTCYCLVNC*

>P1;TvLDH
sequence:TvLDH:      : :      : ::: 0.00: 0.00
RTCMIRREGWGRCLIDTTCAHSCKNKGYIGGNCKGMTKTCYCLVNC*
```

B **Script TvLDHi.pap**

```
 _aln.pos            10        20        30        40
1ti5       RTCMIKKEGWGKCLIDTTCAHSCKNRGYIGGNCKGMTRTCYCLVNC
TvLDH      RTCMIRREGWGRCLIDTTCAHSCKNKGYIGGNCKGMTKTCYCLVNC
 _consrvd  *****  **** ************* *********** ********
```

Fig. 4 (A) TvLDH-1ti5.ali, illustrating the alignment between the template and target sequences. (B) TvLDH-1ti5.pap, highlighting by "*" the conserved residues between the two sequences. Created with the support of Biorender (BioRender.com).

3.1.2.6 Generating theoretical models using model-single.py

(a) After the alignment has been executed by the script align2d.py, open the file model-single.py, which calculates 3D models for the target sequence based on the template structure. The first line of this script sets the environment. The following lines comprise parameter values for the "model" routine. The variable (ALNFILE) sets the name of the alignment file "TvLDH-1ti5.al," the parameter (KNOWNS) defines the known template structure and (SEQUENCE) needs to be assigned as TvLDH, which has the target sequence. At last, the parameters (A.STARTING_MODEL) and (A.ENDING_MODEL) define the number of models that will be calculated. The last line refers to the command to run the "model" routine (Fig. 5A).

A Script model-single.py

```
from modeller import *
from modeller.automodel import *

env = environ()
a = automodel(env, alnfile='TvLDH-1ti5.ali',
              knowns='1ti5', sequence='TvLDH',
              assess_methods=(assess.DOPE, assess.GA341))
a.starting_model = 1
a.ending_model = 100
a.make()
```

B Script model-disulfide.py

```
# Comparative modeling by the automodel class
from modeller import *                # Load standard Modeller classes
from modeller.automodel import *     # Load the automodel class

# Redefine the special_patches routine to include the additional disulfides
# (this routine is empty by default):
class MyModel(automodel):
    def special_patches(self, aln):
        # A disulfide between residues 3 and 46:
        self.patch(residue_type='DISU', residues=(self.residues['3'],
                                                  self.residues['46']))
        # A disulfide between residues 12 and 32:
        self.patch(residue_type='DISU', residues=(self.residues['12'],
                                                  self.residues['32']))

log.verbose()
env = environ()


env.io.atom_files_directory = ['.', '../atom_files']

a = MyModel(env,
            alnfile='TvLDH-1ti5.ali',
            knowns='1ti5', sequence = 'TvLDH',
            assess_methods=(assess.DOPE))
a.starting_model= 1
a.ending_model  = 100

a.make()
```

Fig. 5 Representation of scripts used to run molecular modeling simulations (A) representing model-single.py script used to generate 3D theoretical models for the target sequence based on the template structure. (B) model-disulfide.py script, used to generate models that contain disulfide bonds in their structure. Created with the support of Biorender (BioRender.com).

(b) When modeling a sequence with disulfide bonds, it may be necessary to use a modified version of model-single.py, named model-disulfide.py (https://salilab.org/modeller/manual/node24.html#SECTION:model-disulfide). The restraints to disulfide bonds are added in the script. To generate CHARMM topology it is used the file DISU in the parameter (RESIDUE_TYPE), the residues with the disulfide bonds are inserted in (SELF.RESIDUES["]) parameter. The instruction should be repeated according to the number of disulfide bonds with each specific pair of cysteines (i.e., cys3-cys46 and cys12-cys32) (Fig. 5B).

3.1.2.7 Using model-disulfide.py used to generate 3D theoretical

(a) Command lines to generate 3D theoretical models with unpaired disulfide bonds

i. To generate models without disulfide bonds restrains, run mod10.1 model-single.py

ii. To generate models containing disulfide bonds without coverage in template structure, run mod10.1 model-disulfide.py

iii. Choose the lowest free-energy generated model based on the DOPE score. The DOPE (Discrete Optimized Protein Energy) is added into MODELLER, based on a refined reference state that coincides with non-interacting atoms in a uniform sphere with a sample native structure radius dependent. It was tested with several non-redundant sets of crystallographic structures. Therefore, it is among the best parameters to calculate peptides/proteins free-energy (Shen & Sali, 2006).

(b) The report of the output files can be found in "model-single.log" and "model-disulfide.log," depending on the scripts used. Helpful information about generated models, including report warnings, errors, and input restraints, can be found in .log files. The model can be visualized using the PyMOL software (https://pymol.org/2/).

3.1.2.8 Structural statistics for a given 3D theoretical model

MODELLER evaluates final models using the scoring function DOPE (Shen & Sali, 2006), developed independently based on a conformation analysis of well-folded peptides and proteins. After the selection of the lowest free-energy model, an additional evaluation step is recommended. For that, many servers can be used. General examples include the ProSA-web server (https://prosa.services.came.sbg.ac.at/prosa.php) and QMEAN (https://swissmodel.expasy.org/qmean/) to evaluate the 3D theoretical model fold in comparison with other peptides of similar size and structurally determined by X-ray crystallography or NMR. Additionally, PROCHECK (https://saves.mbi.ucla.edu/) can be used to check the stereochemistry quality of models. High-quality models are expected to present an NMR or X-ray similar *z*-score on ProSA, a modular *z*-score below to 2 on QMEAN and, at least, 90% of the residues in most favored and additionally allowed regions on the Ramachandran Plot.

3.1.3 Molecular dynamics

3.1.3.1 Installing GROMACS

(a) The first step is to download and install GROMACS software at (http://www.gromacs.org/Downloads), distributed for Mac OS, Windows and Unix/Linux system but works natively on a Unix-type system (such as Linux, or Mac OS X).

(b) The installation can be done according to the GROMACS version that the user chose. There are two ways to build GROMACS. We will use a user-friendly installation. For this, the configuration described below is necessary:

(i) Get the latest version of C and C++ compilers;

(ii) Check if the computer has CMake installed (it varies according to GROMACS version);

(iii) Download and unpack the latest version of GROMACS;

(iv) Select a separate build directory and change to it;

(v) Run the command cmake with a path to the source as an argument;

(vi) Run commands make, make check and make install (Fig. 6).

(c) Here, an MD simulation demonstration will be performed using GROMACS version 5.0.4 on Ubuntu 16.04 LTS. For more information about system configuration and performance, please see: (https://manual.gromacs.org/current/user-guide/mdrun-performance.html).

```
tar xfz gromacs-5.0.7.tar.gz
cd gromacs-5.0.7
mkdir build
cd build
cmake .. -DGMX_BUILD_OWN_FFTW=ON -DREGRESSIONTEST_DOWNLOAD=ON
make
make check
sudo make install
source /usr/local/gromacs/bin/GMXRC
```

Fig. 6 Commands to run when executing the quick and dirty installation of GROMACS.

3.1.3.2 Programing the MD simulation

PDB structure

(a) Go to Protein Data Bank and download (https://www.rcsb.org/) a target structure or use your theoretical molecular model generated by comparative modeling. Here cysteine-rich defensin VrD1 (pdb: 1tl5) was used as an example

Checking scripts

(a) Check the necessary scripts that will be used throughout the MD parametrization. The scripts are:
ions.mdp—to generate ions counting to parametrize the system;
minim.mdp—system energy minimization;
nvt.mdp—temperature parametrization;
npt.mdp—pressure parametrization;
md.mdp—to generate the file that will execute md.tpr.

(b) Open the terminal command on Linux/Ubuntu, then source GROMACS using the following command line:

source/usr/local/gromacs/bin/GMXRC.

3.1.3.3 Start programming your MD simulation

(a) First, it is necessary to generate GROMACS coordinate files (.gro) and topology using as input the .pdb file downloaded from PDB or obtained from molecular modeling. The pdb2gmx tool generates a GROMACS atomic coordinate (.gro) from the input .pdb file. To map a disulfide bond, use the flag: -ss (they will be automatically identified). The other parameters refer to the solvent type (e.g., water: flag -water spce), net-charge (neutral; protonation: flag: -inter), N- and C-terminus modifications (flag: -ter), and polar hydrogen addition or removal (flag: -ignh). For more information, access (https://manual.gromacs.org/archive/5.0.4/programs/gmx-pdb2gmx.html).

(b) Type the following command:

gmx pdb2gmx -f 1ti5.pdb -o 1ti5.gro -ignh -inter -ter -water spce -ss

(c) Next, a list of force fields will pop up (Fig. 7). Select the force field that suits best the kind of macromolecules and system you want to simulate. The GROMOS96 43a force field is commonly used for linear and cyclic peptides and, therefore, it was selected for the methodology described here. Select option 9 for GROMOS96 43a

(d) Next, it will be necessary to protonate the peptide/protein. Register how many residues will be protonated. This data will be used in the following steps

(e) The following step includes the simulation box generation. The command (flag: -editconf) converts generic structure from .gro, (flag: -o) to .pdb. To set the geometric center of the system, use (flag: -c), to center the system in the box, use (flag: -d) followed by the measure of the

```
Select the Force Field:
From '/usr/local/gromacs/share/gromacs/top':
 1: AMBER03 protein, nucleic AMBER94 (Duan et al., J. Comp. Chem. 24, 1999-2012, 2003)
 2: AMBER94 force field (Cornell et al., JACS 117, 5179-5197, 1995)
 3: AMBER96 protein, nucleic AMBER94 (Kollman et al., Acc. Chem. Res. 29, 461-469, 1996)
 4: AMBER99 protein, nucleic AMBER94 (Wang et al., J. Comp. Chem. 21, 1049-1074, 2000)
 5: AMBER99SB protein, nucleic AMBER94 (Hornak et al., Proteins 65, 712-725, 2006)
 6: AMBER99SB-ILDN protein, nucleic AMBER94 (Lindorff-Larsen et al., Proteins 78, 1950-58, 2010)
 7: AMBERGS force field (Garcia & Sanbonmatsu, PNAS 99, 2782-2787, 2002)
 8: CHARMM27 all-atom force field (CHARM22 plus CMAP for proteins)
 9: GROMOS96 43a1 force field
10: GROMOS96 43a2 force field (improved alkane dihedrals)
11: GROMOS96 45a3 force field (Schuler JCC 2001 22 1205)
12: GROMOS96 53a5 force field (JCC 2004 vol 25 pag 1656)
13: GROMOS96 53a6 force field (JCC 2004 vol 25 pag 1656)
14: GROMOS96 54a7 force field (Eur. Biophys. J. (2011), 40,, 843-856, DOI: 10.1007/s00249-011-0700-9)
15: OPLS-AA/L all-atom force field (2001 aminoacid dihedrals)
```

Fig. 7 List of force filed that can be chosen depending on the user's needs.

distance from solute and the box (e.g., 1.0). To define the box shape (e.g., cubic, dodecahedral, octahedral, among others), use the command (flag: -bt). For more information, access (https://manual.gromacs.org/archive/5.0.4/programs/gmx-editconf.html).

(f) The final command should be something as described below:

gmx editiconf -f 1ti5.gro -o box.gro -c -d 1.0 -bt cubic

(g) After preparing the simulation box, proceed to the system's solvation. For this, use the flag: -solvate, followed by the flag: -cp to specify the output files from the previous step (editconf). The flag: -cs is used to specify the solvent, which will be the Simple Point Charge water (SPC) using the flag: -spc216.gro. Nevertheless, other co-solvents can also be added to the system, including 2,2,2-trifluoroethanol and methanol

(h) The final command should be something as described below:

gmx solvate -cp box.gro -cs spc216.gro -o solv.gro -p topol.top

(i) The topology file (topol.top) will be updated constantly from this step, as this file contains information about molecule types and the number of molecules

(j) An additional input file with the molecular dynamics parameter file extension (.mdp) will be used to produce a .tpr file with grompp. Grompp assembles specified parameters from the .mdp file with the coordinates and topology information to generate a .tpr file. Type the following command:

gmx grompp -f ions.mdp -c solv.gro -p topol.top -o ions.tpr

(k) With the atomic-level description of the system in the binary file ions.tpr, use the genion command. To add the necessary ions to neutralize the system's net charge by adding the correct number of negative ions (Na+) or positive ions (Cl-), use the flag: -neutral. Moreover, if a given ionic strength is desired, for instance, 0.15 M NaCl, use the flag: -conc 0.15

(l) For more information, please visit (https://manual.gromacs.org/archive/5.0.4/programs/gmx-genion.html).

The final command should be something as described below:

gmx genion -s ions ions.tpr -o solv_ions.gro -p topol.top -neutral or

gmx genion -s ions.tpr -o solv_ions.gro -p topol.top -neutral -conc 0.15

For each ion that is added to the system, one solvent molecule will be removed.

3.1.3.4 MD energy minimization

(a) Energy minimization steps are used to ensure no steric clashes or inappropriate geometry in the system. A process called energy minimization

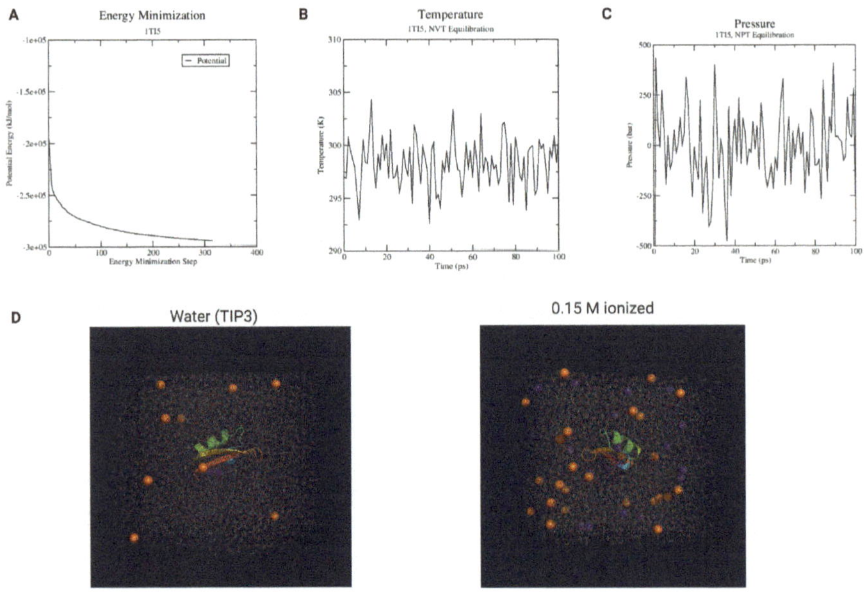

Fig. 8 (A) The plot is represented by an energy potential that results from energy minimization, demonstrating that energy is converging. (B) Plot generated showing the temperature progression of the system according to the number of steps set in the nvt.mdp script. The graph was built using xmgrace. (C) Plot generated using xmgrace shows the system's pressure equilibration, which is usually set at 1 bar. And (D) cubic boxes containing the target cysteine-rich peptide, the TIP3 water molecules and one system only with the necessary ions to neutralize the system's charge (left); whereas the other system (right) shows the same system, however at 0.15 M ionic strength. Both systems had their energies minimized, as described above. Created with the support of Biorender (BioRender.com).

is applied for structure relaxing (Fig. 8A and D). For more information visit: (https://manual.gromacs.org/archive/5.0.4/programs/gmx-grompp.html) and (https://manual.gromacs.org/archive/5.0.4/programs/gmx-mdrun.html).

(b) Run the following command:

gmx grompp -f minim.mdp -c solv_ions.gro -p topol.top -o em.tpr.

gmx mdrun -v -deffnm em

(c) The potential energy graph can be plotted by the following command:

gmx energy -f em.edr -o potential.xvg.

Select the option: Potential.

(d) Then open the plot using xmgrace software. For more information: (https://plasma-gate.weizmann.ac.il/Grace/).

3.1.3.5 MD equilibration

(a) Equilibration is commonly conducted in two steps. The first is conducted under an NVT ensemble (constant Number of particles, Volume, and Temperature). This ensemble can also be referred as "isothermal-isochoric" or "canonical" (Fig. 8B). The system's temperature should reach a plateau at the desired value. For more information, see: (https://manual.gromacs.org/documentation/2018/user-guide/mdp-options.html).

(b) To run temperature equilibration using the nvt.md script type the following commands:

gmx grompp -f nvt.mdp -c em.gro -p topol.top -o nvt.tpr.

gmx mdrun -v -deffnm nvt

(c) To plot the temperature graph, run the following command:

gmx energy -f nvt.edr -o temperature.xvg

Select the Temperature option and open the. xvg output file using xmgrace

(d) Pressure equilibration is the second step conducted under an NPT ensemble, as the number of particles, pressure, and temperature are constant. The ensemble is also called the "isothermal-isobaric" ensemble and closely resembles experimental conditions (Fig. 8C). For more information, please visit (https://manual.gromacs.org/documentation/2018/user-guide/mdp-options.html).

(e) To generate NPT files run the following commands:

gmx grompp -f npt.mdp -c nvt.gro -t nvt.cpt -p topol.top -o npt.tpr

gmx mdrun -v -deffnm npt

(f) To plot the pressure graph, run the following command:

gmx energy -f npt.edr -o pressure.xvg

Select the Pressure option and open the. Xvg output file using xmgrace.

3.1.3.6 MD simulation

(a) After minimization and equilibration steps, the next step is to release the position restraints and run the MD simulation. It is a similar process to the commands described above for NVT and NPT. For this representative simulation, we will run 10 ns of MD simulation. Therefore, it is necessary to edit script md.mdp and alter the value in nsteps parameter (Fig. 9).

(b) Then execute the following commands:

gmx grompp -f md.mdp -c npt.gro -t npt.cpt -p topol.top -o md.tpr.

gmx mdrun -v -deffnm md.tpr

```
; Run parameters
integrator      = md            ; leap-frog integrator
nsteps          = 50000         ; 2 * 500000 = 1000 ps (1 ns)
dt                  = 0.002             ; 2 fs
; Output control
nstxout                 = 5000          ; save coordinates every 10.0 ps
nstvout                 = 5000          ; save velocities every 10.0 ps
nstenergy               = 5000          ; save energies every 10.0 ps
nstlog                  = 5000          ; update log file every 10.0 ps
```

Fig. 9 Representation of the script md.mdp, where the nsteps parameter can be altered according to the time scale desired for a give MD simulation.

(c) The command would be like:

gmx mdrun -v -nice 0 -nb gpu -deffnm md.

3.1.3.7 MD simulation analyzes

(a) A critical analysis to be done to evaluate the peptide's trajectory during the simulation. For this purpose, we can calculate the peptide's root mean square deviation (RMSD).

(b) However, before starting any MD analysis, it is necessary to perform post-processing using the GROMACS tool trjconv.

(c) Run the following command:

gmx trjconv -s md.tpr -f md.xtc -o md_noPBC.xtc -pbc mol -center -ur compact.

To place the peptide at the center of the simulation box, select 1 and 0 (zero). Now the user has the post-processed file md_noPBC.xtc to carry out further analyzes.

(d) To run RMSD analysis execute the following command:

gmx rms -s md.tpr -f md_noPBC.xtc -o rmsd.xvg -tu ns.

Select option 4 (backbone). The resulting .xvg file can be opened using xmgrace for RMSD evaluation (Fig. 10A).

(e) The root-mean-square-fluctuation (RMSF) computes the fluctuation of atoms' positions in the peptide's trajectory. To obtain this result execute the following command:

gmx -rmsf -s md.tpr -f md_noPBC.xtc -o -res.

Select option 4. The resulting .xvg file can be opened using xmgrace for RMSF evaluation (Fig. 10B).

(f) MD simulations can also be evaluated by checking the structures' snapshots during the time scale simulated.

For this, run the following command:

gmxconv -s md.tpr -f md_noPBC.xtc -b 2000 -e 2000 -o pep_2ns.pdb.

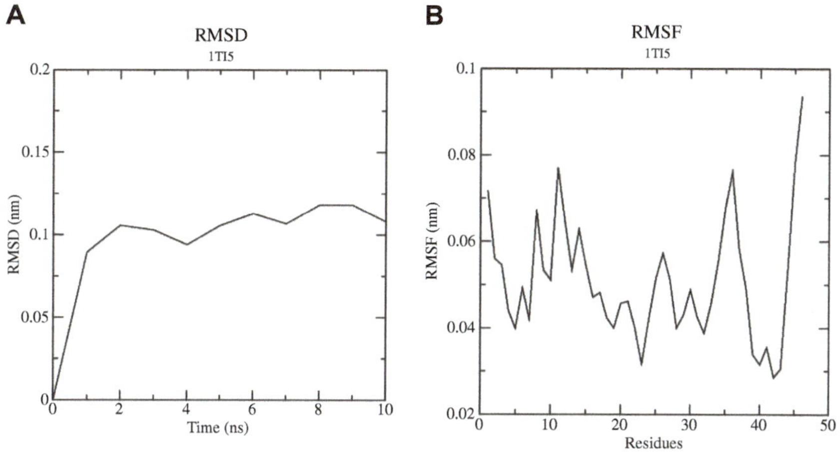

Fig. 10 Demonstrated MD results. (A) RMSD analysis for PDB ID: 1ti5 after 10 ns of MD simulation in water under 0.15 M ionic strength. (B) RMSF plot revealing the fluctuation for each amino acid residue that constitutes the PDB ID 1ti5 structure. Created with the support of Biorender (BioRender.com).

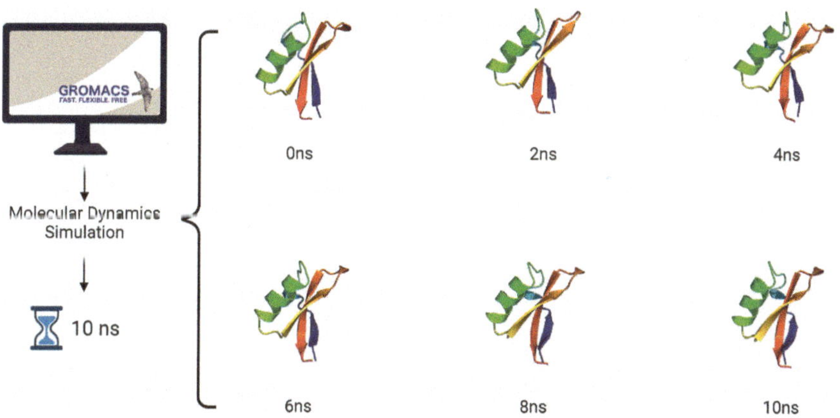

Fig. 11 Structural snapshots of PDB ID: 1ti5 during 10 ns, at 2 ns intervals, generated by molecular dynamics simulation using GROMACS.

The flags -b and -e indicate the initial and final time scale, respectively, from which the 3D structures will be generated. Usually, the flags -b and -e present the same number in picoseconds (ps) to evaluate a specific structural arrangement for a given time point. Different snapshots can be generated to evaluate the peptide's structural modifications at a 3D level (Fig. 11).

For more information on how to program MD simulations with different molecules and solvents, please access: http://www.mdtutorials.com/gmx/lysozyme/06_equil.html.

3.2 Peptide's synthesis

The candidate peptides screened can be synthesized by stepwise solid-phase using the N-9-fluorenylmethyloxycarbonyl (Fmoc) strategy and purified by high-performance liquid chromatography (HPLC), with purity >95% according to Chan and White (1999).

3.3 Activity evaluation *in vitro*

After peptide synthesis, it is necessary to evaluate the biological activities for the lead Cs-AMPs based drug candidates. Here we provide the readers with a description of the methods to evaluated antimicrobial, antifungal and antiviral activity.

3.3.1 Antimicrobial (antibacterial and antifungal) activity evaluation

The methods described below are based on ISO (2019), CLSI (2018), CLSI (2017) and Wiegand, Hilpert, and Hancock (2008).

3.3.1.1 Bacterial cell preparation

Timing: ~3 days

(a) Streak bacterial isolates onto nutrient-rich Mueller–Hinton agar plates to obtain individual colonies, according to Wiegand et al. (2008).

(b) Incubate plates for 18–24 h at 37 °C

(c) Select three to five isolated colonies from the fresh agar and inoculate in one sterile falcon tube containing 5 mL of Mueller–Hinton broth

Note: Use a sterile swab

(d) Incubate the falcon tubes for 18–24 h in a shaker at 200 rpm at 37 °C

(e) Pipette 50 μL of the overnight culture into 4.95 mL

(f) Incubate at 37 °C until the mid-log phase is reached

(g) Adjust bacterial suspension to 1×10^8 CFU mL^{-1}

Note: Adjustment should be made to McFarland Standard 0.5 (absorbance at 600 nm close to 0.08–0.13).

3.3.1.2 Broth microdilution for antibacterial susceptibility testing

Timing: ~ 2 days.

(a) Remove a 96-well sterile microplate from its packing.

Critical: To avoid contamination of the plate, keep the lid closed when not handling it.

(b) Label the microplate (Fig. 12). Use one column for each assay with up to eight different dilutions.

(c) Pipette 50 μL of Mueller–Hinton broth into row B–H of the microplate. Add 100 μL into column 12 (sterility control).

(d) Dilute the concentrated peptide stock solution into Mueller–Hinton broth in microcentrifuge tubes.

(e) Add 100 μL of the twofold concentrated peptide solution into the wells in row A.

(f) Withdraw 50 μL from each well in row A and add this to the wells in row B. This makes row B a two-fold dilution of row A. Mix carefully.

(g) Repeat the procedure down to row H. Discard the withdrawn solution from row H.

(h) Homogenize the bacterial suspension adjusted to 1×10^8 CFU mL^{-1} from Step 1(g) and dilute it at 1:100 to reach a concentration of 1×10^6 CFU mL^{-1}.

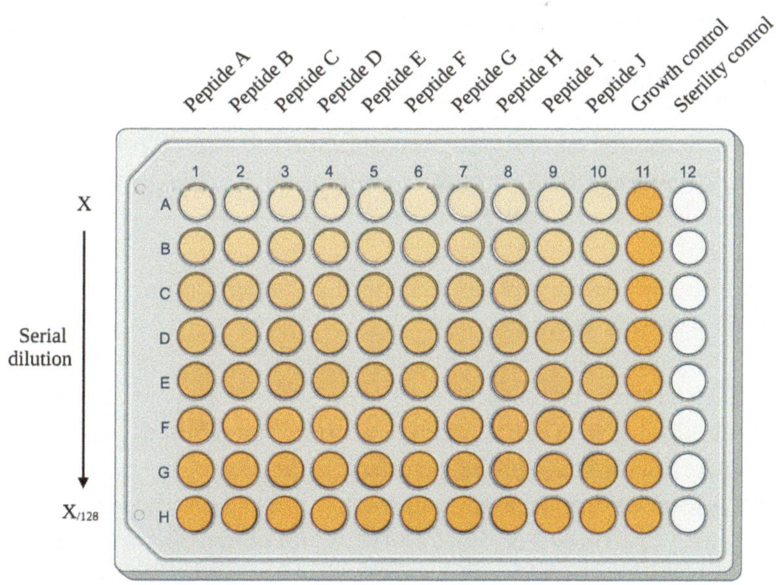

Fig. 12 Outline of the setup of a microplate for antimicrobial susceptibility testing of 10 different antimicrobial peptides. X, the highest concentrations tested; growth control (broth with bacterial inoculum, no peptide); sterility control (broth only). Created with the support of Biorender (BioRender.com).

(i) Inoculate each well containing the peptide solution and the growth control column with 50 μL of the bacterial suspension (row A–H, column 1–11). The final inoculum is 5×10^5 CFU mL^{-1}.
(j) Incubate microplate at 37 °C for 16–20 h, or up to adequate growth is obtained. Critical: Make sure the volume in the wells does not evaporate during incubation.
(k) Check bacterial growth inhibition using a microplate reader at 600 nm.
(l) The minimal inhibitory concentration (MIC) can be described as the lowest concentration of peptide that presents an absence in bacterial growth.

3.3.1.3 Minimal bactericidal concentration assay

Timing: ~ 1 day

(a) After determining the MIC, remove 10 μL from each well with no bacteria growth and plate in Mueller–Hinton agar plates
(b) Incubate plates for 16–20 h at 37 °C
(c) Count colonies, if possible
(d) The minimal bactericidal concentration (MBC) is the concentration at which no colonies (bacterial growth) are observed

3.3.1.4 Time-kill assay

Timing: it will depend on the chosen time points, ranging from 1 to 24 h or more.

(a) After determining MBC in Step 3, dilute the antimicrobial peptide in sterile water (or PBS) until a concentration is achieved that is 10-fold concentrated than MBC
(b) Settings:

1. Monotherapy with peptide at a specific concentration (1 MBC and 4 MBC is usually used);
2. Growth control (bacteria without peptide);
3. Sterility control (broth only).

Note: It is possible to evaluate different peptides and different concentrations.

(c) Prepare falcon tubes per experiment settings
(d) Each tube of monotherapy must contain the desired concentration of peptide (1 MBC or 4 MBC) and a final inoculum of 5×10^5 CFU mL^{-1} (CLSI, 2018). Growth control bacteria only, and sterility control broth only. All tubes must have a final volume of 5 mL
(e) For time point 0 (starting point), withdraw 100 μL from each falcon tube, serially dilute 10-fold with sterile 0.9% saline, and spread out for colony counts on Mueller–Hinton agar plates

(f) Incubate plates for 18–24 h at 37 °C
(g) Incubate all falcon tubes at 37 °C
(h) For each time point chosen, repeat as Step 4 (e–f).
(i) After 18–24 h of incubation, the colonies can be evaluated and values plotted to obtain the growth kinetics as a function of time.

3.3.1.5 Preparation of the fungi suspension

Timing: ~3–4 days.

Note: Specific assay to yeast-like fungi

(a) Streak the fungi isolates onto growth medium potato dextrose agar plates to obtain isolated colonies
(b) Incubate plates for 24–48 h at 28–35 °C

Note: The temperature can vary, depending on the isolate

(c) Select few colonies from the agar plate and inoculate a sterile falcon tube with 5 mL of RPMI-1640 medium

Note: Use sterile material.

(d) Incubate the falcon tubes for 24 h in a shaker at 200 rpm at 28–35 °C
(e) Adjust fungi suspension to 5×10^3 CFU mL^{-1} (CLSI, 2017). This will be the active suspension

Note: Appropriate dilutions are suggested to obtain the right concentration for counting in a cell-counting hemocytometer (Neubauer chamber). Additionally, for some fungi species, it is suggested that the suspension be made in a 0.01–0.1% Tween 80 solution so that the count can be done correctly.

3.3.1.6 Broth microdilution for antifungal susceptibility testing

Timing: ~ 2 days.

Note: Specific assay to yeast-like fungi

(a) Remove a 96-well sterile microplate from its packing

Critical: Keep the lid closed when not handling the plate to prevent contamination.

(b) Label the microplate with the respective peptide dilutions (Fig. 12). Use one column for each test with up to eight different dilutions
(c) Pipette 50 µL of RPMI-1640 medium into row B—H of the microplate. Add 100 µL into column 12 (sterility control).
(d) Dilute the concentrated peptide stock solution into RPMI-1640 medium in microtubes
(e) Add 100 µL of the twofold concentrated peptide solution into the wells in row A

(f) Withdraw 50 μL from each well in row A and add this to the wells in row B. This makes row B a two-fold dilution of row A. Mix carefully

(g) Repeat the procedure down to row H. Discard the withdrawn solution from row H

(h) Inoculate each well containing the peptide solution and the growth control column with 50 μL of the active suspension from Step 5(e) (row A–H, column 1–11). This results in the final desired inoculum of 2.5×10^3 CFU mL^{-1}

(i) Incubate the microplate at 28–35 °C for 16–20 h

Critical: Make sure that the volume in the wells does not evaporate during incubation.

(j) Verify fungi growth inhibition by optical density using a microplate reader at 530 nm

(k) The minimal inhibitory concentration (MIC) can be described as the lowest concentration of peptide that presents an absence in fungi growth (CLSI, 2017).

3.3.1.7 Expected outcome

The MIC can be described as the lowest concentration of the peptide, which can inhibit the growth of the tested isolate (Fig. 13). On the other hand, it is recorded as less than or equal to the lowest concentration when it can interrupt growth in any wells, except growth control. The MBC can be

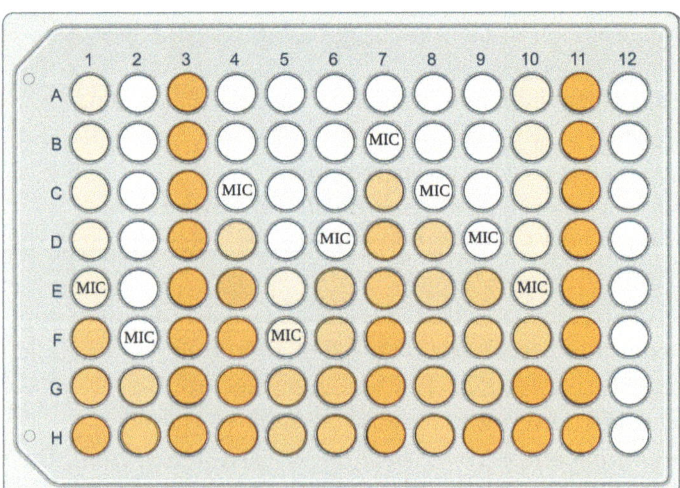

Fig. 13 Outline of possible growth patterns in MIC microplates. Created with the support of Biorender (BioRender.com).

described as the concentration of the antimicrobial substance at which no colonies growth is observed. Determining the MBC it is possible to characterize an antimicrobial substance as bactericidal, where it kills the bacteria, or bacteriostatic, where bacterial cell division is interrupted, but the cells remain viable.

3.3.1.8 Advantages

Antibacterial and antifungal activity evaluation, including MIC, MBC and time-kill assays, can be used to monitor the development of antimicrobial drug resistance. MIC distribution databases are accessible for relevant species–drug associations. Microorganisms that are positively selected and with a particular resistance profile can be easily identified by presenting MIC values higher than the epidemiological cut-off value. Even small changes can become clinically relevant and, therefore, MIC determination is a measurement for resistance surveillance. Indeed, for new drug candidates, including Cs-AMPs, MIC and time-kill assays are the initial steps to assess antimicrobial potential.

3.3.1.9 Limitations

The main limitation of the MIC and time-kill is related to the bacterial load compared to the load found in infection sites since usually about 100 × less is used. In the time-kill, a limitation, in addition to the lower bacterial load, is that the action time can be overestimated or underestimated since, in a more complex system, in which the immune system is involved, this time can be shorter or longer. Furthermore, the MIC does not make it possible to consider more complex biological parameters, such as the possibility of degradation of the peptides before their action on the microorganism.

3.3.2 Antiviral activity evaluation

3.3.2.1 Reagents' preparation: Prior testing, make sure all the reagents are at hand

Timing: ∼ 1 day.

(a) For Crystal violet stain: 1% Crystal Violet in 20% ethanol and distilled H2O.

(b) Formaldehyde solution at 10% for fixation, diluted in distilled H2O.

Critical: Use proper personal protective equipment and ventilation while handling formaldehyde.

(c) Prepare a CMC overlays stock solution at 2% in dH2O. Mix with a sitting bar and autoclave for complete dissolve.

(d) Prepare 500 mL DMEM, diluted in ultrapure H2O, add 3.7 g of sodium bicarbonate, and filter using a 0.2 μm filter. After filtration, supplement with 1% L-Glutamine and 1% penicillin/streptomycin. Keep in the refrigerator.

Note:

1. For cell growth media, supplement with 10% Fetal Bovine Serum (FBS) and ultrapure autoclaved H_2O for 1 × solution.
2. For viral assays, add ultrapure autoclave H_2O for 1 × solution.
3. For CMC overlay media, use a 1:1 mixture of warmed 2 × DMEM and CMC stock solution, supplement 2% FBS.

3.3.2.2 Determination of viral titers based on Reed and Muench method

Timing: ~ 3 days.

(a) On 96-well plates, confirm cells monolayers' confluence (~90%) and viability on the inverted microscope (Fig. 14) before starting the assay.

(b) After confirmed confluence and viability, remove growth media and add 100 μL in each well DMEM 1 × without serum.

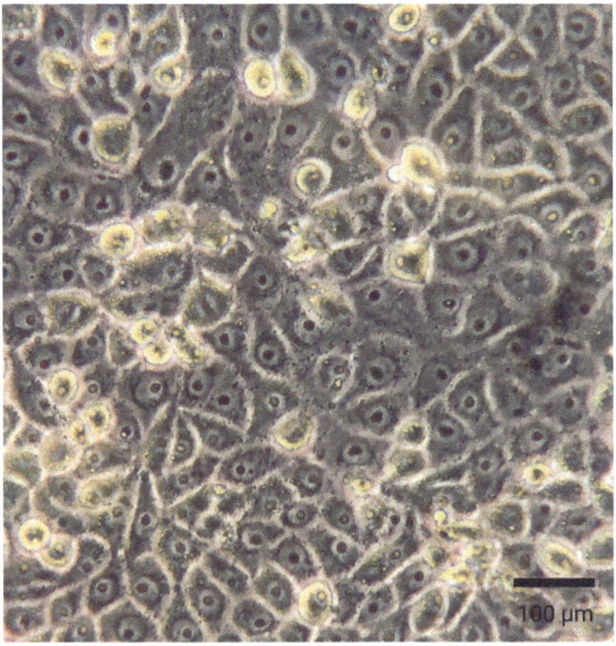

Fig. 14 Vero cells in culture showing over 90% of confluency. Magnification 40 ×.

(c) Prepare a 10-fold serial dilution of the viral sample using the DMEM 1 × without serum as diluent.
(d) In six-well plates, add 25 μL of each viral dilution.
(e) Place the 96-well inside a 5% CO_2 incubator, at 37 °C, for 48 h.
(f) After the incubation period, analyze each well for viral cytopathic effect. If present, read as positive (+). If not, read as negative (−). Subsequently, perform the calculation as proposed by Reed and Muench (1938) to determine the 50% tissue culture infective dose ($TCID_{50}$).

3.3.2.3 Determination of viral titers based on plaque assay

Timing: ~ 15 days

(a) On 6 or 24-well plates, confirm cell monolayers' confluence (~90%) and viability on the inverted microscope before starting the assay. Remove the media before inoculum.
(b) Prepare a 10-fold serial dilution of the viral sample using the DMEM 1 × without serum as a diluent (Fig. 15A).
(c) Add 500 μL (when using six-well plates) or 200 μL (when using 24-well plates) of each viral dilution (in triplicate) and incubate the plates in a

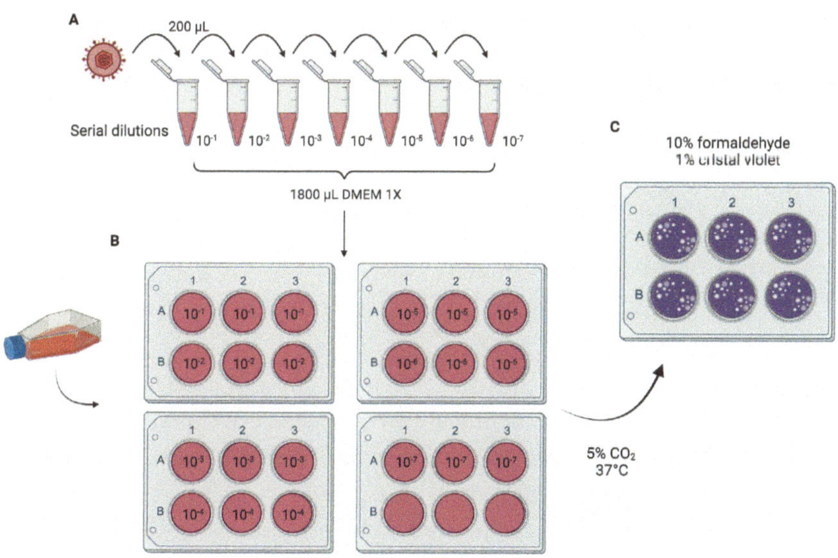

Fig. 15 Determination of viral titers based on plaque assay. (A) Serial dilution of the viral sample using the DMEM 1 ×. (B) Distribution of dilutions in six-well plates and incubation in a 5% CO_2 incubator. (C) Fixation with formaldehyde solution and stained with crystal violet. Created with the support of Biorender (BioRender.com).

5% CO_2 incubator, at 37 °C, for 1 h (Fig. 15B). Move the plates 4–5 times to spread the inoculum during this period and keep the cells from drying.

(d) After the incubation period, remove the inoculum and add, very carefully, 4 mL of CMC overlay media. Incubate again in a 5% CO_2 incubator at 37 °C. The overlay media will prevent the virus from spreading, therefore, forming distinct plaques visually countable. Plaque formation can take 2–14 days, depending on the type of virus.

Critical: Once the overlay media are placed, be careful when moving the plates, as there is a risk of displacing the cell monolayer.

(e) Subsequently, add 3 mL of 10% formaldehyde fixation solution. Keep at room temperature to a minimum of 1 h and 30 min or overnight (Fig. 15C). Remove the fixation solution + CMC overlay media and wash it with water.

(f) Add 600 μL (when using six-well plates) or 100 μL (when using 24-well plates) of crystal violet stain. Keep it moving for 15–20 min for complete coverage. Remove and wash the stain with water.

(g) Once the plates are fixed, stained, and dried, they can be stored indefinitely.

(h) Perform plaque counting of each well, taking the average plaques, and apply them to the following formula. (Note: Make sure the wells with fewer than five or greater than 100 plaques are discarded) (Baer & Kehn-Hall, 2014)

$$PFU/mL = \frac{Avg.of\ plaques}{DxV}$$

D = dilution V = volume of diluted virus added to the plate.

3.3.2.4 Antiviral testing

Timing: ~2 days.

(a) On 24-well plates, confirm cells monolayers' confluence (~90%) and viability on the inverted microscope before starting the assay. Remove the media before inoculum.

(b) Prepare the sample on the desired concentration using the DMEM 1 × without serum as diluent. Add 1 mL of sample in one well.

(c) Prepare viral suspension at 100 $TCID_{50}$ using the DMEM 1 × without serum as diluent. Add 100 μL of the suspension in each well used in the test, including the positive viral control.

(d) Incubate the plate in a 5% CO_2 incubator, at 37 °C, for 48 h. After this period, check the wells for the presence of the viral cytopathic effect. Perform a freeze and thaw cycle to lysate the cells and release all viral particles.

Critical: When performing the freeze and thaw cycle, remember to freeze the plate slowly (4 °C, −20 °C, −80 °C), and thaw rapidly (place the plate in the incubator at 37 °C).

(e) Next, prepare a 10-fold serial dilution of both the positive viral control and the sample tested using the DMEM 1 × without serum as diluent. Perform the plaque assay as described above

(f) To determine the percentage of inhibition (PI) of the sample, Nishimura, Toku, and Fukuyasu (1977) proposed the following equation:

$$1 - \frac{Vd}{Vc} x100$$

where Vd = the number of plaques in the presence of the sample; and Vc = the number of plaques in the absence of the sample (the viral positive control) (Nishimura et al., 1977).

3.3.2.5 Expected outcomes

The correct assessment of antiviral testing relies on the following factors: a well-maintained cell culture to be used as the host, with proper media and growth conditions for both cellular and viral multiplication, and adequate time for viral incubation period to allow distinct and countable plaque formation (Fig. 16). Therefore, achieving quality assays. Furthermore, it is to note that these methods can test only viruses that perform cytopathic effects. Thus, the cytopathic effects are expected on the positive viral control when looked at in the inverted microscope.

3.3.2.6 Advantages

The main advantages of antiviral testing in cell culture are the controlled environment, sample homogeneity and cheaper cost (Baer & Kehn-Hall, 2014). Moreover, the plaque reduction assay relies on fully mature viral particles, making it possible to quantify the number of infective virions (Cooper, 1962; Dulbecco & Vogt, 1953).

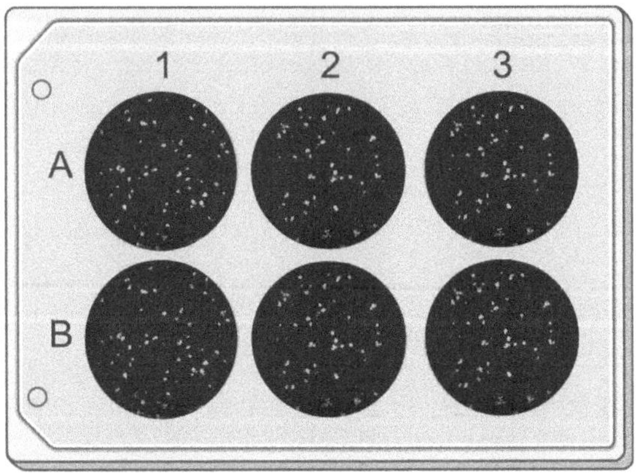

Fig. 16 The expected result from a plaque assay. Plaques should be clear enough to be counted. Created with the support of Biorender (BioRender.com).

3.3.2.7 Limitations

The limitations of antiviral testing in cell culture are that some viruses do not perform cytopathic effects when in culture or do not replicate at all (Dulbecco & Vogt, 1953). Also, *in vitro* testing in cell culture can be time-consuming.

4. Safety consideration

Execute all procedures following the local codes of safe practice. Standard personal protective equipment should follow institutional guidelines but should consist of at least a lab coat and gloves, which should be worn at all times. DMSO is a substance toxic to cells. Thus, its handling must be done carefully using suitable personal protective equipment (PPE).

Acknowledgment

This work was supported by Conselho Nacional de Desenvolvimento Científico e Tecnológico (CNPq), Coordenação de Aperfeiçoamento de Pessoal de Nível Superior (CAPES), Fundação de Apoio a Pesquisa do Estado de Mato Grosso do Sul (FUNDECT), Fundação de Amparo a Pesquisa do Distrito Federal (FAPDF).

References

Baer, A., & Kehn-Hall, K. (2014). Viral concentration determination through plaque assays: Using traditional and novel overlay systems. *Journal of Visualized Experiments: JoVE*, *93*, 1–10.

Bates, P. A., Kelley, L. A., MacCallum, R. M., & Sternberg, M. J. (2001). Enhancement of protein modeling by human intervention in applying the automatic programs 3D-JIGSAW and 3D-PSSM. *Proteins: Structure, Function, and Bioinformatics*, *45*(S5), 39–46.

Boas, L. C. P. V., Campos, M. L., Berlanda, R. L. A., de Carvalho Neves, N., & Franco, O. L. (2019). Antiviral peptides as promising therapeutic drugs. *Cellular and Molecular Life Sciences*, *76*(18), 3525–3542.

Cardoso, M. H., Oshiro, K. G. N., Rezende, S. B., Candido, E. S., & Franco, O. L. (2018). The structure/function relationship in antimicrobial peptides: What can we obtain from structural data? *Advances in Protein Chemistry and Structural Biology*, *112*, 359–384. https://doi.org/10.1016/bs.apcsb.2018.01.008.

Chan, W., & White, P. (1999). *Fmoc solid phase peptide synthesis: A practical approach. Vol. 222*. OUP Oxford.

CLSI. (2017). *Performance standards for antifungal susceptibility testing of yeasts*. Wayne, PA: Clinical and Laboratory Standards Institute.

CLSI. (2018). Methods for dilution antimicrobial susceptibility tests for bacteria that grow aerobically. In *CLSI standard M07* (11th ed.). Wayne, PA: Clinical and Laboratory Standards Institute.

Cock, P. J., Antao, T., Chang, J. T., Chapman, B. A., Cox, C. J., Dalke, A., et al. (2009). Biopython: Freely available Python tools for computational molecular biology and bioinformatics. *Bioinformatics*, *25*(11), 1422–1423. https://doi.org/10.1093/bioinformatics/btp163.

Cooper, P. (1962). *The plaque assay of animal viruses advances in virus research. Vol. 8* (pp. 319–378). Elsevier.

Costa, L. S. M., Pires, A. S., Damaceno, N. B., Rigueiras, P. O., Maximiano, M. R., Franco, O. L., et al. (2020). *In silico* characterization of class II plant defensins from Arabidopsis thaliana. *Phytochemistry*, *179*, 112511. https://doi.org/10.1016/j.phytochem.2020.112511.

de Oliveira Dias, R., & Franco, O. L. (2015). Cysteine-stabilized αβ defensins: From a common fold to antibacterial activity. *Peptides*, *72*, 64–72.

Dulbecco, R., & Vogt, M. (1953). Some problems of animal virology as studied by the plaque technique. In *Paper presented at the Cold Spring Harbor symposia on quantitative biology*.

Fiser, A., & Šali, A. (2003). Modeller: Generation and refinement of homology-based protein structure models. *Methods in Enzymology*, *374*, 461–491.

Fleitas Martínez, O., Cardoso, M. H., Ribeiro, S. M., & Franco, O. L. (2019). Recent advances in anti-virulence therapeutic strategies with a focus on dismantling bacterial membrane microdomains, toxin neutralization, quorum-sensing interference and biofilm inhibition. *Frontiers in Cellular and Infection Microbiology*, *9*, 74.

Huan, Y., Kong, Q., Mou, H., & Yi, H. (2020). Antimicrobial peptides: Classification, design, application and research progress in multiple fields. *Frontiers in Microbiology*, *11*, 582779. https://doi.org/10.3389/fmicb.2020.582779.

ISO. (2019). *Susceptibility testing of infectious agents and evaluation of performance of antimicrobial susceptibility test devices—Part 1: Broth micro-dilution reference method for testing the in vitro activity of antimicrobial agents against rapidly growing aerobic bacteria involved in infectious diseases*. Geneva, Switzerland: International Organization for Standardization.

Maximiano, M. R., & Franco, O. L. (2021). Biotechnological applications of versatile plant lipid transfer proteins (LTPs). *Peptides*, *140*, 170531.

McGuffin, L. J., Adiyaman, R., Maghrabi, A. H., Shuid, A. N., Brackenridge, D. A., Nealon, J. O., et al. (2019). IntFOLD: An integrated web resource for high performance protein structure and function prediction. *Nucleic Acids Research*, *47*(W1), W408–W413.

Nijnik, A., & Hancock, R. (2009). Host defence peptides: Antimicrobial and immunomodulatory activity and potential applications for tackling antibiotic-resistant infections. *Emerg Health Threats J.*, *2*, e1. https://doi.org/10.3134/ehtj.09.001.

Nishimura, T., Toku, H., & Fukuyasu, H. (1977). Antiviral compounds. XII. Antiviral activity of amidinohydrazones of alkoxyphenyl-substituted carbonyl compounds against influenza virus in eggs and in mice. *The Kitasato Archives of Experimental Medicine*, *50*(1–2), 39–46.

Porto, W. F., Pires, A. S., & Franco, O. L. (2017). Computational tools for exploring sequence databases as a resource for antimicrobial peptides. *Biotechnology Advances*, *35*(3), 337–349. https://doi.org/10.1016/j.biotechadv.2017.02.001.

Reed, L. J., & Muench, H. (1938). A simple method of estimating fifty per cent endpoints12. *American Journal of Epidemiology*, *27*(3), 493–497. https://doi.org/10.1093/oxfordjournals.aje.a118408.

Roy, A., Kucukural, A., & Zhang, Y. (2010). I-TASSER: A unified platform for automated protein structure and function prediction. *Nature Protocols*, *5*(4), 725–738.

Schwede, T., Kopp, J., Guex, N., & Peitsch, M. C. (2003). SWISS-MODEL: An automated protein homology-modeling server. *Nucleic Acids Research*, *31*(13), 3381–3385.

Shelenkov, A., Slavokhotova, A., & Odintsova, T. (2020). Predicting antimicrobial and other cysteine-rich peptides in 1267 plant transcriptomes. *Antibiotics (Basel)*, *9*(2). https://doi.org/10.3390/antibiotics9020060.

Shen, M. Y., & Sali, A. (2006). Statistical potential for assessment and prediction of protein structures. *Protein Science*, *15*(11), 2507–2524.

Silverstein, K. A., Moskal, W. A., Jr., Wu, H. C., Underwood, B. A., Graham, M. A., Town, C. D., et al. (2007). Small cysteine-rich peptides resembling antimicrobial peptides have been under-predicted in plants. *The Plant Journal*, *51*(2), 262–280. https://doi.org/10.1111/j.1365-313X.2007.03136.x.

Söding, J., Biegert, A., & Lupas, A. N. (2005). The HHpred interactive server for protein homology detection and structure prediction. *Nucleic Acids Research*, *33*(Suppl. 2), W244–W248.

Song, Y., DiMaio, F., Wang, R. Y.-R., Kim, D., Miles, C., Brunette, T., et al. (2013). High-resolution comparative modeling with RosettaCM. *Structure*, *21*(10), 1735–1742.

Srivastava, S., Dashora, K., Ameta, K. L., Singh, N. P., El-Enshasy, H. A., Pagano, M. C., et al. (2021). Cysteine-rich antimicrobial peptides from plants: The future of antimicrobial therapy. *Phytotherapy Research*, *35*(1), 256–277.

Tomczak, A., Sontheimer, J., Drechsel, D., Hausdorf, R., Gentzel, M., Shevchenko, A., et al. (2012). 3D profile-based approach to proteome-wide discovery of novel human chemokines. *PLoS One*, 7(5), e36151. https://doi.org/10.1371/journal.pone.0036151.

Van Rossum, G. (2020). *The Python library reference, release 3.8. 2. Vol. 16*. Python Software Foundation.

Webb, B., & Sali, A. (2016). Comparative protein structure modeling using MODELLER. *Current Protocols in Bioinformatics*, *54*(1), 5.6.1–5.6.37.

Wiegand, I., Hilpert, K., & Hancock, R. E. (2008). Agar and broth dilution methods to determine the minimal inhibitory concentration (MIC) of antimicrobial substances. *Nature Protocols*, *3*(2), 163–175. https://doi.org/10.1038/nprot.2007.521.

Xu, J., Mcpartlon, M., & Li, J. (2021). Improved protein structure prediction by deep learning irrespective of co-evolution information. *Nature Machine Intelligence*, *3*, 601–609. https://doi.org/10.1038/s42256-021-00348-5.

Zhu, S. (2008). Discovery of six families of fungal defensin-like peptides provides insights into origin and evolution of the CSalphabeta defensins. *Molecular Immunology*, *45*(3), 828–838. https://doi.org/10.1016/j.molimm.2007.06.354.

CHAPTER FIVE

Exploring synergy and its role in antimicrobial peptide biology

Matthew Howell, Antonina K. Wenc, Caroline M. Donaghy, Devon V. Wasche, Izabela Abissi, Marvin D. Naing, Scott Pierce, and Alfredo M. Angeles-Boza*

Department of Chemistry and Institute of Materials Science, University of Connecticut, Storrs, CT, United States

*Corresponding author: e-mail address: alfredo.angeles-boza@uconn.edu

Contents

Methods in Enzymology, Volume 663
ISSN 0076-6879
https://doi.org/10.1016/bs.mie.2021.09.017

Abstract

Antimicrobial peptides will be an essential component in combating the escalating issue of antibiotic resistance. Identifying synergistic combinations of two or more substances will increase the value of these peptides further. Several potential pitfalls in conducting synergy testing with peptides are discussed in detail. As case studies, we describe observations of AMP synergy with peptides, antibiotics, and metal ions as well as some of the mechanistic details that have been uncovered. The Bliss and Loewe models for synergy are presented prior to recommending protocols for conducting checkerboard, minimal inhibitory concentration, and time-kill assays. Establishing mechanisms of action and exploring the potential for resistance will be crucial to translate these studies into the clinic.

1. Introduction

There is a global crisis of antibiotic resistance. Many of these infections are acquired at alarming rates in hospital settings (Rosenthal et al., 2012). In spite of the challenge, many are optimistic that antimicrobial peptides (AMPs) will be an effective tool in responding to this quandary. Micromolar quantities of these peptides have routinely been observed to have potent antimicrobial effects. It has been argued that the affinity between cationic AMPs and anionic bacterial membranes results in a membrane-associated peptide concentration that is four orders of magnitude greater than its initial solution concentration (Melo, Ferre, & Castanho, 2009). These membrane-active peptides (a term that also includes the chemically similar β-amyloids and cell-penetrating peptides) can stabilize short-lived membrane defects, which can lead to pore formation and peptide translocation (Last, Schlamadinger, & Miranker, 2013). AMPs, as well as other peptidomimetics, often present great difficulty to microbes in developing resistance (Pollard et al., 2012).

It is perhaps for the reasons above that no fewer than 27 different AMPs have survived experimental scrutiny and advanced to clinical trials (Koo & Seo, 2019). The interest in using these peptides as therapeutic antimicrobials gained traction as the issue of resistance to conventional antibiotics became appreciated (Koo & Seo, 2019). This explains in part why the number of peptides in clinical development is presently small relative to conventional antibiotics. Additionally, the susceptibility of AMPs to degradation *in vivo* and the cost of synthesizing peptides have been obstacles to more widespread interest (Koo & Seo, 2019). Despite the greater expense relative to conventional antibiotics, peptides are a quickly growing fraction of the drug market

(Koo & Seo, 2019). And later, we will mention how researchers have chemically modified AMPs to improve *in vivo* efficacy.

As with conventional antibiotics, mechanisms of resistance for AMPs have been reported. Prolonged exposure can result in Gram-positive organisms adding lysine residues to membrane lipids (Peschel et al., 2001), while Gram-negative organisms can append aminoarabinose and ethanolamine to Lipid A (Hashemi et al., 2017). These small molecules are positively charged under physiological conditions. Some microbes produce membrane-associated cationic peptides which appear to repel AMPs (Vo, Spahn, Heilemann, & Bode, 2021). In general, it seems that reducing the Coulombic attraction between cationic AMP and anionic membrane decreases efficacy. Furthermore, AMPs can be degraded by endogenous peptidases (Sieprawska-Lupa et al., 2004). These mechanisms underscore the need to rely not solely on the potential of individual AMPs but also on synergy testing, which has been shown to present microbes an even more challenging task in acquiring resistance (Dobson, Purves, Kamysz, & Rolff, 2013).

Synergy is a phenomenon where a combination treatment of two (or more) substances results in an effect that surpasses the individual substances in isolation. We will discuss many of the details below, but one should intuitively appreciate combination treatments that result in more killing in time-kill assays or an equivalent inhibition in checkerboard assays at lower concentrations as practical examples of synergism. Of course, synergy can take place between several classes of compounds, not solely peptides. Below, we will consider synergy between AMPs, antibiotics, and metal ions as case studies. Synergistic combinations can help combat antimicrobial resistance by hitting a common molecular target (e.g., the membrane) more severely, by attacking distinct vulnerabilities, or by having the second substance address the mechanism of resistance that rendered the first substance ineffective. As an example of a synergistic combination in current use, the β-lactamase inhibitor clavulanic acid can be co-administered with a β-lactam antibiotic (e.g., amoxicillin) to overcome the widespread resistance to β-lactams (Girouard, Maclean, Ronald, & Albritton, 1981).

2. Considerations for activity testing in peptides

Since synergy testing is within the scope of assaying peptide activity in general, several factors must be considered in order for synergy testing to be a fruitful endeavor. Regardless of how the peptide is obtained (e.g., chemical synthesis or isolation from the appropriate organism), the researcher has an immediate need for establishing both concentration and purity. For many AMPs, quantification at 214 nm is the most feasible option for determining how much peptide is present (Kuipers & Gruppen, 2007). For peptides containing tryptophan or tyrosine, quantification at 280 nm is an approach that gives less variability (Gill & von Hippel, 1989; Pace, Vajdos, Fee, Grimsley, & Gray, 1995). Our recent work exploring the activity of peptide-ruthenium conjugates (Pierce, Jennings, Juliano, & Angeles-Boza, 2020) has provided an opportunity to quantify at the λ_{max} of the transition metal complex, arguably giving an even less variable measure of concentration. Of course, the gold standard involves hydrolyzing the peptide, derivatizing the resultant amino acids with a small molecule, and chromatographically comparing the derivatives to known standards (Bartolomeo & Maisano, 2006). This allows determination of the actual molar absorptivity, which can deviate from the predicted value. We recommend characterizing a peptide sample both by chromatography and mass spectrometry prior to performing biological assays. Much has been written (e.g., Jensen, Tofteng, & Pedersen, 2013) about the many side reactions that can take place during synthesis and beyond.

Many AMPs are cationic, while agar possesses negatively charged sulfate groups (BeMiller, 2018). Consequently, attempts at using radial diffusion assays to measure activity give small zones of inhibition (Lee, Cho, & Lehrer, 1997). We therefore discourage the use of these assays and the conceptually similar E-test (Doern, 2014) for cationic AMP synergy testing. We will not consider anionic AMPs in detail, and recommend Harris, Dennison, and Phoenix (2009) for a thorough review on that topic.

A further complication is the tendency of cationic peptides to stick to glass and plastic surfaces. Minimizing sample handling, using low-adsorption plastics (e.g., polypropylene), and preparing high concentration peptide stocks are essential strategies for avoiding sample loss (Kristensen, Henriksen, & Andresen, 2015). Some researchers have resorted to making peptide solutions in aqueous 0.01% (v/v) acetic acid and 0.2% (w/v) BSA (Steinberg et al., 1997) to deal both with the sample loss at low concentrations and also the precipitation of peptides at high concentrations in media, which is recommended for standard broth microdilution procedures (Wiegand, Hilpert, & Hancock, 2008).

Our lab has spent considerable effort in characterizing a group of peptides from *Styela clava*, the clavanins (Juliano et al., 2020; Juliano, Pierce, deMayo, Balunas, & Angeles-Boza, 2017). For researchers working with histidine-rich AMPs, it is important to note that their activity will generally be dependent on acidic conditions (Kacprzyk et al., 2007). Not only does the initial media pH concern the researcher, but so too does the pH of the media as the microbial population of interest expands. As an example, as *E. coli* grows in LB broth the media will become alkaline as the oligopeptides are metabolized and ammonia is released (Sezonov, Joseleau-Petit, & D'Ari, 2007). For many microbes, one strategy for maintaining acidic conditions is supplementing the media with glucose (Sánchez-Clemente et al., 2018).

The ionic strength of the medium is another important consideration, as increasing ionic strength has been shown to decrease the efficacy of many peptides (Lee et al., 1997). The ionic strength of MHB is 0.12 (Beggs & Andrews, 1976), while the ionic strengths of LB-Miller (10 g/L NaCl), LB-Lennox (5 g/L NaCl), and LB-Luria (0.5 g/L NaCl) are 0.17, 0.085, and 0.0085, respectively (Walczak et al., 2012). While MHB is most common for antimicrobial susceptibility testing, Yan and Hancock (2001) point out the ionic strength of MHB is sufficient to inhibit the activity of many AMPs. Nevertheless, the ionic strength of MHB is physiologically relevant (Liu, Poolman, & Boersma, 2017) and is likely why it is used for making clinical recommendations.

Our lab recently contributed to a comprehensive review of antimicrobial susceptibility testing for AMPs (Mercer et al., 2020). We direct the curious reader to that publication for a more thorough treatment. Furthermore, we assume the reader is already familiar with the checkerboard and time-kill assays. If this is not the case, see Doern (2014) for an introduction.

3. Synergy between AMPs

Perhaps one of the earliest demonstrations of peptide-peptide synergy is that of basic attacin and cecropin B, both produced by *Hyalophora cecropia* (Engström, Carlsson, Engström, Tao, & Bennich, 1984). More extensive studies include the work of Yan and Hancock (2001) who applied different pairings of bovine indolicidin, porcine protegrin 1, human LL-37, and bovine bactenecin. Yu, Baeder, Regoes, and Rolff (2016) took a combinatorial approach in investigating the potency of two- and three-agent combinations of an LL-37 derivative, indolicidin, pexiganan (a synthetic analog of magainin 2), and insect peptides cecropin A, melittin, and apidaecin. Experiments such as these have established that a plethora of synergistic combinations of AMPs exist.

In recent years, there has been a pivot from merely documenting the presence of synergy towards probing mechanistic details. As one example, we consider two peptides present in the skin of *Xenopus laevis*, PGLa and magainin 2. It has been shown recently that both Gly-7 and Gly-11 in PGLa are necessary for the observed synergy with magainin 2 (Zerweck et al., 2017). The authors propose that the key residues allow formation of a synergistic complex consisting of two PGLa and two magainin 2 peptides. This may relate to the work of Zou, Zhu, Tu, Wu, and Landry (2018), which argues that optimal peptide aggregation is necessary for potency. As aggregates become larger, the energy cost of membrane transposition increases and is accompanied by a decrease in antimicrobial activity. Dynamic light scattering measurements suggest that optimal aggregation is responsible for the enhanced activity of combining magainin 2 with the horseshoe crab peptide, tachyplesin 1 (Remington et al., 2020). A similar phenomenon has been observed with the temporins, isolated from *Rana temporaria*. Temporin L synergizes with both temporins A and B by preventing their lipopolysaccharide-dependent oligomerization, which likely facilitates permeation into the cytoplasm (Rosenfield, Barra, Simmaco, Shai, & Mangoni, 2006).

Rahnamaeian, Cytrynska, Zdybicka-Barabas, and Vilcinskas (2016) observed a synergistic effect of pore-forming peptides cecropin A and stomoxyn with the bumblebee peptide, abaecin, which associates with intracellular *E. coli* DNaK. This led the authors to speculate that a general strategy for developing synergistic combinations was pairing a pore-forming peptide with an intracellularly-acting peptide. The results of He, Starr, and Wimley (2015), which failed to document synergy between membrane permeabilizing AMPs and antibiotics, cast doubt on this generalization. We will address the experimental design of He et al. (2015) later, but concede that there is perhaps no general pairing that reliably predicts synergy.

Hanson et al. (2019) used CRISPR to generate a series of *Drosophila melanogaster* mutants lacking one or more AMPs prior to infection with different microbes. These powerful *in vivo* studies demonstrated synergistic relationships between AMPs directed against particular microbes. Furthermore, these results suggest that individual AMPs are essential to resisting infection against specific microbes. Research groups with the ability to utilize genetic tools to explore *in vivo* consequences of removing AMPs should look to this work as a template. An underappreciated emphasis on the functional specialization of individual AMPs is echoed in a recent review by Lazzaro, Zasloff, and Rolff (2020).

4. Synergy between AMPs and antibiotics

Synergy between peptides and clinically relevant antibiotics has likewise been of great interest to researchers. An extensive study by Giacometti et al. (2000) examined multiple peptide-antibiotics pairings applied against clinical strains of *Stenotrophomonas maltophilia*. They considered three peptides: buforin 2 (produced by the toad, *Bufo gargarizans*), magainin 2, and cecropin P1 (produced by the nematode, *Ascaris suum*). They considered 10 antibiotics: several β-lactams (piperacillin, ceftazidime, and meropenem), the macrolide clarithromycin, the translation-inhibitor chloramphenicol, the tetracycline doxycycline, the transcription-inhibitor rifampin, the fluoroquinolone oflaxacin, the aminoglycoside netilmicin, and the cyclic peptide polymyxin E (also known as colistin). All three of the peptides were strongly synergistic with the β-lactams, clarithromycin, and polymyxin E. This study did not attempt to uncover why certain pairings were synergistic and others were not. Zharkova et al. (2019) observed a pattern, though not a rigid rule, between membrane-permeabilizing peptides and antibiotics with intracellular targets.

The convenience of reporting the results of incubating bacteria with antimicrobials is one reason why most synergy studies are *in vitro* or potentially *ex vivo*. In an example of the latter case, combination treatment of nisin (delivered in a biogel) and the antiseptic chlorhexidine was superior to other treatments in the inhibition of biofilm formation of *Staphylococcus aureus* isolates collected from patients with diabetic foot infections (Santos, Ruza, Cunha, Tavares, & Oliveira, 2019). There are a few successful *in vivo* studies that demonstrate the power of synergistic drug treatment. For example, Cirioni et al. (2018) used a combination treatment of pexiganan and tigecycline informed by checkerboard and time-kill assays to significantly reduce bacterial load in a mouse model infected with clinical isolates of *Pseudomonas aeruginosa*. Studies such as these will be instrumental in more combination treatments entering the clinical trial phase.

As with AMP-AMP synergy studies, more recent efforts have attempted to establish mechanistic insights. Gupta, Singh, and van Hoek (2015) examined AMPs capable of membrane permeabilization and depolarization in *Mycobacterium smegmatis*, which is closely related to *Mycobacterium tuberculosis*. They found synergistic pairings with antibiotics that acted either extracellularly or intracellularly, though no overarching narrative was presented. Rishi, Preet, Bharrhan, and Verma (2011) studied the effect of combination treatment of ampicillin and membrane-permeabilizing cryptdin 2 against *Salmonella enterica* serovar Typhimurium, another microbe capable of infecting and surviving within macrophages. After establishing *in vitro* synergy between the two agents using time-kill and checkerboard assays, *ex vivo* studies of infected macrophages showed that combination treatment potentiated killing of intracellular *Salmonella* and also decreased the lipid peroxidation which may be an early biomarker of tissue damage following the macrophage's attempt to eliminate the pathogen. Additionally, *Salmonella*-infected mice receiving combination treatment exhibited significantly lower bacterial loads in examined organs.

5. Synergy between AMPs and metal ions

Metal ions are a physiologically relevant candidate for synergism. Invertebrate cytoplasmic granules accumulate a plethora of different ions (Deb & Fukushima, 1999). *E. coli*, and likely other microbes as well, possesses a supply of copper ions (Outten & Munson, 2013) that copper-synergizing peptides could use to their advantage. Macrophages accumulate copper ions in response to inflammation signals that follow infection (White, Lee, Kambe, Fritsche, & Petris, 2009). Building off of this, Libardo et al. (2018) demonstrated that a peptide, DAB-10, could utilize this phagosomal copper to kill intracellular *M. tuberculosis*. And more generally, Walkenhorst, Sundrud, and Laviolette (2014) found synergy between the peptide *ARVA and either zinc or copper when applied to *P. aeruginosa* or *S. aureus*.

The tunicate peptide heavily studied by our lab, Clavanin A (ClavA), exhibits synergy with Zn^{2+} against *E. coli* MG1655 (Juliano et al., 2017) and a host of other strains (Juliano et al., 2020). Furthermore, the Zn^{2+}-binding capacity of the clavanins has been supported by the work of Miller, Matera-Witkiewicz, Mikołajczyk, Wieczorek, and Rowińska-Żyrek (2021). The ClavA-Zn^{2+} complex has nuclease activity which depends on the presence of the metal (Juliano et al., 2020). Zn^{2+} enhancement depends on both His-17 and His-21 (Juliano et al., 2017). Researchers should consider establishing essential residues for synergism, particularly in well-studied cases.

The amino terminal Cu^{2+}- and Ni^{2+}-binding (ATCUN) motif can facilitate high-affinity metal-binding, DNA cleavage, and generation of reactive oxygen species (Harford & Sarkar, 1997). Alternatively termed the XXH motif (due to the relaxed constraint of two amide nitrogens between the N-terminal amine and a histidine residue), it can be found in a variety of naturally-occurring and synthetic peptides and proteins. This motif is essential for the oxidative activity of the tick peptide ixosin as well as its ability to synergize with ixosin B (Libardo, Gorbatyuk, & Angelez-Boza, 2016). Our lab demonstrated that incorporation of an ATCUN motif into the wasp peptide anoplin significantly increased its activity by conferring copper-dependent reactive oxygen species generation (Libardo, Nagella, Lugo, Pierce, & Angelez-Boza, 2015). However, this strategy is not infallible, as shown by Agbale et al. (2019). Bouraguba et al. (2020) points out that most ATCUN-containing peptides cannot as efficiently bind Cu^{1+} following intracellular reduction of Cu^{2+}. Therefore, ATCUN-containing AMPs may instead help deliver copper into the cell where it can then act in an independent manner. Considering the susceptibility of cysteine-containing species towards oxidation (Paulsen & Carroll, 2013) and thiols more generally, we raise the possibility that biomolecule oxidation has been underexplored as a mechanism of copper-AMP synergy.

Mismetallation occurs when the native metal of a metal-binding compound is replaced by another metal. Copper is particularly adept at displacing other ions, as predicted by its position in the Irving-Williams series (Irving & Williams, 1953). Nickel and zinc typically occupy the second and third position in displacement potential, swapping places depending on the ligand under consideration. In *E. coli*, zinc can replace iron following oxidative stress, impairing the activity of mononuclear iron enzymes (Imlay, 2014). Furthermore, it has been shown that mismetallation (Macomber & Imlay, 2009), and not the generation of reactive oxygen species (Macomber, Rensing, & Imlay, 2007), is the major mechanism of copper toxicity in *E. coli*. We hypothesize that mismetallation following peptide-mediated delivery is an underappreciated mechanism of synergy between peptides and metal ions.

Our lab has recently taken an interest in photodynamic therapy, in which light is utilized to generate reactive oxygen species. Inspired by the work of Johnson et al. (2014) in conjugating a small molecule to a peptide, we fused a ruthenium-based transition metal complex to buforin II (Pierce et al., 2020). These agents show great promise in treating topical infections (e.g., MRSA) and degrading biofilms in medical equipment. However, many of these transition-metal complexes are activated by blue light, which over extended exposure times can be toxic to human cells (Godley et al., 2005). In addition to avoiding these undesired side reactions, developing red-shifted transition metal complexes will allow deeper tissue penetration for treatment of internal infections.

For those interested in studying synergy between metal ions and peptides, the media of choice's contents must be considered. For example, ICP-MS has been used to show that one commercially available batch of LB contained ~12.2 μM zinc (Takahasi et al., 2015) while different brands of MHB contained between an undetectable amount of zinc and ~9 μM (Girardello, Bispo, Yamanaka, & Gales, 2012). Some have taken the approach of first passaging bacteria in minimal media in order to dilute the metals present on agar plates. Subsequently, the peptide's efficacy in the presence and absence of metals in minimal media is examined (Libardo et al., 2017). The potential of peptides synergizing with metals *in vivo* is worthy of demonstration. If metals are not required as components of a drug formulation, this will reduce regulatory hurdles in getting the combination treatment to market.

6. A primer on Bliss and Loewe

Researchers may be unaware that many synergy analyses rely on one of two perspectives on the nature of the phenomenon. We first introduce the Bliss perspective and then discuss the Loewe perspective.

In the Bliss model of synergy, independent joint action is synonymous with noninteraction or independence (Bliss, 1939). As an illustration analogous to the one made by Chou (2010), consider a situation where one drug gives 50% inhibition at a certain dose while a second drug gives 70% inhibition at its respective dose. Clearly, the sum of the individual inhibitions cannot predict the combination treatment. Based in probability theory, one would predict that for two independently acting agents that the cumulative effect would be 0.5 + 0.7–0.5(0.7) or 85% inhibition. Of course, this is equivalent to (0.5)(1–0.7) or 15% noninhibition. And on that note, we affirm the mathematical equivalence of the formulation of Bliss independence as presented in Remington et al. (2020) and Tang, Wennerberg, and Aittokallio (2015). The former focuses on the proportion of bacteria that grow when exposed to an antimicrobial agent, while the latter considers the proportion of bacteria that do not grow. The time-kill assay pairs nicely with the Bliss model, as four cultures using specific doses can be simultaneously compared (i.e., negative control, drug A, drug B, combination treatment).

The work of Loewe (1953) is the basis for the Loewe model of synergy. The formulation of noninteraction (i.e., what is expected under similar joint action or additivity) is predicated on the notion that a substance cannot synergize with itself (Hall, Middleton, & Westmacott, 1983). When two substances are applied in combination, one can calculate a fractional inhibitory concentration (FIC) as $[A]_{combo} \div [A]_{alone}$ where each concentration represents the amount needed to achieve an equivalent biological effect (e.g., complete inhibition of growth by visual inspection). Once an FIC has been calculated for both agents, their sum can be used to quantify the extent of synergy. Different authors have stated standards of variable stringency. We advocate the use of synergism for $\Sigma FIC \leq 0.5$, noninteraction/indifference/additivity for $0.5 < \Sigma FIC < 4.0$, and antagonism $\Sigma FIC \geq 4.0$. Checkerboard assays lend themselves to the Loewe model, and a synergistic combination can be intuitively understood as one that lowers each minimum inhibitory concentration (MIC) by a factor of 4 or more. A hypothetical combination where one drug had a threefold MIC reduction and the second drug had a sixfold MIC reduction would be an equivalent hypothetical case to give $\Sigma FIC = 0.5$.

As a concrete example, against *E. coli* MG1655 cultured in LB-Lennox (pH 5.5) the MIC of Zn^{2+} was 1024 μM and the MIC of ClavA was 64 μM (Juliano et al., 2017). When combined in a 2:1 metal: peptide ratio, the MIC for ClavA under the same conditions was 4 μM. We can calculate ΣFIC as $4 \div 64 + 8 \div 1024$ or 0.07. Using a very stringent definition, this is an unambiguous example of very strong synergy.

Many organisms produce a family of similar peptides. For example, *Styela clava* produces at least six clavanins and five styelins (Lehrer, Tincu, Taylor, Menzel, & Waring, 2003). There appears to be a trend in observing stronger synergy in three-agent combination treatments (Yu et al., 2016) which of course has clinical implications. Additionally, examining more sophisticated combinations than two-agent pairings may be necessary in establishing biological function. After all, one explanation why an organism expends energy in making similar peptides could be that multiple peptides work together in accomplishing a certain function. The computation for three agents extends naturally to $([A]_{combo} \div MIC_A) + ([B]_{combo} \div MIC_B) + ([C]_{combo} \div MIC_C)$ and the interpretation of the sum is synergistic if less than 1 (Berenbaum, 1978). For those insisting on stringent definitions of synergy, a sum less than 0.75 seems appropriately analogous to the two-agent case. We point the mathematically inclined reader to Tabashnik (1992) for a more rigorous exposition on the theory of similar joint action.

For completeness, we acknowledge that several other definitions of synergy can be found in the literature (Greco, Bravo, & Parsons, 1995). In our view, the Loewe and Bliss perspectives predominate and lend themselves organically to the checkerboard and time-kill assays, respectively.

An alternative that appears to have some widespread adoption is the Chou-Talalay method (Chou, 2010). This approach is rooted in the median-effect equation that underlies the Henderson-Hasselbach, Hill, Michaelis-Menten, and Scatchard equations. Chou and Talalay's approach to quantifying synergy is more computationally involved, and we commend the zealous academic that wishes to incorporate this perspective in addition to Bliss and Loewe. Despite the lack of a uniform terminology to discuss the phenomenon of synergy, the researcher can greatly aid the field by explicitly mentioning in their publications precisely what *they* mean when they use the term.

Tang et al. (2015) propose an "interaction barometer" for categorizing the strength of two-agent interaction. This requires knowledge of the individual drug effects as well as the Loewe and Bliss combined action expected for additivity and independence, respectively. The authors acknowledge that their proposed analysis does not advise the researcher how to practically carry out experiments. Indeed, it would be quite cumbersome to apply Bliss to checkerboard assays or Loewe to time-kill assays. We instead offer a pragmatist's solution by seeking agreement between the checkerboard assay using Loewe and the time-kill assay using Bliss.

7. Interpreting checkerboard assays

Consider a hypothetical checkerboard assay where there are no errors in sample preparation or serial dilution. Assume that drug A has MIC = 128 μM, drug B has MIC = 64 μM, and ΣFIC = 0.4 (Fig. 1). Bonapace, Bosso, Friedrich, and White (2002) points out that the same 96-well plate can be interpreted in at least four different ways, which may be one contributing factor to confusion in the synergy literature. Averaging the nine turbid wells at the growth-inhibition interface, one would report ΣFIC = 0.55. Selecting the well that suggests the strongest synergy (32 μM drug A and 16 μM drug B), one would conclude ΣFIC = 0.50. Choosing the row concentration (64 μM drug A) and column concentration (32 μM drug B) that gives complete inhibition to inform the calculation, one would find ΣFIC = 1.00. Lastly, a qualitative approach would conclude the presence of synergy between the two drugs due to the absence of growth at (2 × MIC drug A, 2 × MIC drug B) and (0.25 × MIC drug A, 0.25 × MIC drug B). Since we prefer a stringent definition of synergy ($\Sigma FIC \leq 0.5$), we recommend reporting the minimum observed FIC. In this example, it is the only quantitative approach that would correctly conclude that synergy existed between the two drugs according to a conservative standard. We

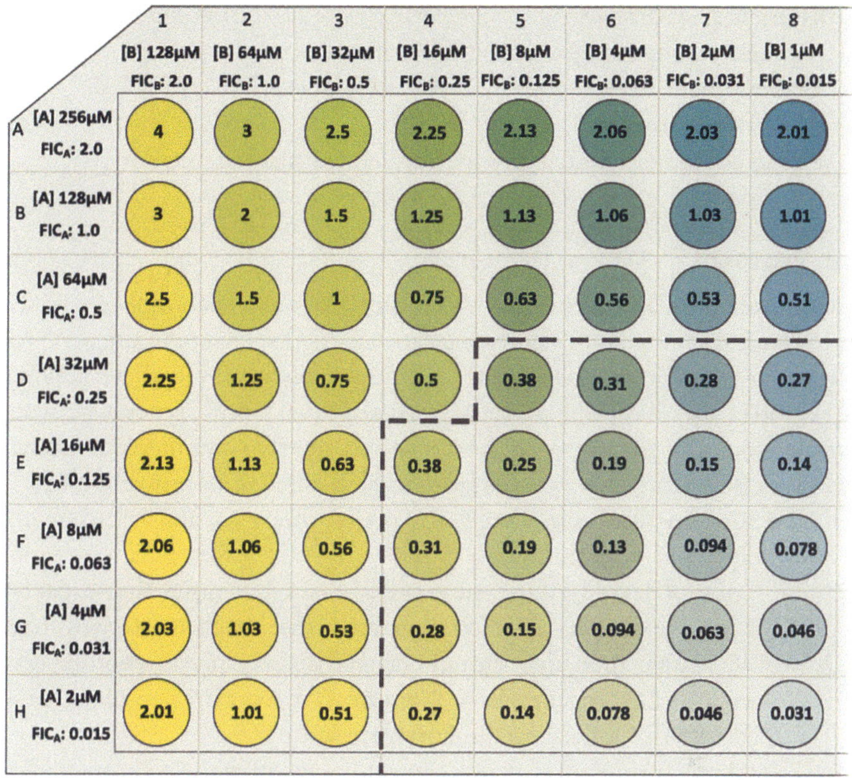

Fig. 1 Error-free checkerboard assay demonstrating different interpretation methods. The dotted line represents the growth-inhibition interface. Yellow and blue shading illustrates the concentration gradient explored with the checkerboard assay. Numbers within wells represent ΣFIC. Sums greater than 0.4 in this example should result in no growth observed, while sums less than 0.4 should result in turbid wells. Note that an additional column and row would be required to make independent MIC determinations of both drugs.

advocate constructing the checkerboard assay in a way that allows independent MIC determination for both substances, which serves as an internal control against experimental error.

8. Checkerboard assay

8.1 Materials/reagents

- Low-adsorption polypropylene tubes
- Stock solutions of antimicrobial agents
- Biological media
- Polypropylene 96-well plates

- Multichannel pipette
- Agar plate containing microbe of interest
- Shaking incubator
- Spectrophotometer
- Plastic container and damp paper towels for plate incubation

8.2 Protocol

This protocol is adapted from Schwalbe, Steele-Moore, and Goodwin (2007) and Wiegand et al. (2008). For testing cationic AMPs, be sure to use polypropylene 96-well plates. Polystyrene will nonspecifically bind these peptides (Wiegand et al., 2008). The protocol is written for exploring synergy between two peptides, but of course can be applied to any two antimicrobial agents.

1. To a polypropylene 96-well plate (which has 12 columns and 8 rows), add 50 μL media to Column 2–8. To Column 9–12, add 100 μL media to help avoid evaporation.
2. To Column 1, add 100 μL of peptide A at eight times the suspected MIC. For example, if it is suspected that peptide A will have MIC = 32 μM, then one should add 100 μL of 256 μM peptide A to Column 1. This should allow Column 2 to contain peptide A at its suspected MIC following the complete setup.
3. Take 50 μL from each well in Column 1 and add to the adjacent well in Column 2. Pipette up and down at least six times to ensure adequate mixing. Now take 50 μL from each well in Column 2 and add to the adjacent well in Column 3, mixing thoroughly as before. Repeat this process through Column 7, discarding the final 50 μL after mixing well. Since Column 8 will allow for MIC determination of peptide B and Column 9–12 will serve as a sterility control, none of these wells will receive the serially diluted peptide A.
4. To a second polypropylene 96-well plate, add 50 μL media to Row B-H.
5. To Row A, add 100 μL of peptide B at eight times the suspected MIC. This should allow Row B to contain peptide B at its MIC after the complete setup.
6. As before, serially dilute peptide B from Row A-G. Row H will allow for MIC determination of peptide A.
7. Now take 50 μL from Row A on the second polypropylene plate and add this to the corresponding wells in Row A on the first polypropylene plate. Repeat for Row B-H. Each well should now be diluted by a factor of 2.

8. Inoculate between two and five colonies of the microbe of interest in 5 mL of the media of choice (e.g., MHB). Incubate in a shaking incubator at 37 °C until a mid-log phase OD_{600} is achieved. For *E. coli*, this is approximately between 0.4 and 0.6.
9. Dilute the bacteria into 1.5 mL of media using the formula $OD_{final} V_{final} = OD_{initial} V_{initial}$, to attain a cell count of 1×10^6 CFU/mL. The researcher may first be required to perform growth kinetic assays to determine the cell count that corresponds to a given optical density. For *E. coli*, an OD_{600} of 0.1 corresponds to this cell count.
10. Add 100 μL bacteria to wells in the 8 × 8 grid (i.e., Row A-H & Column 1–8). Each well should now be diluted by an additional factor of 2.
11. Place the 96-well plate in a plastic holder (e.g., a Ziploc bag) and include a damp paper towel to help avoid evaporation. Incubate plate at 37 °C for 16–20 h or until sufficient growth is observed.

9. A modified MIC assay for synergy testing

Of course, errors in sample preparation and serial dilutions do routinely occur. Furthermore, reporting the minimum FIC derived from a checkerboard assay only gives one datum for analysis. Inspired by the approach of He et al. (2015) to enhance reproducibility and statistical power, we modify the protocol to first use one or more checkerboard assays to derive an informed guess at the optimum ratio of concentrations. In the example above on interpreting checkerboard assays, the optimum ratio was 2:1 A:B. Note that this protocol allows one to perform four replicates of substance A (Column 1–4), substance B (Column 5–8), and the combination treatment (Column 9–12) on a single plate. The protocol is written for exploring synergy between two peptides, but of course can be applied to any two antimicrobial agents.

10. Modified MIC assay

10.1 Materials/reagents

- Low-adsorption polypropylene tubes
- Stock solutions of antimicrobial agents
- Biological media
- Polypropylene 96-well plate
- Multichannel pipette
- Petri dish containing microbe of interest

- Shaking incubator
- Spectrophotometer
- Plastic container and damp paper towels for plate incubation

10.2 Protocol

1. To Row B-G of a polypropylene 96-well plate, add 50 μL media. To Row H, add 100 μL media. Row H will serve as a sterility control.
2. To Column 1–4 in Row A, add 100 μL peptide A at twice the desired initial concentration. Serially dilute 50 μL peptide A from Row A-G, discarding the final 50 μL. At each well, pipette up and down at least six times to ensure adequate mixing.
3. To Column 5–8 in Row A, add 100 μL peptide B at twice the desired initial concentration. In the same manner as before, serially dilute 50 μL peptide B from Row A-G, discarding the final 50 μL.
4. To Column 9–12 in Row A, add 100 μL of the combination treatment at twice the desired initial concentrations. Serially dilute in like manner 50 μL from Row A-G, discarding the final 50 μL.
5. Inoculate between two and five colonies of the microbe of interest in 5 mL of the media of choice (e.g., MHB). Incubate in a shaking incubator at 37 °C until a mid-log phase OD_{600} is achieved. For *E. coli*, this is approximately between 0.4 and 0.6.
6. Dilute the bacteria into 1.5 mL of media using the formula $OD_{final} V_{final} = OD_{initial} V_{initial}$, to attain a cell count of 1×10^6 CFU/mL.
7. Add 50 μL of bacteria to all wells of Row A-G.
8. Place the 96-well plate in a plastic holder (e.g., a Ziploc bag) and include a damp paper towel to help avoid evaporation. Incubate plate at 37 °C for 16–20 h or until sufficient growth is observed.

11. Interpreting time-kill assays

Consider a time-kill assay where 90% of bacteria are killed when exposed to drug A at dose x. Furthermore, one observes that 90% of bacteria are killed when exposed to drug B at dose y. In each case, 10% of the bacteria remain when exposed to the antimicrobial agent in isolation. According to the Bliss model, in the absence of synergy one would expect as few as 1% of bacteria to survive if exposed simultaneously to drug A at dose × and drug B at dose y. This represents a 2-$\log_{10}$ reduction relative to a bacterial culture not exposed to the drugs. A further decrease in survival would be indicative of synergy. However, according to the National Committee on Clinical

Laboratory Standards protocol (1999) synergy is characterized by a 2-$\log_{10}$ reduction relative to the most active agent. In this example, fewer than 0.1% of bacteria surviving would be indicative of synergy. Our lab has personally experienced a case where this arbitrary standard would have forced us to conclude the absence of synergy between two peptides that had been shown in both checkerboard assays and MIC assays to be synergistic (unpublished observations). Because of this, we recommend using the Bliss approach in interpreting time-kill assays. The researcher should feel confident that synergy has been correctly pronounced if the recommended approaches to carrying out checkerboard assays, MIC assays, and time-kill assays all concur. Additionally, given the resource- and labor-intensive nature of the time-kill assay, we advise first performing MIC and checkerboard assays to arrive at informed decisions for optimal concentrations to use in the time-kill assay.

12. Time-kill assay

12.1 Materials/reagents

- Low-adsorption polypropylene tubes
- Stock solutions of antimicrobial agents
- Biological media
- Agar plates for enumeration
- Borosilicate glass beads or plate spreader
- Multichannel pipette
- Petri dish containing microbe of interest
- Shaking incubator
- Spectrophotometer

12.2 Protocol

This protocol is adapted from the one reported in Portelinha, Heilemann, Jin, and Angeles-Boza (2021).

1. Inoculate between two and five colonies of the microbe of interest in 10mL of the media of choice (e.g., MHB). Incubate in a shaking incubator at 37°C until a mid-log phase OD_{600} is achieved. For *E. coli*, this is approximately between 0.4 and 0.6.
2. Dilute the bacteria into 10mL of media using the formula $OD_{final} V_{final} = OD_{initial} V_{initial}$, to attain a cell count of 1×10^6 CFU/mL. The researcher may first be required to perform growth kinetic assays to

determine the cell count that corresponds to a given optical density. For *E. coli*, an OD_{600} of 0.1 corresponds to this cell count.

3. Prepare all peptide solutions in 350 μL media at twice the desired initial concentration. For examining synergy between two peptides, three peptide solutions should be prepared: peptide A alone at dose x, peptide B alone at dose y, and peptide A at dose x + peptide B at dose y. As an example, suppose that the final concentration of each peptide will be 10 μM. Therefore, one would prepare solutions of 20 μM peptide A, 20 μM peptide B, and 20 μM peptide A + 20 μM peptide B prior to dilution in the next step.
4. Into a low-adsorption polypropylene vessel, add 350 μL diluted bacteria to 350 μL of the peptide solution. This exposes 5×10^5 CFU/mL to the peptide at the desired dose. A growth control should be simultaneously prepared, consisting of 350 μL of the diluted bacteria and 350 μL of media. Thus, for testing synergy between two peptides, there should be a total of four cultures. After mixing well, the $t = 0$ plates should be immediately prepared to verify the initial CFU count. Incubate cultures in a shaking incubator at 37 °C.
5. At each time point of interest, take 10 μL from each culture and prepare dilutions for serial plating. We typically prepare dilutions of the culture up to 10^4, especially for time points after additional growth has taken place. Add 100 μL of diluted bacteria to plates before distributing with borosilicate glass beads or a plate spreader. Incubate all plates at 37 °C for 18–20 h. Ensure that an excess of plates is available, as many plates will be consumed. The first trial should err on the side of covering many dilution factors, but subsequent iterations can omit the statistically uncountable plates.
6. When enumerating bacteria, select plates that have between 20 and 200 colonies to achieve a reliable count. Volumes in steps 4 and 5 should be scaled up if needed to achieve these counts. Calculate cell count as follows: CFU/mL = (# of colonies)(10) ÷ (dilution factor). For example, 53 colonies counted on a 10^{-3} plate (which received 0.1 mL inoculum) corresponds to 5.3×10^5 CFU/mL.

12.3 Synergy analysis

7. For each time point, calculate the drug-treated sample cell count relative to the growth control. For example, if after 1 h the drug A-treated sample had a cell count of 5×10^5 CFU/mL and the growth control had a cell

count of 5×10^6 CFU/mL, this calculation would reflect the fact that the actual cell count for the drug-treated sample is 10% of the growth control. We denote these normalized cell counts for the peptides as G_A, G_B, and G_{A+B}, respectively.

8. Determine the Bliss model's expected cell count of the combination treatment as $G_A \times G_B$.
9. Compare the actual growth of the combination treatment (i.e., G_{A+B}) to $G_A \times G_B$. Less growth than expected supports the pronouncement of synergy between the two peptides. The user is commended if supporting this inference with statistical analysis.

13. Mechanistic studies

A necessary first step in anticipating resistance is establishing mechanisms of action for AMPs and their synergistic partners. Fortunately for the researcher, much has been written on the activities of AMPs to inform initial exploration (Brogden, 2005; Hancock, Haney, & Gill, 2016; Lazzaro et al., 2020). Biophysical studies have allowed categorization of membrane poration mechanisms as toroidal (Henzler-Wildman, Lee, & Ramamoorthy, 2003), carpet (Yamaguchi et al., 2001), and barrel-stave (Yang, Harroun, Weiss, Ding, & Huang, 2001). However, Last et al. (2013) cautions the researcher not to put too much stock in associating a single poration mechanism with a peptide, as different conditions can give rise to distinct mechanisms.

AMPs have been shown to act on several intracellular targets, in addition to their membrane-associated roles. Indolicidin has been shown to significantly inhibit the replication of DNA and moderately inhibit the transcription of RNA (Subbalakshmi & Sitaram, 1998). Buforin 2 binds to DNA and RNA after passing through the microbial membrane, rapidly leading to cell death (Park, Kim, & Kim, 1998). Mersacidin binds to Lipid II, preventing peptidoglycan biosynthesis (Brötz, Bierbaum, Leopold, Reynolds, & Sahl, 1998). Aspartate-rich anionic peptides present in ovine pulmonary surfactant can flocculate cytoplasmic components (Brogden, De Lucca, Bland, & Elliott, 1996). Recent publications (Chongsiriwatana et al., 2017; Zhu, Mohapatra, & Weisshaar, 2019) argue that cationic AMPs in general have intracellular mechanisms of action and that membrane activity in many cases is a means to an end.

Finally, AMPs have been shown to serve a variety of immunomodulatory roles, such as inducing the expression of cytokines in target cells

(Hancock et al., 2016). Because AMPs serve a broader set of physiological functions beyond antimicrobial activities, Hancock advocates adoption of the phrase host defense peptide (HDP). Since most studies have examined the effects of AMPs on isolated microbes, there is likely much more to be learned concerning the role of AMPs in the host-pathogen interface.

Conjugation of small-molecule fluorophores to AMPs can facilitate microscopy studies. Verifying that the biological activity is not affected by this modification is necessary, but not sufficient, for quality control. Researchers should be aware that the fluorophore may artificially direct the subcellular localization of the peptide (Zhao et al., 2016).

Note that membrane depolarization and membrane poration are two distinct processes. While many assays examine the entrance of membrane-impermeable dyes, voltage sensitive dyes can be used to measure changes in the membrane potential (Winkel, Gray, Seistrup, Hamoen, & Strahl, 2016).

Examining the activity of mirror-image peptides has been used to elucidate mechanistic details. It had been thought that if the activities of the native AMP and enantiomer were identical, a membrane-associated mechanism of action was likely. Differences in activity were thought to suggest interactions with proteins (i.e., a chiral interaction). Henriques, Peacock, Benfield, Wang, and Craik (2019) demonstrated the oversimplicity in this view by showing that membrane interactions can in fact be chiral. Using enantiomeric biomolecules that are the putative targets of AMPs present another opportunity for establishing AMP functions.

14. Proactive assessment of resistance

Why are AMPs superior to conventional antibiotics in their low tendency to lead to acquired resistance? This is an unsettled question, but there have been credible explanations put forward. Recall that micromolar concentrations of AMPs can reach millimolar concentrations at the membrane surface due to preferential partitioning (Melo et al., 2009). While this presents a high selective pressure, He et al. (2015) suggests that a global remodeling of the membrane to eliminate the nonspecific interaction imposes a very significant fitness cost. This is in contrast to conventional antibiotics, which have a specific cellular target, which allows simpler circumvention. In an analogous fashion, micromolar amounts of cationic AMPs entering the cell to exert an intracellular mechanism of action can achieve a millimolar local concentration, due to the cell's small volume (Zhu et al., 2019). Chongsiriwatana et al. (2017) argues that there is a widespread and underappreciated phenomenon of cationic AMPs interacting with polyanionic species (e.g., ribosomes and nucleic acids) in a cytotoxic manner. Individual peptides can vary in their affinities for different polyanions, and the fact that organisms produce multiple peptides suggests that a lack of consistent selection makes it difficult for resistance to develop (Chongsiriwatana et al., 2017).

As done by Rishi et al. (2011), researchers should additionally report fractional bactericidal concentrations (FBCs) to verify that combination treatment enhances killing and not merely inhibition. After all, failure to eliminate the microbial population by combination treatment can select for resistant microbes that render the treatment moot (Pena-Miller et al., 2013). This metric can be calculated by replacing MICs in the FIC calculation with minimum bactericidal concentrations (MBCs). After performing a checkerboard assay (which will allow MIC and FIC calculations), one can take the remaining media from wells of interest and inoculate agar plates for MBC/FBC determination. *In vitro* results can then inform follow-up experiments in animal models.

Researchers are encouraged to perform serial passaging experiments, whereby microbes are exposed to incrementally higher concentrations of antimicrobials in order to generate resistant strains. Whole-genome sequencing of replicate analyses can identify conserved mutations. While some groups have found differences in the mutations that arise depending on the selection regime, Jahn, Munck, Ellabaan, and Sommer (2017) documented a consistency of adaptations using four distinct methodologies. A more detailed discussion of factors governing the onset of resistance can be found in Andersson, Hughes, and Kubicek-Sutherland (2016).

An interesting result was obtained in serial passaging experiments reported in Pollard et al. (2012). *P. aeruginosa* that was serially exposed to colistin indeed acquired resistance to this drug. However, serial exposure of a ceragenin (a peptidomimetic) not only led to resistance against this compound, but also a 14-fold increase in colistin resistance, compared to the parent strain. This raises the possibility that acquiring resistance to one drug, and subsequently acquiring resistance to a synergistic partner might entail a different mechanism of resistance than that associated with simultaneous exposure to the two drugs (not to mention the mechanism involved if the order of exposure is inverted). For those interested in seeing synergy studies make additional headway into the clinic, this is yet another avenue that deserves exploration. Note that in some cases resistance can be transient, meaning that serial passaging in the absence of antimicrobial will lead the resistant strain to revert back to the parental susceptibility (Pollard et al., 2012). In the case of Pollard et al. (2012), the observed reversion was consistent with the hypothesis that global membrane remodeling imposes too great a fitness cost to maintain in the absence of AMP/peptidomimetic.

15. Overcoming the challenge of scale

Simply demonstrating the existence of synergy with *in vitro* or *ex vivo* studies can consume large quantities of peptide. Performing the recommended mechanistic and serial passaging experiments, as well as *in vivo* studies in animal models, will require even greater stockpiles. However, recent advances in the field have mitigated this issue. Broadly speaking, one can approach the problem from either a molecular biology or synthetic chemistry perspective.

In the former case, one can utilize recombinant DNA techniques. For example, one group designed an overexpression plasmid encoding calmodulin fused to the peptide of interest, separated by a cleavable linker sequence (Ishida, Nguyen, Gopal, Aizawa, & Vogel, 2016). Utilization of calmodulin protects both the peptide of interest from degradation as well as the expression host from the toxic action of the AMP. Following expression, purification, and protease treatment directed at the linker, the peptide could be isolated chromatographically. Optimization of this procedure using minimal media, N-15 ammonium chloride, and C-13 glucose may provide a cheaper synthetic route for NMR or MS studies where complete incorporation of stable isotopes is desired. Using genetic code expansion techniques (Xie & Schultz, 2005) can allow incorporation of unnatural amino acids at specific residues, making this approach just as sensible as chemical synthesis depending on the researcher's application.

The second approach involves solid phase peptide synthesis of the desired compounds (Amblard, Fehrentz, Martinez, & Subra, 2006). While the entire procedure can be carried out at room temperature, synthesizers that use heating blocks and/or microwave radiation can drastically cut the required time. Combining the efficiency conferred by automation with ligation techniques to stitch together peptide fragments (e.g., native chemical ligation followed by desulfurization), even large proteins containing multiple post-translational modifications (PTMs) have been synthesized (Premdjee, Andersen, Larance, Conde-Frieboes, & Payne, 2021). Considering a peptide or protein where multiple PTMs have been documented, chemical synthesis may offer a simple solution for investigating the impact of individual PTMs in isolation. Commercially available synthesizers also allow for extensive parallel peptide synthesis. This enables rapid exploration of the consequence of mutagenizing individual peptide positions (Amara, Foe, Onguka, Garland, & Bogyo, 2019). We cited an example above where key residues were discovered that were essential for synergy between PGLa and magainin 2. Alanine scanning (i.e., replacing individual residues with alanine) for two peptides known to synergize would be a manageable task for a synthesizer capable of parallel synthesis. This would arguably be a faster route to generating peptides for synergy testing, as compared to recombinant techniques. The same could be said about attempts to quickly determine which residues were key for metal-binding, as mentioned in the example of ClavA and Zn^{2+}. In general, chemical synthesis may be preferable to recombinant techniques when selective incorporation of residues, stable isotopes, or PTMs is needed. Of course, both approaches have great utility and should be viewed as complementary techniques.

16. Conclusions

The interest in AMPs will continue to grow as a more complete range of applications are enumerated. Hosts may use AMPs to shape symbiotic microbial communities (Loudon et al., 2020). The anionic surface of cancer cells may allow broad repurposing of cationic AMPs in developing future anti-cancer agents (Aronson et al., 2020). A recent mechanistic demonstration of synergy between AMPs and histone H2A (Doolin et al., 2020) raises the possibility that AMPs work together with additional biomolecules within their native biological contexts.

Some have taken a cynical view of the potential value of synergy testing for clinical applications. Herein we reject that pessimism and lay out a framework for an informed pursuit of synergy. Examples in the literature such as White, Burgess, Manduru, and Bosso (1996) suggest that different methods of exploring synergy have unacceptably low levels of agreement. We have argued that a number of missteps can cause individual assays to fail in confirming synergism between substances. He et al. (2015) failed to detect any synergistic combinations between three membrane-permeabilizing peptides and four commonly used antibiotics (streptomycin, ampicillin, ciprofloxacin, and vancomycin). But the authors only examined one ratio of concentrations for the combination treatment, and they themselves demonstrated that their procedure likely overestimates the actual FIC. A clinical trial found no improved outcome for cystic fibrosis (*CF*) patients who had antibiotic regimens determined through multiple combination bactericidal testing compared to conventional antibacterial susceptibility testing (Aaron et al., 2005). This study in part led the Cystic Fibrosis Foundation to advise against synergy testing for *CF* patients (Flume et al., 2009). Finding combination treatments that in both checkerboard and time-kill assays agreed on synergistic potential, and additionally applying scrutiny in tests with animal models, would have very likely produced superior therapeutic regimens.

The potential of AMPs to combat antibiotic resistance has expanded with additional innovation. Examples include stapled AMPs (Mourtada et al., 2019), peptides containing non-proteinogenic amino acids (Lizio et al., 2021), peptoids (Bicker & Cobb, 2020), and D-peptides (McGrath et al., 2013). These endeavors will yield reduced toxicity, less degradation, and a larger synthetic space. *In silico* analyses of large AMP databases (Shi et al., 2021) will almost certainly reveal previously undetected structural and functional trends. Synergy testing will expand this horizon even further.

Acknowledgments

This material is based upon work supported by the National Science Foundation (Grant No. DGE-1747453 to M.H. and C.M.D. and Grant No. MCB1715494 to A.M.A.-B.). Any opinions, findings, and conclusions or recommendations expressed in this material are those of the authors and do not necessarily reflect the views of the National Science Foundation.

References

Aaron, S. D., Vandemheen, K. L., Ferris, W., Fergusson, D., Tullis, E., Haase, D., et al. (2005). Combination antibiotic susceptibility testing to treat exacerbations of cystic fibrosis associated with multiresistant bacteria: A randomised, double-blind, controlled clinical trial. *Lancet*, *366*(9484), 463–471. https://doi.org/10.1016/S0140-6736(05)67060-2.

Agbale, C. M., Sarfo, J. K., Galyuon, I. K., Juliano, S. A., Silva, G. G. O., Buccini, D. F., et al. (2019). Antimicrobial and antibiofilm activities of helical antimicrobial peptide sequences incorporating metal-binding motifs. *Biochemistry*, *58*(36), 3802–3812. https://doi.org/10.1021/acs.biochem.9b00440.

Amara, N., Foe, I. T., Onguka, O., Garland, M., & Bogyo, M. (2019). Synthetic fluorogenic peptides reveal dynamic substrate specificity of depalmitoylases. *Cell Chemical Biology*, *26*(1), 35–47. https://doi.org/10.1016/j.chembiol.2018.10.005.

Amblard, M., Fehrentz, J., Martinez, J., & Subra, G. (2006). Methods and protocols of modern solid phase peptide synthesis. *Molecular Biotechnology*, *33*(3), 239–254. https://doi.org/10.1385/MB:33:3:239.

Andersson, D. I., Hughes, D., & Kubicek-Sutherland, J. Z. (2016). Mechanisms and consequences of bacterial resistance to antimicrobial peptides. *Drug Resistance Updates*, *26*, 43–57. https://doi.org/10.1016/j.drup.2016.04.002.

Aronson, M. R., Dahl, E. S., Halle, J. A., Simonson, A. W., Gogal, R. A., Glick, A. B., et al. (2020). Re-engineering antimicrobial peptides into oncolytics targeting drug-resistant ovarian cancers. *Cellular and Molecular Bioengineering*, *13*(5), 447–461. https://doi.org/10.1007/s12195-020-00626-z.

Bartolomeo, M. P., & Maisano, F. (2006). Validation of a reversed-phase HPLC method for quantitative amino acid analysis. *Journal of Biomolecular Techniques*, *17*(2), 131–137.

Beggs, W. H., & Andrews, F. A. (1976). Role of ionic strength in salt antagonism of aminoglycoside action on *Escherichia coli* and *Pseudomonas aeruginosa*. *The Journal of Infectious Diseases*, *134*(5), 500–504. https://doi.org/10.1093/infdis/134.5.500.

BeMiller, J. N. (2018). *Carbohydrate chemistry for food scientists* (3rd ed.). Elsevier. https://doi.org/10.1016/C2016-0-01960-5.

Berenbaum, M. C. (1978). A method for testing for synergy with any number of agents. *The Journal of Infectious Diseases*, *137*(2), 122–130. https://doi.org/10.1093/infdis/137.2.122.

Bicker, K. L., & Cobb, S. L. (2020). Recent advances in the development of anti-infective peptoids. *Chemical Communications*, *56*(76), 11158–11168. https://doi.org/10.1039/D0CC04704J.

Bliss, C. I. (1939). The toxicity of poisons applied jointly. *Annals of Applied Biology*, *26*(3), 585–615. https://doi.org/10.1111/j.1744-7348.1939.tb06990.x.

Bonapace, C. R., Bosso, J. A., Friedrich, L. V., & White, R. L. (2002). Comparison of methods of interpretation of checkerboard synergy testing. *Diagnostic Microbiology and Infectious Disease*, *44*(4), 363–366. https://doi.org/10.1016/S0732-8893(02)00473-X.

Bouraguba, M., Glattard, E., Naudé, M., Pelletier, R., Aisenbrey, C., Bechinger, B., et al. (2020). Copper-binding motifs Xxx-His or Xxx-Zzz-His (ATCUN) linked to an antimicrobial peptide: Cu-binding, antimicrobial activity and ROS production. *Journal of Inorganic Chemistry*, *213*, 111255. https://doi.org/10.1016/j.jinorgbio.2020.111255.

Brogden, K. A. (2005). Antimicrobial peptides: Pore formers or metabolic inhibitors in bacteria? *Nature Reviews Microbiology*, *3*(3), 238–250. https://doi.org/10.1038/nrmicro1098.

Brogden, K. A., De Lucca, A. J., Bland, J., & Elliott, S. (1996). Isolation of an ovine pulmonary surfactant-associated anionic peptide bactericidal for *Pasteurella haemolytica*. *Proceedings of the National Academy of Sciences of the United States of America*, *93*(1), 412–416. https://doi.org/10.1073/pnas.93.1.412.

Brötz, H., Bierbaum, G., Leopold, K., Reynolds, P. E., & Sahl, H. (1998). The lantibiotic mersacidin inhibits peptidoglycan synthesis by targeting lipid II. *Antimicrobial Agents and Chemotherapy*, *42*(1), 154–160. https://doi.org/10.1128/AAC.42.1.154.

Chongsiriwatana, N. P., Lin, J. S., Kapoor, R., Wetzler, M., Rea, J. A. C., Didwania, M. K., et al. (2017). Intracellular biomass flocculation as a key mechanism of rapid bacterial killing by cationic, amphipathic antimicrobial peptides and peptoids. *Scientific Reports*, 7, 16718. https://doi.org/10.1038/s41598-017-16180-0.

Chou, T. (2010). Drug combination studies and their synergy quantification using the Chou-Talalay method. *Cancer Research*, *70*(2), 440–446. https://doi.org/10.1158/0008-5472.CAN-09-1947.

Cirioni, O., Simonetti, O., Morroni, G., Brescini, L., Kamysz, W., Kamysz, E., et al. (2018). Efficacy of pexiganan combination with tigecycline in a mouse model of Pseudomonas aeruginosa sepsis. *Current Topics in Medicinal Chemistry*, *18*(24), 2127–2132. https://doi.org/10.2174/1568026619666181219123431.

Deb, S. C., & Fukushima, T. (1999). Metals in aquatic ecosystems: Mechanisms of uptake, accumulation and release— Ecotoxicological perspectives. *International Journal of Environmental Studies*, *56*(3), 385–417. https://doi.org/10.1080/00207239908711212.

Dobson, A. J., Purves, J., Kamysz, W., & Rolff, J. (2013). Comparing selection on S. aureus between antimicrobial peptides and common antibiotics. *PLoS One*, *8*(10), e76521. https://doi.org/10.1371/journal.pone.0076521.

Doern, C. D. (2014). When does 2 plus 2 equal 5? A review of antimicrobial synergy testing. *Journal of Clinical Microbiology*, *52*(12), 4124–4128. https://doi.org/10.1128/JCM.01121-14.

Doolin, T., Amir, H. M., Duong, L., Rosenzweig, R., Urban, L. A., Bosch, M., et al. (2020). Mammalian histones facilitate antimicrobial synergy by disrupting the bacterial proton gradient and chromosome organization. *Nature Communications*, *11*(1), 3888. https://doi.org/10.1038/s41467-020-17699-z.

Engström, P., Carlsson, A., Engström, A., Tao, Z. J., & Bennich, H. (1984). The antibacterial effect of attacins from the silk moth *Hyalophora cecropia* is directed against the outer membrane of *Escherichia coli*. *The EMBO Journal*, *3*(13), 3347–3351. https://doi.org/10.1002/j.1460-2075.1984.tb02302.x.

Flume, P. A., Mogayzel, P. J., Robinson, K. A., Goss, C. H., Rosenblatt, R. L., Kuhn, R. J., et al. (2009). Cystic fibrosis pulmonary guidelines: Treatment of pulmonary exacerbations. *American Journal of Respiratory and Critical Care Medicine*, *180*(9), 802–808. https://doi.org/10.1164/rccm.200812-1845PP.

Giacometti, A., Cirioni, O., Del Prete, M. S., Barchiesi, F., Fortuna, M., Drenaggi, D., et al. (2000). In vitro activities of membrane-active peptides alone and in combination with clinically used antimicrobial agents against *Stenotrophomonas maltophilia*. *Antimicrobial Agents and Chemotherapy*, *44*(6), 1716–1719. https://doi.org/10.1128/AAC.44.6.1716-1719.2000.

Gill, S. C., & von Hippel, P. H. (1989). Calculation of protein extinction coefficients from amino acid sequence data. *Analytical Biochemistry*, *182*(2), 319–326. https://doi.org/10.1016/0003-2697(89)90602-7.

Girardello, R., Bispo, P. J. M., Yamanaka, T. M., & Gales, A. C. (2012). Cation concentration variability of four distinct Mueller-Hinton agar brands influences polymyxin B susceptibility results. *Journal of Clinical Microbiology*, *50*(7), 2414–2418. https://doi.org/10.1128/JCM.06686-11.

Girouard, Y. C., Maclean, I. W., Ronald, A. R., & Albritton, W. L. (1981). Synergistic antibacterial activity of clavulanic acid and amoxicillin against β-lactamase-producing strains of *Haemophilus ducreyi*. *Antimicrobial Agents and Chemotherapy*, *20*(1), 144–145. https://doi.org/10.1128/AAC.20.1.144.

Godley, B. F., Shamsi, F. A., Liang, F., Jarrett, S. G., Davies, S., & Boulton, M. (2005). Blue light induces mitochondrial DNA damage and free radical production in epithelial cells. *Journal of Biological Chemistry*, *280*(22), 21061–21066. https://doi.org/10.1074/jbc.M502194200.

Greco, W. R., Bravo, G., & Parsons, J. C. (1995). The search for synergy: A critical review from a response surface perspective. *Pharmacological Reviews*, *47*(2), 331–385.

Gupta, K., Singh, S., & van Hoek, M. L. (2015). Short, synthetic cationic peptides have antibacterial activity against *Mycobacterium smegmatis* by forming pores in membrane and synergizing with antibiotics. *Antibiotics*, *4*(3), 358–378. https://doi.org/10.3390/antibiotics4030358.

Hall, M. J., Middleton, R. F., & Westmacott, D. (1983). The fractional inhibitory concentration (FIC) index as a measure of synergy. *Journal of Antimicrobial Chemotherapy*, *11*(5), 427–433. https://doi.org/10.1093/jac/11.5.427.

Hancock, R. E. W., Haney, E. F., & Gill, E. E. (2016). The immunology of host defence peptides: Beyond antimicrobial activity. *Nature Reviews Immunology*, *16*(5), 321–334. https://doi.org/10.1038/nri.2016.29.

Hanson, M. A., Dostálová, A., Ceroni, C., Poidevin, M., Kondo, S., & Lemaitre, B. (2019). Synergy and remarkable specificity of antimicrobial peptides in vivo using a systematic knockout approach. *eLife*, *8*, e44341. https://doi.org/10.7554/eLife.44341.001.

Harford, C., & Sarkar, B. (1997). Amino terminal Cu(II)- and Ni(II)-binding (ATCUN) motif of proteins and peptides: Metal binding, DNA cleavage, and other properties. *Accounts of Chemical Research*, *30*(3), 123–130. https://doi.org/10.1021/ar9501535.

Harris, F., Dennison, S. R., & Phoenix, D. A. (2009). Anionic antimicrobial peptides from eukaryotic organisms. *Current Protein & Peptide Science*, *10*(6), 585–606. https://doi.org/10.2174/138920309789630589.

Hashemi, M. M., Rovig, J., Weber, S., Hilton, B., Forouzan, M. M., & Savage, P. B. (2017). Susceptibility of colistin-resistant, Gram-negative bacteria to antimicrobial peptides and ceragenins. *Antimicrobial Agents and Chemotherapy*, *61*. e00292–17 https://doi.org/10.1128/AAC.00292-17.

He, J., Starr, C. G., & Wimley, W. C. (2015). A lack of synergy between membrane-permeabilizing cationic antimicrobial peptides and conventional antibiotics. *Biochimica et Biophysica Acta*, *1848*(1), 8–15. https://doi.org/10.1016/j.bbamem.2014.09.010.

Henriques, S. T., Peacock, H., Benfield, A. H., Wang, C. K., & Craik, D. J. (2019). Is the mirror image a true reflection? Intrinsic membrane chirality modulates peptide binding. *Journal of the American Chemical Society*, *141*(51), 20460–20469. https://doi.org/10.1021/jacs.9b11194.

Henzler-Wildman, K. A., Lee, D., & Ramamoorthy, A. (2003). Mechanism of lipid bilayer disruption by the human antimicrobial peptide, LL-37. *Biochemistry*, *42*(21), 6545–6558. https://doi.org/10.1021/bi0273563.

Imlay, J. A. (2014). The mismetallation of enzymes during oxidative stress. *Journal of Biological Chemistry*, *289*(41), 28121–28128. https://doi.org/10.1074/jbc.R114.588814.

Irving, H., & Williams, R. J. P. (1953). The stability of transition-metal complexes. *Journal of the Chemistry Society (Resumed)*, 3192–3210. https://doi.org/10.1039/JR9530003192.

Ishida, H., Nguyen, L. T., Gopal, R., Aizawa, T., & Vogel, H. J. (2016). Overexpression of antimicrobial, anticancer, and transmembrane peptides in Escherichia coli through a calmodulin-peptide fusion system. *Journal of the American Chemical Society*, *138*(35), 11318–11326. https://doi.org/10.1021/jacs.6b06781.

Jahn, L. J., Munck, C., Ellabaan, M. M. H., & Sommer, M. O. A. (2017). Adaptive laboratory evolution of antibiotic resistance using different selection regimes lead to similar phenotypes and genotypes. *Frontiers in Microbiology*, *8*, 816. https://doi.org/10.3389/fmicb.2017.00816.

Jensen, K. J., Tofteng, P., & Pedersen, S. L. (Eds.). (2013). *Peptide synthesis and applications* (2nd ed.). Humana Press. https://doi.org/10.1007/978-1-62703-544-6.

Johnson, G. A., Ellis, E. A., Kim, H., Muthukrishnan, N., Snavely, T., & Pellois, J. (2014). Photoinduced membrane damage of E. coli and S. aureus by the photosensitizer-antimicrobial peptide conjugate eosin-(KLAKLAK)2. *PLoS One*, *9*(3), e91220. https://doi.org/10.1371/journal.pone.0091220.

Juliano, S. A., Pierce, S., deMayo, J. A., Balunas, M. J., & Angeles-Boza, A. M. (2017). Exploration of the innate immune system of Styela clava: Zn^{2+} binding enhances the antimicrobial activity of the tunicate peptide Clavanin A. *Biochemistry*, *56*(10), 1403–1414. https://doi.org/10.1021/acs.biochem.6b01046.

Juliano, S. A., Serafim, L. F., Duay, S. S., Chavez, M. H., Sharma, G., Rooney, M., et al. (2020). A potent host defense peptide triggers DNA damage and is active against multidrug-resistant Gram-negative pathogens. *ACS Infectious Disease*, *6*(5), 1250–1263. https://doi.org/10.1021/acsinfecdis.0c00051.

Kacprzyk, L., Rydengård, V., Mörgelin, M., Davoudi, M., Pasupuleti, M., Malmsten, M., et al. (2007). Antimicrobial activity of histidine-rich peptides is dependent on acidic conditions. *Biochimica et Biophysica Acta*, *1768*(11), 2667–2680. https://doi.org/10.1016/j.bbamem.2007.06.020.

Koo, H. B., & Seo, J. (2019). Antimicrobial peptides under clinical investigation. *Peptide Science*, *111*, e24122. https://doi.org/10.1002/pep2.24122.

Kristensen, K., Henriksen, J. R., & Andresen, T. L. (2015). Adsorption of cationic peptides to solid surfaces of glass and plastic. *PLoS One*, *10*(5), e0122419. https://doi.org/10.1371/journal.pone.0122419.

Kuipers, B. J. H., & Gruppen, H. (2007). Prediction of molar extinction coefficients of proteins and peptides using UV absorption of the constituent amino acids at 214 nm to enable quantitative reverse phase high-performance liquid chromatography-mass spectrometry analysis. *Journal of Agricultural and Food Chemistry*, *55*(14), 5445–5451. https://doi.org/10.1021/jf070337l.

Last, N. B., Schlamadinger, D. E., & Miranker, A. D. (2013). A common landscape for membrane-active peptides. *Protein Science*, *22*(7), 870–882. https://doi.org/10.1002/pro.2274.

Lazzaro, B. P., Zasloff, M., & Rolff, J. (2020). Antimicrobial peptides: Application informed by evolution. *Science*, *368*(6490), eaau5480. https://doi.org/10.1126/science.aau5480.

Lee, I. H., Cho, Y., & Lehrer, R. I. (1997). Effects of pH and salinity on the antimicrobial properties of clavanins. *Infection and Immunity*, *65*(7), 2898–2903. https://doi.org/10.1128/iai.65.7.2898-2903.1997.

Lehrer, R. I., Tincu, J. A., Taylor, S. W., Menzel, L. P., & Waring, A. J. (2003). Natural peptide antibiotics from tunicates: Structures, functions, and potential uses. *Integrative & Comparative Biology*, *43*(2), 313–322. https://doi.org/10.1093/icb/43.2.313.

Libardo, M. D. J., Bahar, A. A., Ma, B., Fu, R., McCormick, L. E., Zhao, J., et al. (2017). Nuclease activity gives an edge to host-defense peptide piscidin 3 over piscidin 1, rendering it more effective against persisters and biofilms. *The FEBS Journal*, *284*(21), 3662–3683. https://doi.org/10.1111/febs.14263.

Libardo, M. D. J., de la Fuente-Nuñez, C., Anand, K., Krishnamoorthy, G., Kaiser, P., Pringle, S. C., et al. (2018). Phagosomal copper-promoted oxidative attack on intracellular *Mycobacterium tuberculosis*. *ACS Infectious Diseases*, *4*(11), 1623–1634. https://doi.org/10.1021/acsinfecdis.8b00171.

Libardo, M. D. J., Gorbatyuk, V. Y., & Angelez-Boza, A. M. (2016). Central role of the copper-binding motif in the complex mechanism of action of ixosin: Enhancing oxidative damage and promoting synergy with ixosin B. *ACS Infectious Diseases*, *2*(1), 71–81. https://doi.org/10.1021/acsinfecdis.5b00140.

Libardo, M. D. J., Nagella, S., Lugo, A., Pierce, S., & Angelez-Boza, A. M. (2015). Copper-binding tripeptide motif increases potency of the antimicrobial peptide anoplin via reactive oxygen species generation. *Biochemical and Biophysical Research Communications*, *456*(1), 446–451. https://doi.org/10.1016/j.bbrc.2014.11.104.

Liu, B., Poolman, B., & Boersma, A. J. (2017). Ionic strength sensing in living cells. *ACS Chemical Biology*, *12*(10), 2510–2514. https://doi.org/10.1021/acschembio.7b00348.

Lizio, M. G., Campana, M., De Poli, M., Jefferies, D. F., Cullen, W., Andrushchenko, V., et al. (2021). Insight into the mechanism of action and peptide-membrane interactions of Aib-rich peptides: Multitechnique experimental and theoretical analysis. *Chembiochem*, *22*(9), 1656–1667. https://doi.org/10.1002/cbic.202000834.

Loewe, S. (1953). The problem of synergism and antagonism of combined drugs. *Arzneimittel-Forschung*, *3*, 285–290.

Loudon, A. H., Kurtz, A., Esposito, E., Umile, T. P., Minbiole, K. P. C., Parfrey, L. W., et al. (2020). Columbia spotted frogs (*Rana luteiventris*) have characteristic skin microbiota that may be shaped by cutaneous skin peptides and the environment. *FEMS Microbiology Ecology*, *96*(10), fiaa168. https://doi.org/10.1093/femsec/fiaa168.

Macomber, L., & Imlay, J. A. (2009). The iron-sulfur clusters of dehydratases are primary intracellular targets of copper toxicity. *Proceedings of the National Academy of Sciences of the United States of America*, *106*(20), 8344–8349. https://doi.org/10.1073/pnas.0812808106.

Macomber, L., Rensing, C., & Imlay, J. A. (2007). Intracellular copper does not catalyze the formation of oxidative DNA damage in Escherichia coli. *Journal of Bacteriology*, *189*(5), 1616–1626. https://doi.org/10.1128/JB.01357-06.

McGrath, D. M., Barbu, E. M., Driessen, W. H. P., Lasco, T. M., Tarrand, J. J., Okhuysen, P. C., et al. (2013). Mechanism of action and initial evaluation of a membrane active all-D-enantiomer antimicrobial peptidomimetic. *Proceedings of the National Academy of Sciences of the United States of America*, *110*(9). 3477–3472 https://doi.org/10.1073/pnas.1221924110.

Melo, M. N., Ferre, R., & Castanho, M. A. R. B. (2009). Antimicrobial peptides: Linking partition, activity and high membrane-bound concentrations. *Nature Reviews Microbiology*, 7, 245–250. https://doi.org/10.1038/nrmicro2095.

Mercer, D. K., Torres, M. D. T., Duay, S. S., Lovie, E., Simpson, L., von Köckritz-Blickwede, M., et al. (2020). Antimicrobial susceptibility testing of antimicrobial peptides to better predict efficacy. *Frontiers in Cellular and Infection Microbiology*, *10*, 326. https://doi.org/10.3389/fcimb.2020.00326.

Miller, A., Matera-Witkiewicz, A., Mikołajczyk, A., Wieczorek, R., & Rowińska-Żyrek, M. (2021). Chemical "butterfly effect" explaining the coordination chemistry and antimicrobial properties of clavanin complexes. *Inorganic Chemistry*, *60*(17), 12730–12734. https://doi.org/10.1021/acs.inorgchem.1c02101.

Mourtada, R., Herce, H. D., Yin, D. J., Moroco, J. A., Wales, T. E., Engen, J. R., et al. (2019). Design of stapled antimicrobial peptides that are stable, nontoxic and kill antibiotic-resistant bacteria in mice. *Nature Biotechnology*, *37*(10), 1186–1197. https://doi.org/10.1038/s41587-019-0222-z.

National Committee on Clinical Laboratory Standards. (1999). *Methods for determining bactericidal activity of antimicrobial agents; approved guideline*. Document M26-A National Committee on Clinical Laboratory Standards.

Outten, F. W., & Munson, G. P. (2013). Lability and liability of endogenous copper pools. *Journal of Bacteriology*, *195*(20), 4553–4555. https://doi.org/10.1128/JB.00891-13.

Pace, C. N., Vajdos, F., Fee, L., Grimsley, G., & Gray, T. (1995). How to measure and predict the molar absorption coefficient of a protein. *Protein Science*, *4*(11), 2411–2423. https://doi.org/10.1002/pro.5560041120.

Park, C. B., Kim, H. S., & Kim, S. C. (1998). Mechanism of action of the antimicrobial peptide buforin II: Buforin II kills microorganisms by penetrating the cell membrane and inhibiting cellular functions. *Biochemical and Biophysical Research Communications*, *244*(1), 253–257. https://doi.org/10.1006/bbrc.1998.8159.

Paulsen, C. E., & Carroll, K. S. (2013). Cysteine-mediated redox signaling: Chemistry, biology, and tools for discovery. *Chemical Reviews*, *113*(7), 4633–4679. https://doi.org/10.1021/cr300163e.

Pena-Miller, R., Laehnemann, D., Jansen, G., Fuentes-Hernandez, A., Rosenstiel, P., Schulenburg, H., et al. (2013). When the most potent combination of antibiotics selects for the greatest bacterial load: The smile-frown transition. *PLoS Biology*, *11*(4), e1001540. https://doi.org/10.1371/journal.pbio.1001540.

Peschel, A., Jack, R. W., Otto, M., Collins, L. V., Staubitz, P., Nicholson, G., et al. (2001). Staphylococcus aureus resistance to human defensins and evasion of neutrophil killing via the novel virulence factor MprF is based on modification of membrane lipids with L-lysine. *Journal of Experimental Medicine*, *193*(9), 1067–1076. https://doi.org/10.1084/jem.193.9.1067.

Pierce, S., Jennings, M. P., Juliano, S. A., & Angeles-Boza, A. M. (2020). Peptide-ruthenium conjugate as an efficient photosensitizer for the inactivation of multidrug-resistant bacteria. *Inorganic Chemistry*, *59*(20), 14866–14870. https://doi.org/10.1021/acs.inorgchem.0c02491.

Pollard, J. E., Snarr, J., Chaudhary, V., Jennings, J. D., Shaw, H., Christiansen, B., et al. (2012). In vitro evaluation of the potential for resistance development to ceragenin CSA-13. *Journal of Antimicrobial Chemotherapy*, *67*(11), 2665–2672. https://doi.org/10.1093/jac/dks276.

Portelinha, J., Heilemann, K., Jin, J., & Angeles-Boza, A. M. (2021). Unraveling the implications of multiple histidine residues in the potent antimicrobial peptide gaduscidin-1. *Journal of Inorganic Biochemistry*, *219*, 111391. https://doi.org/10.1016/j.jinorgbio.2021.111391.

Premdjee, B., Andersen, A. S., Larance, M., Conde-Frieboes, K. W., & Payne, R. J. (2021). Chemical synthesis of phosphorylated insulin-like growth factor binding protein 2. *Journal of the American Chemical Society*, *143*(14), 5336–5342. https://doi.org/10.1021/jacs.1c02280.

Rahnamaeian, M., Cytrynska, M., Zdybicka-Barabas, A., & Vilcinskas, A. (2016). The functional interaction between abaecin and pore-forming peptides indicates a general mechanism of antibacterial potentiation. *Peptides*, *78*, 17–23. https://doi.org/10.1016/j.peptides.2016.01.016.

Remington, J. M., Liao, C., Sharafi, M., Ste Marie, E. J., Ferrell, J. B., Hondal, R. J., et al. (2020). Aggregation state of synergistic antimicrobial peptides. *The Journal of Physical Chemistry Letters*, *11*(21), 9501–9506. https://doi.org/10.1021/acs.jpclett.0c02094.

Rishi, P., Preet, S., Bharrhan, S., & Verma, I. (2011). In vitro and in vivo synergistic effects of cryptdin 2 and ampicillin against Salmonella. *Antimicrobial Agents and Chemotherapy*, *55*(9), 4176–4182. https://doi.org/10.1128/AAC.00273-11.

Rosenfield, Y., Barra, D., Simmaco, M., Shai, Y., & Mangoni, M. L. (2006). A synergism between temporins towards Gram-negative bacteria overcomes resistance imposed by the lipopolysaccharide protective layer. *The Journal of Biological Chemistry*, *281*(39), 28565–28574. https://doi.org/10.1074/jbc.M606031200.

Rosenthal, V. D., Bijie, H., Maki, D. G., Mehta, Y., Apisarnthanarak, A., Medeiros, E. A., et al. (2012). International Nosocomial Infection Control Consortium (INICC) report, data summary of 36 countries, for 2004-2009. *American Journal of Infection Control*, *40*(5), 396–407. https://doi.org/10.1016/j.ajic.2011.05.020.

Sánchez-Clemente, R., Igeño, M. I., Población, A. G., Guijo, M. I., Merchán, F., & Blasco, R. (2018). Study of pH changes in media during bacterial growth of several environmental strains. *Proceedings*, *2*(20), 1297. https://doi.org/10.3390/proceedings2201297.

Santos, R., Ruza, D., Cunha, E., Tavares, L., & Oliveira, M. (2019). Diabetic foot infections: Application of a nisin-biogel to complement the activity of conventional antibiotics and antiseptics against Staphylococcus aureus biofilms. *PLoS One*, *14*(7), e0220000. https://doi.org/10.1371/journal.pone.0220000.

Schwalbe, R., Steele-Moore, L., & Goodwin, A. C. (Eds.). (2007). *Antimicrobial susceptibility testing protocols* (1st ed.). CRC Press. https://doi.org/10.1201/9781420014495.

Sezonov, G., Joseleau-Petit, D., & D'Ari, R. (2007). *Escherichia coli* physiology in Luria-Bertani broth. *Journal of Bacteriology*, *189*(23), 8746–8749. https://doi.org/10.1128/JB.01368-07.

Shi, G., Kang, X., Dong, F., Liu, Y., Zhu, N., Hu, Y., et al. (2021). DRAMP 3.0: An enhanced comprehensive data repository of antimicrobial peptides. *Nucleic Acids Research*. https://doi.org/10.1093/nar/gkab651.

Sieprawska-Lupa, M., Mydel, P., Krawczyk, K., Wójcik, K., Puklo, M., Lupa, B., et al. (2004). Degradation of human antimicrobial peptide LL-37 by Staphylococcus aureus-derived proteinases. *Antimicrobial Agents and Chemotherapy*, *48*(12), 4673–4679. https://doi.org/10.1128/AAC.48.12.4673-4679.2004.

Steinberg, D. A., Hurst, M. A., Fujii, C. A., Kung, A. H. C., Ho, J. F., Cheng, F. C., et al. (1997). Protegrin-1: A broad-spectrum, rapidly microbicidal peptide with in vivo activity. *Antimicrobial Agents and Chemotherapy*, *41*(8), 1738–1742. https://doi.org/10.1128/AAC.41.8.1738.

Subbalakshmi, C., & Sitaram, N. (1998). Mechanism of antimicrobial action of indolicidin. *FEMS Microbiology Letters*, *160*(1), 91–96. https://doi.org/10.1111/j.1574-6968.1998.tb12896.x.

Tabashnik, B. E. (1992). Evaluation of synergism among *Bacillus thuringiensis* toxins. *Applied and Environmental Microbiology*, *58*(10), 3343–3346. https://doi.org/10.1128/aem.58.10.3343-3346.1992.

Takahasi, H., Oshima, T., Hobman, J. L., Doherty, N., Clayton, S. R., Iqbal, M., et al. (2015). The dynamic balance of import and export of zinc in Escherichia coli suggests a heterogeneous population response to stress. *Journal of the Royal Society Interface*, *12*(106), 20150069. https://doi.org/10.1098/rsif.2015.0069.

Tang, J., Wennerberg, K., & Aittokallio, T. (2015). What is synergy? The Saariselkä agreement revisited. *Frontiers in Pharmacology*, *6*, 181. https://doi.org/10.3389/fphar.2015.00181.

Vo, T. D., Spahn, C., Heilemann, M., & Bode, H. B. (2021). Microbial cationic peptides as a natural defense mechanism against insect antimicrobial peptides. *ACS Chemical Biology*, *16*(3), 447–451. https://doi.org/10.1021/acschembio.0c00794.

Walczak, J. J., Wang, L., Bardy, S. L., Feriancikova, L., Li, J., & Xu, S. (2012). The effects of starvation on the transport of Escherichia coli in saturated porous media are dependent on pH and ionic strength. *Colloids and Surfaces B: Biointerfaces*, *90*, 129–136. https://doi.org/10.1016/j.colsurfb.2011.10.010.

Walkenhorst, W. F., Sundrud, J. N., & Laviolette, J. M. (2014). Additivity and synergy between an antimicrobial peptide and inhibitory ions. *Biochimica et Biophysica Acta*, *1838*(9), 2234–2242. https://doi.org/10.1016/j.bbamem.2014.05.005.

White, R. L., Burgess, D. S., Manduru, M., & Bosso, J. A. (1996). Comparison of three different in vitro methods of detecting synergy: Time-kill, checkerboard, and E test. *Antimicrobial Agents and Chemotherapy*, *40*(8), 1914–1918. https://doi.org/10.1128/AAC.40.8.1914.

White, C., Lee, J., Kambe, T., Fritsche, K., & Petris, M. J. (2009). A role for the ATP7A copper-transporting ATPase in macrophage bactericidal activity. *Journal of Biological Chemistry*, *284*(49), 33949–33956. https://doi.org/10.1074/jbc.M109.070201.

Wiegand, I., Hilpert, K., & Hancock, R. E. W. (2008). Agar and broth dilution methods to determine the minimal inhibitory concentration (MIC) of antimicrobial substances. *Nature Protocols*, *3*(2), 163–175. https://doi.org/10.1038/nprot.2007.521.

Winkel, J. D. T., Gray, D. A., Seistrup, K. H., Hamoen, L. W., & Strahl, H. (2016). Analysis of antimicrobial-triggered membrane depolarization using voltage sensitive dyes. *Frontiers in Cell and Developmental Biology*, *4*, 29. https://doi.org/10.3389/fcell.2016.00029.

Xie, J., & Schultz, P. G. (2005). Adding amino acids to the genetic repertoire. *Current Opinion in Chemical Biology*, *9*(6), 548–554. https://doi.org/10.1016/j.cbpa.2005.10.011.

Yamaguchi, S., Huster, D., Waring, A., Lehrer, R. I., Kearney, W., Tack, B. F., et al. (2001). Orientation and dynamics of an antimicrobial peptide in the lipid bilayer by solid-state NMR spectroscopy. *Biophysical Journal*, *81*(4), 2203–2214. https://doi.org/10.1016/S0006-3495(01)75868-7.

Yan, H., & Hancock, R. E. W. (2001). Synergistic interactions between mammalian antimicrobial defense peptides. *Antimicrobial Agents and Chemotherapy*, *45*(5), 1558–1560. https://doi.org/10.1128/AAC.45.5.1558-1560.2001.

Yang, L., Harroun, T. A., Weiss, T. M., Ding, L., & Huang, H. W. (2001). Barrel-stave model or toroidal model? A case study on melittin pores. *Biophysical Journal*, *81*(3), 1475–1485. https://doi.org/10.1016/S0006-3495(01)75802-X.

Yu, G., Baeder, D. Y., Regoes, R. R., & Rolff, J. (2016). Combination effects of antimicrobial peptides. *Antimicrobial Agents and Chemotherapy*, *60*(3), 1717–1724. https://doi.org/10.1128/AAC.02434-15.

Zerweck, J., Strandberg, E., Kukharenko, O., Reichert, J., Bürck, J., Wadhwani, P., et al. (2017). Molecular mechanism of synergy between the antimicrobial peptides PGLa and magainin 2. *Scientific Reports*, *7*, 13153. https://doi.org/10.1038/s41598-017-12599-7.

Zhao, C., Fernandez, A., Avlonitis, N., Velde, G. V., Bradley, M., Read, N. D., et al. (2016). Searching for the optimal fluorophore to label antimicrobial peptides. *ACS Combinatorial Science*, *18*(11), 689–696. https://doi.org/10.1021/acscombsci.6b00081.

Zharkova, M. S., Orlov, D. S., Golubeva, O. Y., Chakchir, O. B., Eliseev, I. E., Grinchuk, T. M., et al. (2019). Application of antimicrobial peptides of the innate immune system in combination with conventional antibiotics—A novel way to combat antibiotic resistance? *Frontiers in Cellular and Infection Microbiology*, *9*, 128. https://doi.org/10.3389/fcimb.2019.00128.

Zhu, Y., Mohapatra, S., & Weisshaar, J. C. (2019). Rigification of the *Escherichia coli* cytoplasm by the human antimicrobial peptide LL-37 revealed by superresolution fluorescence microscopy. *Proceedings of the National Academy of Sciences of the United States of America*, *116*(3), 1017–1026. https://doi.org/10.1073/pnas.1814924116.

Zou, R., Zhu, X., Tu, Y., Wu, J., & Landry, M. P. (2018). Activity of antimicrobial peptide aggregates decreases with increased cell membrane embedding free energy cost. *Biochemistry*, *57*(18), 2606–2610. https://doi.org/10.1021/acs.biochem.8b00052.

CHAPTER SIX

Colorimetric assays for the rapid and high-throughput screening of antimicrobial peptide activity against diverse bacterial pathogens

Jessie L. Allen, Sarah J. Kennedy, and Lindsey N. Shaw*
University of South Florida, Tampa, FL, United States
*Corresponding author: e-mail address: shaw@usf.edu

Contents

Methods in Enzymology, Volume 663
ISSN 0076-6879
https://doi.org/10.1016/bs.mie.2021.10.008

Abstract

With the post-antibiotic era rapidly approaching, naturally-sourced antimicrobial peptides (AMPs) are a prime resource for restocking our antibiotic medicine cupboard. The efficiency of identification requires high-throughput screens that can identify bioactive peptides present within abundant natural-products chemical-space. While there are multiple amenable and high sensitivity colorimetric-based screening approaches available, resazurin-based assays are cost-effective, peptide compatible, and expedient, allowing one to screen a multitude of AMPs in a high-throughput fashion. Herein, we provide a detailed protocol for the optimization and use of resazurin assays for AMP testing, providing key experimental insight, and highlight pitfalls to be avoided.

1. Introduction

Antimicrobial peptides (AMPs) are continually being investigated for the therapeutic potential of targeting multi-drug resistant microorganisms. Every domain of life produces some form of AMP, leaving vast opportunities for the discovery of clinically-relevant peptide structures. While these naturally-derived structures may not always be bioactive, the scaffolds can inspire semi-or entirely synthetic follow-up discoveries with improved bioactivity. To date, there are 3,273 AMPs within the antimicrobial peptide database (Wang, Li, & Wang, 2016), however, only seven AMPs have FDA approval for the treatment of bacterial infections (Chen & Lu, 2020). This discrepancy highlights the need to continue AMP development efforts on a high-throughput scale to find those with the greatest therapeutic potential. AMPs are different from standard small molecules because they are typically cationic in nature and tend to have short half-lives. Because of their short half-life, bioassays have to be adjusted to include minimizing freeze-thaw cycles and decreasing assay times. As a result, standard overnight minimum inhibitory concentration (MIC) methods are not always ideal for AMP testing because potentially active peptides can go unnoticed due to stability issues. Importantly, a feature of FDA approved AMPs is their stability in vivo; which is problematic if stability is already an issue in vitro. With that said, in the event an unstable peptide is found to be active, its clinical potential can be remediated through targeted structural modifications to increase stability while maintaining bioactivity.

With the threat of a post-antibiotic age, and an underrepresentation of clinically approved AMPs, there exists a need for an easy-to-use, peptide-compatible, high-throughput system to parallelly screen and guarantee same-day results. Resazurin, also called Alamar Blue, diazo-rescorcinol, azoresorcin, or resazoin, is a non-toxic and water-soluble colorimetric indicator of metabolic activity in a variety of organisms. Bacterial NAD(P)H-dependent oxidoreductases and dehydrogenases convert resazurin from its oxidized form to a reduced form, causing fluorescence and color changes (Braissant, Astasov-Frauenhoffer, Waltimo, & Bonkat, 2020). Such changes are easily visualized and can be analyzed qualitatively or quantitatively. As suggested, standard overnight assays may underestimate bioactivity due to a peptide's short half-life. Metabolic-dye based approaches, however, indirectly monitor bioactivity through bacterial viability and therefore can assess both bactericidal and bacteriostatic AMP activity while limiting assay time to compensate for shortened half-lives.

In addition to being high-throughput, resazurin assays are also remarkably inexpensive and can be performed in relatively short experimental timeframes. Herein, we detail the benefits and pitfalls of using resazurin to assess AMP bioactivity and describe an assay optimization protocol that is easy to use and can be adapted to a variety of different bacterial species.

2. AMP drug discovery

Due to the physicochemical properties of AMPs, there are constraints with regard to their testing and development. Overall, AMPs are characterized as small, cationic, amphipathic, and generally unstable. If the tested AMP amino acid sequences are known and similar enough to existing peptides, then *in silico* predictions of physicochemical stability and half-life can generally identify bioassay limitations and guide experimental designs (Santos, Pujols, Pallarès, Iglesias, & Ventura, 2020). However, for rapid and high-throughput screens, the sequence identities are not always known, especially for novel chemical scaffolds derived from natural products. This uncertainty means there are generalized considerations that need to be addressed when designing experiments for antibacterial screening.

2.1 Stability, charge, solubility, and aggregation of AMPs

It is important to acknowledge the cationic nature of AMPs when screening for bioactivity. The positive charge of peptides can cause off-target binding before AMPs are ever introduced to a tested pathogen; resulting in missed bioactivity. This off-target, or non-pathogen, binding can include to the experimental materials, compound vials, and even the pipette tips used to transfer the material. Consequently, the handling and preparation of AMPs should account for the limitations inherent to AMP physiochemistry. The plastic or glassware most commonly used in laboratories (polypropylene, polystyrene, borosilicate glass) all have literature precedence for AMP-binding (Chico, Given, & Miller, 2003; Kristensen, Henriksen, & Andresen, 2015). This off-target binding to assay materials is even demonstrated during AMP-filtration for sterilization. As an example of off-target binding that may occur, commonly used (and negatively-charged) cellulose acetate filters can bind positively-charged AMPs (Wiegand, Hilpert, & Hancock, 2008). These AMP physicochemical limitations are inherent to working with peptides, and experimental design can only minimize (and not eliminate) sample attrition. Furthermore, when supplementing AMP solutions with DMSO or other diluents to minimize off-target binding, it is important to consider that such additions can contribute their own confounding biological interactions to the bioassay, perhaps resulting in misleading MIC values (Mercer et al., 2020). For example, it is now known that while polysorbate 80 (Tween 80) can decrease off-target binding, it also contains inherent antibacterial properties that can synergistically interact with the tested peptide to decrease MICs (Mercer et al., 2020). When designing AMP bioactivity screens, the design must incorporate approaches that address and overcome AMP limitations. For example, when preparing AMP solutions, one could increase the concentrations of AMPs to account for sample loss to off-target interactions, minimize excessive freeze-thaw cycles, and pursue same-day test results if pathogen growth characteristics allow for it.

Consequently, once the preparation of an AMP solution is complete, a thorough consideration of peptide interaction with experimental materials within the bioassay should be performed. Such assays are perhaps best performed using broth microdilutions, as cationic AMPs have been shown to interact with the negatively-charged components of agar during disk diffusion approaches (Lehrer, Rosenman, Harwig, Jackson, & Eisenhauer, 1991). Media and nutrient conditions must accommodate the tested pathogen; however, one should also optimize the media to account for AMP physiochemistry. Nutrient media optimizations should limit AMP insolubility, aggregation, or precipitation since this inactivation of AMPs could otherwise mask or mislead bioassay results. Excessive amounts of media salts are known to "salt-out" AMP aggregates (Zai et al., 2021), and literature does show precedence for improved AMP bioactivity upon removal of excessive anions that precipitate cationic AMPs (Turner, Cho, Dinh, Waring Alan, & Lehrer Robert, 1998). Toward this point, Mueller Hinton broth (MHB) is generally accepted for AMP bioassays, despite it poorly mimicking human in vivo conditions (Mercer et al., 2020). Various MHB supplements can accommodate more fastidious organisms (Mercer et al., 2020; Wiegand et al., 2008); this flexibility provides an almost-internationally standard media important for communicating and reproducing bioassay results. Toward the point of reproducibility, however, one should avoid standard turbidity-based broth microdilutions. The lack of quick and informative bioassay results of a standard turbidity experiment cannot accommodate the short-lived AMP half-life, and turbidity results do not necessarily equate to viable cells. If an AMP is inactivated during testing, the tested pathogen may recover and therefore mask potential bioactivity. For quick and rapid results required for overcoming AMP instability, one should instead consider reporter dyes that monitor bacterial viability via metabolism.

2.2 The advantages of using resazurin as compared to other metabolic dyes

Many fluorescent metabolic reporter dyes have been used for a variety of purposes when studying organisms from all three domains of life. While there are clear advantages to using a metabolic-based approach to AMP discovery, there are also factors to account for in experimental design. When considering the many fluorescent dyes (MTT, XTT, INT, LDH, etc), resazurin is perhaps the best choice, having been used since 1929 (Rampersad, 2012) for the detection of metabolism in bacteria, fungi, and eukaryotic cell lines alike. One advantage to choosing resazurin is that it is non-toxic and can be used on bacteria that are still viable, while tetrazolium-based dyes (MTT, XTT, INT) require an end-point assessment and cell lysis prior to data acquisition. The reason for this latter point is that the formation and accumulation of toxic formazan crystals within cells does cause a color change, however, the accurate quantification of this necessitates access to the cytoplasm for proper solubilization of crystals. Another advantage is that resazurin is water-soluble, while the other dyes referenced are commonly solubilized in organic solvents; these solvents can be toxic to bacteria and may confound bioactivity determinations. Furthermore, resazurin acts as an intermediate electron acceptor in the electron transport chain without interfering with the normal transfer of electrons. Resazurin can be reduced by: NADPH, $FADH_2$, FMNH, NADH as well as cytochromes while many other dyes are unable to be metabolized by cytochromes (Rampersad, 2012). Toward this point, resazurin has been used in a variety of bacterial culture media and with an array of taxonomy. There is also literature precedence for antibacterial AMP assays utilizing resazurin for high-throughput screens. Table 1 summarizes selected literature examples (Barreras-Serrano et al., 2012; Chuah et al., 2014; Coban et al., 2005; Farkas et al., 2018; Jia et al., 2020; Kim & Jang, 2017; Kirkpatrick et al., 2017; Nagarajan et al., 2018, 2019; Rangel et al., 2020; Sala et al., 2018; Sarker et al., 2007; Sivakumar et al., 2010; Travnickova et al., 2019).

2.3 Overcoming the difficulties of assaying various pathogens using the same assay

To identify AMPs that are bioactive in a high-throughput manner, one should consider parallelly screening multiple pathogen species, such as the ESKAPE (*Enterococcus faecium, Staphylococcus aureus, Klebsiella pneumoniae, Acinetobacter baumannii, Pseudomonas aeruginosa, Enterobacter cloacae*) pathogen set. Although it is beneficial for expanding the potential for discovering novel bioactivity, examining multi-drug resistant ESKAPE variants against a library of antimicrobials poses several obstacles. Two of the organisms are Gram-positive, while the remaining four are Gram-negative. In addition, the generation times for each organism are unique. Conversely, whilst the pathogens may have their differences, they all share the same form of respiration in the conversion of NADH to NAD+. This shared metabolic pathway makes resazurin a prime candidate to use for each organism, however, some optimization is needed for organism-specific conditions.

Prior to starting the screen, it is ideal to determine three things: (1) the doubling time of each bacterial species, (2) the OD_{600} value that corresponds to an ideal starting CFU/mL, (3) the rate at which resazurin is metabolized by the bacterial species being tested.

When designing an experiment to investigate bioactivity against multiple pathogens, it is important to consider a generalized media amenable to the tested compounds, reporter dye, and pathogen nutritional needs. For this purpose, the Mueller Hinton broth (MHB) discussed above has literature precedence for non-fastidious species, especially for AMP-resazurin bioassays (Wiegand et al., 2008).

Table 1 Literature precedence of resazurin-based screening of AMPs. From left to right: the genera, species, media, inocula (CFU/mL), starting optical density (OD in nm; grey shading indicates value not reported), incubation, results determination, tested compound/peptide type, and reference.

Genus	Species	Media	Inoculum (CFU/mL)	Optical density before dilutions (OD, nm)	Incubation	Determination	Peptide type	Reference
Enterococcus								
	Clinical spp. (control ATCC 29212)	caMHB and isosensitest	5 x 10^5	0.5 McFarland Standard	35 °C: 5 h bacterial incubation, then 5 h with 30 μL of 0.01% resazurin	Endpoint: visual color change	Vancomycin (glycopeptide)	Coban et al. (2005)
		caMHB and isosensitest	5 x 10^6	0.5 McFarland Standard	35 °C: 4 h bacterial incubation, then 5–10 min with 30 μL of 0.01% resazurin	Endpoint: visual color change	Vancomycin (glycopeptide)	(Coban et al., 2013)
	faecalis (CCM 4224)	TSB	1–3 x 10^5	OD_{600} = 1.0	37 °C: 4 h bacterial incubation with membrane, then 5 min with 20 μmol/L resazurin and membranes removed	Intervals of 30 min for 720 min: 520 nm ex, 590 nm em	Not a peptide; antimicrobial electrospun nanofiber filtration membrane	Travnickova et al. (2019)
	faecalis (clinical spp.)	MHB	5 x 10^5–5 x 10^4	0.5 McFarland Standard	37 °C: 12 h bacterial incubation, then 12 h with 30 μL of 0.02% resazurin	Endpoint: 530 nm ex, 590 nm em	*In silico*-guided synthetic peptides	Nagarajan et al. (2019)
Staphylococcus								
	aureus (NCTC 10788, NCTC 11940)	Isosentitest	5 x 10^5		37 °C: 18–24 h bacterial incubation with 10 μL of 270 mg/40 mL solubilized resazurin tablet	Endpoint: visual color change	Not a peptide; purified and crude plant extracts	Sarker, Nahar, and Kumarasamy (2007)

Continued

Table 1 Literature precedence of resazurin-based screening of AMPs. From left to right: the genera, species, media, inocula (CFU/mL), starting optical density (OD in nm; grey shading indicates value not reported), incubation, results determination, tested compound/peptide type, and reference.—cont'd

Genus	Species	Media	Inoculum (CFU/mL)	Optical density before dilutions (OD, nm)	Incubation	Determination	Peptide type	Reference
	aureus (ATCC 25923, ATCC 700699)	MHB	5×10^5	0.5 McFarland Standard: $OD_{625} = 0.08–0.13$	37 °C: 16–18 h bacterial incubation, then 2 h with 30 µL of 0.01% resazurin	Endpoint: visual color change	Crude plant extracts	Chuah, Zakaria, Suhaili, Abu Bakar, and Mohd Desa (2014)
	aureus (ATCC 33591)	caMHB	5×10^5	0.5 McFarland Standard	37 °C: 16 h bacterial incubation, then 1 h with 30 µL of 0.01% resazurin	Endpoint: visual color change	Canine-derived scaffold, synthetic K9CATH AMP	Barreras-Serrano et al. (2012)
	aureus (clinical spp.)	MHB	5×10^5–5×10^4	0.5 McFarland Standard	37 °C: 12 h bacterial incubation, then 12 h with 30 µL of 0.02% (w/v) resazurin	Endpoint: 530 nm ex, 590 nm em	*In silico*-guided synthetic peptides	Nagarajan et al. (2019)
	aureus (NCIM 5021)	MHB	5×10^5		37 °C: 18 h bacterial incubation, then 2 h with 10 µL of 0.01% resazurin	Endpoint: visual color change	Not a peptide; cyclical compounds	Sivakumar, Ganesan, Veluchamy, and Doble (2010)
Klebsiella								
	aerogenes (NCTC 9528)	Isosentitest	5×10^5		37 °C: 18–24 h bacterial incubation with 10 µL of 270 mg/40 mL solubilized resazurin tablet	Endpoint: visual color change	Not a peptide; purified and crude plant extracts	Sarker et al. (2007)
	pneumoniae (ATCC 13883)	LB	3×10^4	$OD_{600} = 0.0001$	37 °C: 6 h bacterial incubation, then 1 h RT incubation with 5 µL of 0.125 mg/mL resazurin	Endpoint: 530 nm ex, 590 nm em	Colistin, gentamicin, tetracycline, imipenem	Kim and Jang (2017)

pneumoniae (clinical spp.)	MHB	5×10^5–5×10^4	0.5 McFarland Standard	37 °C: 12 h bacterial incubation, then 12 h with 30 μL of 0.02% (w/v) resazurin	Endpoint: 530 nm ex, 590 nm em	*In silico*-guided synthetic peptides	Nagarajan et al. (2019)
Acinetobacter							
baumannii MDR-clinical strains (ATCC 19606 control)	caMHB, pH 7.3	1×10^6	0.5 McFarland Standard	37 °C: 24 h bacterial incubation, then 1 h with 0.02% resazurin	Endpoint: visual color change	Bee venom-derived melittin	Rangel et al. (2020)
baumannii (clinical spp.)	MHB	5×10^5–5×10^4	0.5 McFarland Standard	37 °C: 12 h bacterial incubation, then 12 h with 30 μL of 0.02% (w/v) resazurin	Endpoint: 530 nm ex, 590 nm em	*In silico*-guided synthetic peptides	Nagarajan et al. (2019)
baumannii (clinical spp.)	MHB	1×10^9	3.5 McFarland Standard	35 ± 2 °C: 3 h bacterial incubation, then 22 μL of resazurin PrestoBlue	Intervals of 15 min for 1 h: visual color change	Colistin	Jia et al. (2020)
Pseudomonas							
aeruginosa (clinical spp.)	MHB	5×10^5–5×10^4	0.5 McFarland Standard	37 °C: 12 h bacterial incubation, then 12 h with 30 μL of 0.02% (w/v) resazurin	Endpoint: 530 nm ex, 590 nm em	*In silico*-guided synthetic peptides	Nagarajan et al. (2019)
aeruginosa (NCTC 6750)	Isosentitest	5×10^5		37 °C: 18–24 h bacterial incubation with 10 μL of 270 mg/40 mL solubilized resazurin tablet	Endpoint: visual color change	Not a peptide; purified and crude plant extracts	Sarker et al. (2007)
aerguinosa (clinical spp. and ATCC 27853 control)	MHB	1×10^8	3.5 McFarland Standard	35 ± 2 °C: 3 h bacterial incubation, then 22 μL of resazurin PrestoBlue	Intervals of 15 min for 1 h: visual color change	Colistin	Jia et al. (2020)

Continued

Table 1 Literature precedence of resazurin-based screening of AMPs. From left to right: the genera, species, media, inocula (CFU/mL), starting optical density (OD in nm; grey shading indicates value not reported), incubation, results determination, tested compound/peptide type, and reference.—cont'd

Genus	Species	Media	Inoculum (CFU/mL)	Optical density before dilutions (OD, nm)	Incubation	Determination	Peptide type	Reference
	aeruginosa (NCIM 5029)	MHB	5×10^5		37 °C: 18 h bacterial incubation, then 2 h with 10 μL of 0.01% resazurin	Endpoint: visual color change	Not a peptide; cyclical compounds	Sivakumar et al. (2010)
Enterobacter								
	aerogenes (NCIM 5139)	MHB	5×10^5		37 °C: 18 h bacterial incubation, then 2 h with 10 μL of 0.01% resazurin	Endpoint: visual color change	Not a peptide; cyclical compounds	Sivakumar et al. (2010)
	cloacae (clinical sp.)	MHB	1×10^7	$OD_{600} = 0.5$	37 °C: 1.5–4 h bacterial incubation, then 1 h with 5 μL of 1.19 mM resazurin	Endpoint: 544 nm ex, 590 nm em	Plant peptides	Kirkpatrick et al. (2017)
Escherichia								
	coli (clinical spp.)	MHB	5×10^5— 5×10^4	0.5 McFarland Standard	37 °C: 12 h bacterial incubation, then 12 h with 30 μL of 0.02% (w/v) resazurin	Endpoint: 530 nm ex, 590 nm em	*In silico*-guided synthetic peptides	Nagarajan et al. (2019)
	coli (ATCC 8739)	MHB, MHBII, modified low-salt medium	5×10^5	$OD_{600} = 0.5–1.0$	37 °C: 20 h incubation of bacteria, peptides, and 10 μL of 0.1% resazurin	Endpoint: visual color change	Synthesized derivatives of plant peptides	Farkas, Pap, Kondorosi, and Maróti (2018)

coli (NCTC 8110)	Isosentitest	5×10^5		37 °C: 18–24 h bacterial incubation with 10 μL of 270 mg/40 mL solubilized resazurin tablet	Endpoint: visual color change	Not a peptide; purified and crude plant extracts	Sarker et al. (2007)
coli (CCM 3954)	TSB	$1–3 \times 10^5$	$OD_{600} = 1.0$	37 °C: 4 h bacterial incubation with membrane, then 5 min with 20 μmol/L resazurin and membranes removed	Intervals of 30 min for 720 min: 520 nm ex, 590 nm em	Not a peptide; antimicrobial electrospun nanofiber filtration membrane	Travnickova et al. (2019)
coli (clinical spp. and control ATCC 25922)	MHB	1×10^8	3.5 McFarland Standard	35 ± 2 °C: 3 h bacterial incubation, then 22 μL of resazurin PrestoBlue	Intervals of 15 min for 1 h: visual color change	Colistin	Jia et al. (2020)
Other							
Bacillus subtilis (NCIM 2718)	MHB	5×10^5		37 °C: 18 h bacterial incubation, then 2 h with 10 μL of 0.01% resazurin	Endpoint: visual color change	Not a peptide; cyclical compounds	Sivakumar et al. (2010)
Bacillus subtilis (ATCC 11774)	MHB, MHBII, modified low-salt medium	5×10^5	$OD_{600} = 0.5–1.0$	37 °C: 20 h incubation of bacteria, peptides, and 10 μL of 0.1% resazurin	Endpoint: visual color change	Synthesized derivatives of plant peptides	Farkas et al. (2018)
Bacillus cereus (NCTC 9689)	Isosentitest	5×10^5		37 °C: 18–24 h bacterial incubation with 10 μL of 270 mg/40 mL solubilized resazurin tablet	Endpoint: visual color change	Not a peptide; purified and crude plant extracts	Sarker et al. (2007)

Continued

Table 1 Literature precedence of resazurin-based screening of AMPs. From left to right: the genera, species, media, inocula (CFU/mL), starting optical density (OD in nm; grey shading indicates value not reported), incubation, results determination, tested compound/peptide type, and reference.—cont'd

Genus	Species	Media	Inoculum (CFU/mL)	Optical density before dilutions (OD, nm)	Incubation	Determination	Peptide type	Reference
	Lactobacillus plantarum (NCTC 6376)	Isosentitest	5×10^5		37 °C: 18–24 h bacterial incubation with 10 μL of 270 mg/40 mL solubilized resazurin tablet	Endpoint: visual color change	Not a peptide; purified and crude plant extracts	Sarker et al. (2007)
	Proteus vulgaris (NCIM 2813)	MHB	5×10^5		37 °C: 18 h bacterial incubation, then 2 h with 10 μL of 0.01% resazurin	Endpoint: visual color change	Not a peptide; cyclical compounds	Sivakumar et al. (2010)
	Salmonella typhi (NCIM 2501)	MHB	5×10^5		37 °C: 18 h bacterial incubation, then 2 h with 10 μL of 0.01% resazurin	Endpoint: visual color change	Not a peptide; cyclical compounds	Sivakumar et al. (2010)
	Neisseria mucosa (MTCC 1772), *Corynebacterium pseudotuberculosis* (MTCC 3158), *Micrococcus luteus* (MTCC 425), *Mycobacterium smegmatis* (MTCC 2155)	MHB	5×10^5– 5×10^4	0.5 McFarland Standard	37 °C: 12 h bacterial incubation, then 12 h with 30 μL of 0.02% (w/v) resazurin	Endpoint: 530 nm ex, 590 nm em	*In silico*-guided synthetic peptides	Nagarajan et al. (2018)
	Mycobacterium fortuitum (ATCC 6841), *Mycobacterium smegmatis* (ATCC 19420)	Middlebrook 7H9 broth +/- EDTA	5×10^5	0.5 McFarland Standard	37 °C: 72 h bacterial incubation, then 24 h with 20 μL of 0.01% (w/v) resazurin	Endpoint: visual color change	Derivatives of natural product from cobra venom	Sala et al. (2018)

2.4 Limitations of resazurin-based assays and how to overcome them

As resazurin measures cell viability based on metabolic activity, care must be taken to optimize assays to maintain inocula within limits of detection. If cell densities are too high for the experimental design, then the reporter dye may be too rapidly reduced beyond levels of detection for reporting informative data. For example, *Escherichia coli* and *Salmonella enterica* serovar Typhimurium have an upper limit of detection of 10^{10} CFU/mL (Costa, Gomes, Braz, Pereira, & Almeida, 2021). Upon reaching this upper limit, the dye no longer fluoresces and leads to an underestimation of bacterial metabolic activity. These limitations, along with the issue of light sensitivity, require experimental approaches to limit incubation times to avoid dye overexposure. With that in mind, obtaining species-specific optimal inocula based on generational doubling time, and tracking resazurin fluorescence over time, are the safest ways to remain within detection limits. Unfortunately, the majority of resazurin-based AMP screens rely on extended incubations and endpoint determinations. While many factors contribute to the infrequency of AMP drug discovery, this disregard for AMP instability and short half-lives within experimental design undoubtedly causes a wealth of potential peptides to remain undiscovered. With this AMP limitation in mind, we propose a modified approach to high-throughput AMP-resazurin drug discovery screens, focusing on the initial monitoring of resazurin metabolism over time.

Whilst this method focuses on AMP drug discovery in a high-throughput manner, we must consider the most critical experimental variable: bacterial density. Before any peptide-specific experimental considerations, pathogen-specific cellular densities and doubling rates must be determined. Once bacterial inocula have been optimized, any number of different AMPs can be screened in a high-throughput resazurin screen. Without optimum bacterial density, AMP bioactivity could either be under- or over-estimated. To this point, we preface our proposed resazurin assay by first presenting bacterial inocula optimizations.

3. Using resazurin to assess antibacterial activity

3.1 Equipment

- Citation 5 Plate Reader (BioTek)
- UV-Visible Spectrometer (BioMate 3S, ThermoFisher)

- 1 mL Cuvettes compatible with Spectrometer
- 125 mL Erlenmeyer Flasks (beveled if available)
- Multichannel Pipettes: P10, P300 12-well
- Media Reservoirs
- Micropipettes: P1000, P200, P20, P10
- Test Tubes
- Shaking Incubator, 37 °C
- Static Incubator, 37 °C
- Polypropylene clear, flat-bottom 96-well Plate
- Polystyrene clear, flat-bottom 96-well Plates
- Petri-dishes
- 1.5 mL tubes
- 50 mL conical vials
- 15 mL conical vials
- Autoclave

Alternatives: Any plate reader that has the capacity to read in the range of 530–660 nm is suitable for the detection of resazurin. If not testing AMPs specifically, it is appropriate to use standard polystyrene plates. Further, this can be adapted to using 384-well plates with modifications to the volumes per well. Erlenmeyer flasks (250 mL) containing 100 mL of media for synchronization are also appropriate.

3.2 Reagents

- Resazurin powder
- Phosphate buffered saline (PBS)
- Media of choice
- Agar form of media of choice

4. Protocol: Bacterial inocula optimization

4.1 Dilution plate and tubes set up

Firstly, it is critical to establish the range of bacterial inocula within the spectrophotometer quantification threshold. If bacterial densities are too high, the resazurin-monitored metabolism may rapidly exceed quantification limits, preventing informative bioassay monitoring. Similarly, if bacterial densities are too low, metabolic activities may not be rapid enough for unstable and short-lived AMPs to act upon a bacterial population.

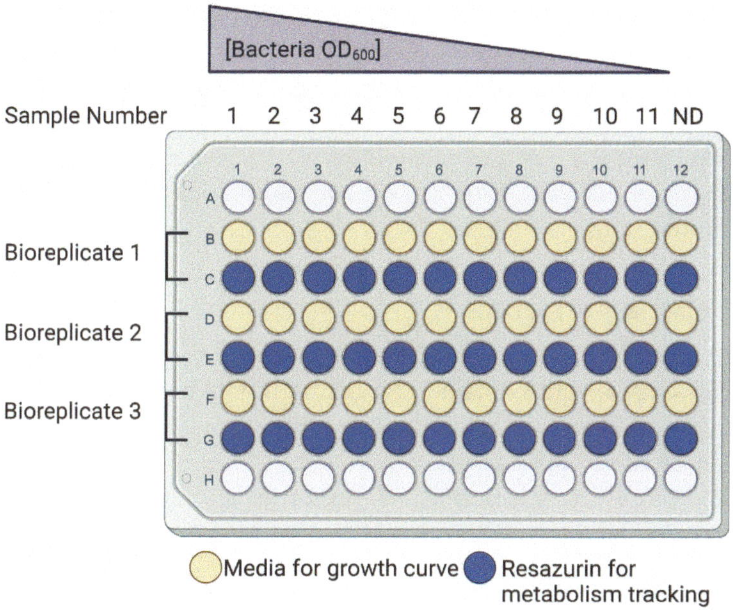

Fig. 1 Polypropylene plate layout for optimization protocol. ND: no drug or media only control. *Created with BioRender.com.*

To this end, we propose an optimization protocol to establish a range of pathogen-specific optical density 600 nm (OD_{600}) values. The range should start high (i.e. OD_{600} of 2.0) and span at least 11 dilutions. The typical 96-well plate format allows for 11 different dilutions of upwards of 8 replicates and allows for the inclusion of proper controls. Controls should include uninoculated media and solvent only containing media.

To adapt the protocol for high-throughput screening, a color-coding system can be employed to differentiate pathogens, biological replicates, ODs, and dilutions. From the above 96-well plate design for a single test pathogen in biological triplicate (Fig. 1), one would transfer the sample identifiers to thirty-three 1.5 mL tubes. The identifiers should indicate the tested OD_{600} and specific dilution per replicate pathogen. For simplicity, within one pathogen biological replicate, we will refer to the dilution samples as #1-#11 per tested OD_{600} with sample #1 signifying the highest OD_{600} to be tested.

4.2 Preparing for bacterial synchronization and spot plating

Prepare 3x 125 mL flasks containing 50 mL of media of choice to be used for synchronizing each biological replicate. Pour and dry 66 Petri dishes with the media/agar of choice. Each biological replicate will need to be plated on agar for CFU/mL using 22 plates. Label each plate with six quadrants (similar to a pie chart with six equal spaces) and label with dilutions $10^{-7} \rightarrow 10^{-2}$ in duplicate (10^{-7}, 10^{-6}, 10^{-5} x2 on one plate and 10^{-4}, 10^{-3}, 10^{-2} x2 on another). Each sample will be plated from each dilution in technical duplicate, labeled with the unique sample identifier following the above instructions.

4.3 Preparing resazurin

Prepare a resazurin stock solution at 10x desired final concentration in a light-blocking/opaque container. Resazurin should be solubilized per manufacturer guidelines (typically in water or phosphate buffered saline (PBS)) and filter-sterilized prior to use.

Note: Store stock solution of resazurin at -20 °C between uses. Freeze-thaw should be minimized by aliquoting multiple concentrated, smaller-volume samples.

4.4 Final optimization preparations

Prepare extra volume of sterile media to use for diluting stocks and overnight culture preparation. Make sure 3 test tubes are sterilized.

Prepare PBS for serial dilutions and spot plating.

For one pathogen tested in biological triplicate: inoculate single colonies into three test tubes containing 5 mL of the media of choice and incubate overnight (15–18 h), shaking at 37 °C.

Materials

(1) 10x Resazurin stock
  - **(a)** 20% stock made in PBS

(2) MHBI media (or media of choice)

(3) MHBI agar (or agar of choice) – 66 Petri dishes

(4) 1.5 mL tubes x33

(5) Polystyrene 96-well plates x3 for dilutions

(6) Polypropylene flat-bottom plate for bioassay (in the plate reader)

(7) Three 5 mL tubes of media for overnight cultures

Establish a range of OD_{600} values to be tested. This range should start high and span 11 dilutions, starting with the highest OD_{600} (~2.0) to ensure the tested range contains the bacterial density required to establish the upper quantification threshold of bacterial metabolism. Similarly, a high number of dilutions will ensure that a lower threshold where resazurin cannot be metabolized is also identified. Within a typical 96-well plate format, 11 different dilutions are possible while including proper controls such as uninoculated media. This procedure can be adapted to include 5–8 samples for the OD_{600} range to generate a reliable correlation graph.

5. Step-by-step method details

Optimization to find ideal parameters for organism of interest.

5.1 Synchronizing cultures to exponential growth

Timing: 3–3.5 h

1. Dilute overnight cultures 1:100 into 125 mL Erlenmeyer flasks containing 50 mL of fresh media.
   (a) 500 μL overnight culture into 50 mL of media.
2. Incubate at 37 °C with shaking to reach exponential phase.
   (b) For organisms such as *S. aureus, E. coli,* or *A. baumannii,* 3 hours is sufficient.

Note: For testing one strain of bacteria, three separate cultures should be used to perform the procedure in biological triplicate.

5.2 Prepare while cultures are synchronizing

Timing: 1–2 h

3. Prepare spectrophotometer cuvettes by filling with 900 μL of fresh media and a blank containing 1000 μL of media.
   (a) Each biological replicate should have its own cuvette.
4. Label three 50 mL conical vials and three 15 mL conical vials corresponding to each of the synchronized cultures (biological replicates).
   (b) The 50 mL vials will be used to concentrate bacterial cultures, and the 15 mL vials will serve as stock cultures for diluting and seeding assay plates.

5.3 Prepare OD_{600} samples and assay plate

Timing: 30 min–1 h

5. Following incubation, transfer synchronized samples into corresponding 50 mL conical vials and centrifuge for 10–15 min at 5000 x *g* to concentrate cells.
6. Discard supernatant and resuspend in 2 mL of fresh media, transferring to 15 mL conical stock vials.

7. Add 100 μL of each bacterial stock solution into cuvettes with mixing until homogenous. Use media only cuvette as a blank and measure samples using a UV-spectrophotometer at OD_{600} nm.
8. Record OD_{600} values while accounting for the 1:10 dilution of each read. Use these results to standardize each bacterial culture to the predetermined starting OD_{600} (the highest desired OD_{600} value, sample #1).

   Critical note: The starting OD_{600} should be determined for each organism individually. For example, if working with *S. aureus*, starting with a highest OD_{600} of 2.0 would be sufficient. It is essential that each organism is treated differently since variations in cell size and shape can influence their OD_{600} values (refer to Table 1 for examples).
   - **(a)** (OD_{600} recorded/desired OD_{600}) x 1000 μL = volume of stock solution to add to 1.5 mL sample vial #1.
   - **(b)** 1000—volume of stock solution calculated in 8a = volume of fresh media to add to 1.5 mL sample vial #1.
9. Add 500 μL of media to remaining 1.5 mL sample vials #2–#11.
10. Perform 1:2 dilutions. Vortex tube #1 or mix until homogenous, then transfer 500 μL from tube #1 to tube #2.
11. Vortex or mix tube #2 and transfer 500 μL to tube #3.
12. Repeat the dilution series until tube #11 is diluted by tube #10.

Note: It is important to discard and use a new micropipette tip in between samples. If the samples are to sit for more than 30 min, it would be ideal to resuspend stocks in PBS at 10x the desired starting concentration instead of media to prevent growth prior to recording on a plate reader. If this is the case, prepare the polypropylene plate with 180 μL of media and add 20 μL of each diluted sample per well.

5.4 Defining plate reader parameters

Timing: 5 min

13. Create a protocol using plate reader software to take an OD_{600} measurement as a baseline, then set parameters to incubate the plate at 37 °C, reading OD_{600} every 15 min followed by a fluorescence measurement of 544 nm excitation and 590 nm emission. Shaking in between steps is optional.
    - **(a)** Turn on plate reader to ensure it is warm and to the correct temperature when samples are ready.

Note: Using these parameters will allow for a growth curve to be performed by taking OD_{600} reads while simultaneously evaluating the reaction time of resazurin with the organism under investigation by measuring fluorescence. If one does not have a plate reader with capabilities to measure OD_{600} and fluorescence concurrently, running the experiment for growth on an alternate day is acceptable.

Alternative: Using an excitation of 530 nm is also widely used and accurate. It is most important that the parameters are not changed once chosen. If 544 nm is used during optimization protocols, it must be used for the remainder of experiments.

5.5 Preparing the polypropylene plate to assay over time in plate reader

Timing: 10–15 min

14. With stock cultures diluted and ready, next prepare the optimization plate that will be measured using the plate reader. Each sample should be seeded within the polypropylene 96-well plate in biological duplicate with 200 µL per well.
   (a) Adding water or PBS to wells A1-12 and H1-12 will minimize evaporation.

15. Biological replicate 1 samples will be added to wells B1-11 and C1-11. Biological replicate #2 samples should be added to wells D1-11 and E1-11, biological replicate #3 samples should be added to wells F1-11 and G1-11 (Fig. 1).
   (a) Each diluted sample will be seeded in duplicate. For example, sample tube #1 for replicate 1 containing the highest OD_{600} will be added to wells B1 and C1. Sample tube #2 will be added to B2 and C2 and so on.

16. Add media-only controls (200 µL of fresh media) to wells B12–G12.

17. Take an immediate measurement of the inoculated plate using the plate reader at OD_{600} to get a baseline and ensure duplicate samples were added to each well with accuracy.

18. Prepare the stock of resazurin using a reservoir and quickly add 20 µL of 10x stock to rows B, D, and F using a multichannel pipette.
   (b) One replicate of each sample will be used to record growth while the other will be subject to resazurin metabolism tracking to determine the ideal end OD_{600} for bioassay.

Critical step: Make sure that resazurin is not exposed to light for an extended period of time, especially when bacterial samples within the polypropylene plate are exposed to light. It is recommended that the stock and polypropylene plate are covered with aluminum foil while not actively being pipetted and when the assay plate is transferred from the benchtop, into the plate reader. We recommend limiting exposure to less than three minutes during the addition of resazurin to samples and less than 30 s when transferring between the foil wrap and the plate reader.

19. Start the plate reader protocol and let it run for at least 4 h. Running the program up to 8 h will give more information but may result in evaporation.

Note: Evaporation typically occurs at the corners of the assay plate first and influences the exterior wells more than the internal wells. Visual differences in liquid levels when viewing the assay plate from the side are clear indications of evaporation. In addition, the OD_{600} will be altered when significant evaporation has occurred. For example, if two wells have the same amount of bacteria, the one that has experienced evaporation will have a higher OD_{600} due to concentrating of the bacteria per total volume within the well. The inclusion of a water moat will help alleviate this, as shown in Fig. 1.

5.6 Using dilutions to generate CFU/mL counts corresponding to OD_{600} samples prepared

Timing: 1–2 h

20. Prepare dilution plates for each sample to be spot plated (samples 1–11 in biological triplicate). Add 270 μL of PBS to wells B1–11 through H1–11 of three 96-well polystyrene plates. Each 96-well polystyrene plate will be used for one biological replicate.

- **(a)** Using 11 samples per replicate with three biological replicates will require three 96-well plates. These can be standard polystyrene plates.
- **(b)** Well A will correspond to the 10^0 dilution, B will contain 10^1, C 10^2, D 10^3, E 10^4, F 10^5, G 10^6, H 10^7.

21. Add 100 μL of sample #1 to well A1 and 30 μL of sample #1 to well B1 mixing thoroughly. Continue this for the remaining samples (sample #2 will be added to A2 and B2).

22. Using a multi-channel pipette, serial dilute 1:10 for the remaining wells by transferring 30 μL from row B to row C, row C to row D, and so on. Remember to change micropipette tips between each dilution step.

23. Pipette 30 μL of each sample at each dilution in duplicate onto agar plates at dilutions corresponding to the biological replicate, OD_{600} or sample number, and dilution factor.
    (a) In total, 66 plates will be used for 3 biological replicates each with 11 dilutions.
24. Allow spot plates to dry, then invert, and incubate in a static incubator at 37 °C for 12–15 h.

5.7 Counting colonies and plotting CFU/mL to OD_{600}

Timing: 30 min

25. The following day, count and record colonies for each sample and use these values to determine the CFU/mL using the following equation:

$$\frac{\text{CFU}}{\text{mL}} = \left(\frac{\text{CFU of sample}}{\text{Dilution factor}}\right) \div 0.03\,\text{mL}$$

26. Generate a graph including the calculated CFU/mL of each sample on the Y-axis and the corresponding OD_{600} on the X-axis.
27. Use a linear regression fit or power function to get a $Y = mX + b$ equation corresponding to the dataset.
    (a) To calculate CFU/mL from OD 600 nm use: OD 600 nm $X = (Y/m)$-b.

6. Expected outcomes

The CFU/mL and corresponding optical densities (ODs) should follow a trend of linearity if dilutions were performed correctly. From the plate reader analysis, a number of growth curves can be compared to allow for a "starting" and a "time to reach" OD to be determined. The "starting" OD_{600} will be the inoculum that each well will be seeded at when the antimicrobial peptide to be tested is added. The "time to reach" OD_{600} will indicate the amount of time the samples should be incubated in the presence of antimicrobial peptide prior to adding resazurin. The resazurin data should be used to determine the final OD_{600} (the end OD_{600} of "time to reach") based on how long it takes the OD_{600} to reach maximum intensity.

7. Quantification and statistical analysis

Utilizing the graph and associated linear equation will allow one to determine which OD_{600} corresponds to the desired "starting" OD_{600} for preparing a bioassay.

The plate reader data should first be evaluated only by looking at the time it takes each OD_{600} to reach a maximum intensity of relative fluorescent units (RFU). Typically, the RFU for each sample will increase over time until a threshold is reached and the peak amount of resofurin is produced. After this time, dihydroresofurin is produced and a decrease in RFU is observed as over-reduction occurs. It is important to determine the OD_{600} that corresponds to a sample that reaches a maximum RFU after approximately 1 h. This should serve as the final OD_{600} that will be reached by the no-treatment controls during the bioassay containing peptides to be screened and will signal that it is time to add resazurin.

Note: Choosing a final OD_{600} that takes 1 h to reach maximum RFU without treatment ensures there is an appropriate buffer between resazurin reduction and over-reduction to dihydroresofurin that results in false positive results. Some biological samples (AMPs) can be stimulants of growth for microorganisms instead of inhibitors, resulting in rapid bacterial growth, increased respiration and thus over-reduction of dihydroresofurin. In addition, having some form of buffer will ensure that peptide samples that may be metabolized and induce growth themselves will be recorded accurately and not result in over-reduction. It is considered best to choose a final OD_{600} that corresponds to a CFU/mL that does not exceed 1×10^9. This is based on extensive experimental experience working with the ESKAPE pathogens and understanding their thresholds for resazurin reduction.

Once a final OD_{600} is established, the starting OD_{600} can be determined based on growth curve data. It is important to find a starting OD_{600} value that ensures the target final OD_{600} value can be reached within 2-4 h, limiting the amount of time the peptide is exposed to high temperature and increasing the number of assays that can be performed within an experimental time window. Using growth curves at varying dilutions of OD_{600} will allow one to determine which starting OD_{600} reaches the now determined final OD_{600} within this time frame. This can then be back-calculated to a corresponding CFU/mL using the equation above.

To screen the desired AMP or panel of AMPs, bacterial cultures should first be synchronized and standardized to the now established starting OD_{600}. The assay plate will be seeded with bacteria at this concentration, the proper controls, and the AMP being tested. Controls without peptide (no-treatment controls) will be included and used to track growth, serving to indicate when resazurin should be added. These wells will also act as a baseline for calculating the inhibitory effects of the peptide and will represent 0% inhibition. Additionally, adding a control that represents 100% inhibition such as a commercial antibiotic is strongly encouraged. The 96-well

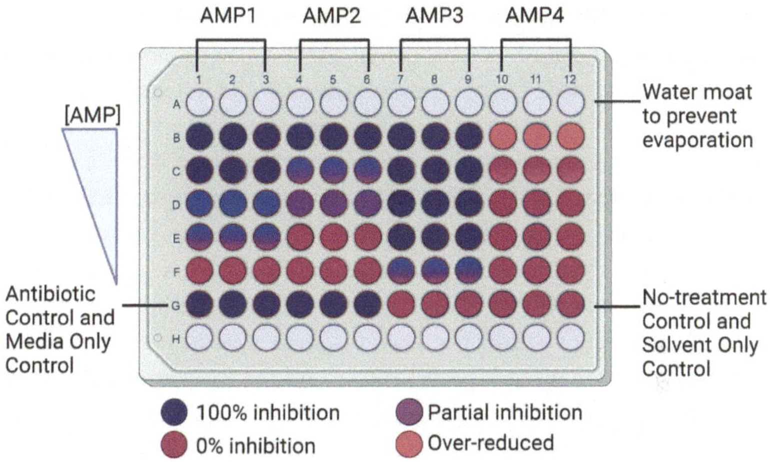

Fig. 2 Bioassay plate layout example. Includes optional layout for screening four separate AMPs against a single organism. AMP, antimicrobial peptide. The AMPs are shown where each is tested in technical triplicate starting with a high concentration (row B) and resulting in two fold dilutions (rows C-F). AMP1 and 3 demonstrate the expected visual result for AMPs that have inhibitory activity at low concentrations. AMP2 demonstrates complete inhibition at the highest concentration tested only. AMP4 provides an example of the test organism metabolizing the AMP, which leads to growth promotion and over-reduction of resazurin. Controls are included to ensure the proper calculations of inhibition are possible. *Created with BioRender.com.*

plate should be incubated for the requisite time determined from optimization protocols to reach the final OD_{600} (2–4h). Once no-treatment controls reach this final OD_{600}, the absorbance should be measured using a plate reader. Following this, aliquots of the 10x resazurin stock can be added to each of the sample wells. The 96-well plate should then be incubated in the dark for one hour. Final fluorescence measurements can be taken using the plate reader set to record at 590nm emission. These values can be used to calculate %inhibition using the following equation:

$$\left(1 - \frac{\text{AMP sample} - \text{antibiotic control}}{\text{average no treatment control} - \text{antibiotic control}}\right) \times 100$$

An example of a bioassay plate setup is shown in Fig. 2.

8. Summary: Considerations when implementing high-throughput assays with optimized parameters

When designing the plate layout for bioassay in a 96-well format, using a final volume of 200 μL is ideal, but the entire procedure can be optimized for a final volume of 100 μL. It will be necessary to include a peptide that will take up less than 5% of the total volume to avoid toxicity or nutrient depravation for the bacteria being challenged. It is suggested that the assay plate be prepared with media and compound leaving space for 10% of the volume to be filled with bacteria standardized to an OD_{600} 10x that of the starting OD_{600}:CFU/mL determined in the optimization assay.

When optimizing assays it is important to determine a starting and final OD_{600}:CFU/mL that is within a range that allows for a buffer between upper and lower thresholds. For example, some natural product peptides may serve as an energy source for a test organism and induce growth instead of inhibiting it. By having a final OD_{600}:CFU/mL that is well below the level of over-reduction, this scenario can be accurately tracked and recorded. Alternatively, starting with an OD_{600}:CFU/mL that is too low could result in a time to reach the final OD_{600}:CFU/mL that falls within the half-life of the peptide being tested.

When designing assays, multiple controls should be included: (1) no treatment controls, (2) no treatment controls with solvent, (3) wells containing a control commercial antibiotic to represent 100% inhibition, (4) media only controls to serve as a baseline for normalization and assurance that no contamination is present within the experiment.

It is important to optimize this approach for each organism. In fact, optimizing each strain would be ideal as some variability between strains of the same species has been observed. For example, it is common for *Pseudomonas aeruginosa* isolates to display unique antibiotic resistance profiles, and growth rates. Thus strains that grow more slowly have respiration rates that vary compared to isolates that grow more rapidly, ultimately resulting in differential rates of resazurin reduction. However, if the high-throughput screening component is a large number of bacterial isolates against one peptide it may be necessary to choose a standard starting and ending OD_{600}:CFU/mL for initial screens. Because peptides with high potential against various organisms will require rounds of confirmatory testing, this approach can be utilized. However, it should again be noted, that it is most beneficial to optimize this screening approach for each bacterial strain.

Further, if the goal is to test multiple different species against a single peptide library, it is ideal to find the OD_{600} specific to each organism that corresponds to the same starting and final CFU/mL. Only the incubation times should vary to some degree. Once the optimization protocol is performed, the experimental parameters can be implemented to test an unlimited number of peptides with consistent results that can be directly compared.

Acknowledgments

This study was supported by grant AI124458 (L.N.S.) from the National Institute of Allergy and Infectious Diseases.

References

Barreras-Serrano, A., Tamayo-Sosa, A. R., Del Villar-Pérez, V. M., Olivas Valdez, J.Á., Rentería-Evangelista, T. B., Pujol-Manríquez, L. C., et al. (2012). Evaluation of the antimicrobial susceptibility of Staphylococcus aureus to the K9CATH peptide by the resazurin microtiter method and the reference broth microdilution method. *Scientific Journal of Veterinary Advances*, *1*(6), 137–141. Retrieved from http://sjournals.com/index.php/sjva/article/view/1480.

Braissant, O., Astasov-Frauenhoffer, M., Waltimo, T., & Bonkat, G. (2020). A review of methods to determine viability, vitality, and metabolic rates in microbiology. *Frontiers in Microbiology*, *11*(2726). https://doi.org/10.3389/fmicb.2020.547458.

Chen, C. H., & Lu, T. K. (2020). Development and challenges of antimicrobial peptides for therapeutic applications. *Antibiotics (Basel)*, *9*(1). https://doi.org/10.3390/antibiotics9010024.

Chico, D. E., Given, R. L., & Miller, B. T. (2003). Binding of cationic cell-permeable peptides to plastic and glass. *Peptides*, *24*(1), 3–9. https://doi.org/10.1016/S0196-9781(02)00270-X.

Chuah, E., Zakaria, Z. A., Suhaili, Z., Abu Bakar, S., & Mohd Desa, M. N. (2014). Antimicrobial activities of plant extracts against methicillin-susceptible and methicillin-resistant Staphylococcus aureus. *Journal of Microbiology Research*, *4*, 6–13. https://doi.org/10.5923/j.microbiology.20140401.02.

Coban, A. Y., Darka, O., Fisgin, N. T., Cihan, C. C., Bilgin, K., Akgunes, A., et al. (2005). The resazurin microplate method for rapid detection of vancomycin resistance in Enterococci. *Journal of Chemotherapy*, *17*(4), 361–366. https://doi.org/10.1179/joc.2005.17.4.361.

Costa, P., Gomes, A. T. P. C., Braz, M., Pereira, C., & Almeida, A. (2021). Application of the resazurin cell viability assay to monitor Escherichia coli and Salmonella typhimurium inactivation mediated by phages. *Antibiotics*, *10*(8). https://doi.org/10.3390/antibiotics10080974.

Farkas, A., Pap, B., Kondorosi, É., & Maróti, G. (2018). Antimicrobial activity of NCR plant peptides strongly depends on the test assays. *Frontiers in Microbiology*, *9*(2600). https://doi.org/10.3389/fmicb.2018.02600.

Jia, H., Fang, R., Lin, J., Tian, X., Zhao, Y., Chen, L., et al. (2020). Evaluation of resazurin-based assay for rapid detection of polymyxin-resistant gram-negative bacteria. *BMC Microbiology*, *20*(1), 7. https://doi.org/10.1186/s12866-019-1692-3.

Kim, H. J., & Jang, S. (2017). Optimization of a resazurin-based microplate assay for large-scale compound screenings against Klebsiella pneumoniae. *3 Biotech*, *8*(1), 3. https://doi.org/10.1007/s13205-017-1034-9.

Kirkpatrick, C. L., Broberg, C. A., McCool, E. N., Lee, W. J., Chao, A., McConnell, E. W., et al. (2017). The "PepSAVI-MS" pipeline for natural product bioactive peptide discovery. *Analytical Chemistry*, *89*(2), 1194–1201. https://doi.org/10.1021/acs.analchem.6b03625.

Kristensen, K., Henriksen, J. R., & Andresen, T. L. (2015). Adsorption of cationic peptides to solid surfaces of glass and plastic. *PLoS One*, *10*(5), e0122419. https://doi.org/10.1371/journal.pone.0122419.

Lehrer, R. I., Rosenman, M., Harwig, S. S., Jackson, R., & Eisenhauer, P. (1991). Ultrasensitive assays for endogenous antimicrobial polypeptides. *Journal of Immunological Methods*, *137*(2), 167–173. https://doi.org/10.1016/0022-1759(91)90021-7.

Mercer, D. K., Torres, M. D. T., Duay, S. S., Lovie, E., Simpson, L., von Köckritz-Blickwede, M., et al. (2020). Antimicrobial susceptibility testing of antimicrobial peptides to better predict efficacy. *Frontiers in Cellular and Infection Microbiology, 10*-(326). https://doi.org/10.3389/fcimb.2020.00326.

Nagarajan, D., Nagarajan, T., Roy, N., Kulkarni, O., Ravichandran, S., Mishra, M., et al. (2018). Computational antimicrobial peptide design and evaluation against multidrug-resistant clinical isolates of bacteria. *Journal of Biological Chemistry, 293*(10), 3492–3509. https://doi.org/10.1074/jbc.M117.805499.

Nagarajan, D., Roy, N., Kulkarni, O., Nanajkar, N., Datey, A., Ravichandran, S., et al. (2019). Ω76: A designed antimicrobial peptide to combat carbapenem- and tigecycline-resistant Acinetobacter baumannii. *Science Advances, 5*(7), eaax1946. https://doi.org/10.1126/sciadv.aax1946.

Rampersad, S. N. (2012). Multiple applications of alamar blue as an indicator of metabolic function and cellular health in cell viability bioassays. *Sensors (Basel, Switzerland), 12*(9), 12347–12360. https://doi.org/10.3390/s120912347.

Rangel, K., Curty Lechuga, G., Almeida Souza, A. L., da Silva, R., Carvalho, J. P., Simões Villas Bôas, M. H., et al. (2020). Pan-drug resistant Acinetobacter baumannii, but not other strains, are resistant to the bee venom peptide melittin. *Antibiotics, 9*(4), 178. Retrieved from https://www.mdpi.com/2079-6382/9/4/178.

Sala, A., Cabassi, C. S., Santospirito, D., Polverini, E., Flisi, S., Cavirani, S., et al. (2018). Novel Naja atra cardiotoxin 1 (CTX-1) derived antimicrobial peptides with broad spectrum activity. *PLoS One, 13*(1), e0190778 https://doi.org/10.1371/journal.pone.0190778.

Santos, J., Pujols, J., Pallarès, I., Iglesias, V., & Ventura, S. (2020). Computational prediction of protein aggregation: Advances in proteomics, conformation-specific algorithms and biotechnological applications. *Computational and Structural Biotechnology Journal, 18*, 1403–1413. https://doi.org/10.1016/j.csbj.2020.05.026.

Sarker, S. D., Nahar, L., & Kumarasamy, Y. (2007). Microtitre plate-based antibacterial assay incorporating resazurin as an indicator of cell growth, and its application in the in vitro antibacterial screening of phytochemicals. *Methods, 42*(4), 321–324. https://doi.org/10.1016/j.ymeth.2007.01.006.

Sivakumar, P. M., Ganesan, S., Veluchamy, P., & Doble, M. (2010). Novel chalcones and 1,3,5-triphenyl-2-pyrazoline derivatives as antibacterial agents. *Chemical Biology & Drug Design, 76*(5), 407–411. https://doi.org/10.1111/j.1747-0285.2010.01020.x.

Travnickova, E., Mikula, P., Oprsal, J., Bohacova, M., Kubac, L., Kimmer, D., et al. (2019). Resazurin assay for assessment of antimicrobial properties of electrospun nanofiber filtration membranes. *AMB Express, 9*(1), 183. https://doi.org/10.1186/s13568-019-0909-z.

Turner, J., Cho, Y., Dinh, N.-N., Waring Alan, J., & Lehrer Robert, I. (1998). Activities of LL-37, a cathelin-associated antimicrobial peptide of human neutrophils. *Antimicrobial Agents and Chemotherapy, 42*(9), 2206–2214. https://doi.org/10.1128/AAC.42.9.2206.

Wang, G., Li, X., & Wang, Z. (2016). APD3: the antimicrobial peptide database as a tool for research and education. *Nucleic Acids Research, 44*(D1), D1087–D1093. https://doi.org/10.1093/nar/gkv1278.

Wiegand, I., Hilpert, K., & Hancock, R. E. W. (2008). Agar and broth dilution methods to determine the minimal inhibitory concentration (MIC) of antimicrobial substances. *Nature Protocols, 3*(2), 163–175. https://doi.org/10.1038/nprot.2007.521.

Zai, Y., Xi, X., Ye, Z., Ma, C., Zhou, M., Chen, X., et al. (2021). Aggregation and its influence on the bioactivities of a novel antimicrobial peptide, temporin-PF, and its analogues. *International Journal of Molecular Sciences, 22*(9). https://doi.org/10.3390/ijms22094509.

CHAPTER SEVEN

In silico prediction and mass spectrometric characterization of botanical antimicrobial peptides

Kevin D. Culver and Leslie M. Hicks*
Department of Chemistry, University of North Carolina at Chapel Hill, Chapel Hill, NC, United States
*Corresponding author: e-mail address: lmhicks@unc.edu

Contents

Methods in Enzymology, Volume 663
ISSN 0076-6879
https://doi.org/10.1016/bs.mie.2021.11.002

Abstract

Antimicrobial peptides (AMPs) are promising compounds for the treatment of antibiotic-resistant bacteria and are found across all organisms, including plants. Unlike most antibiotics, AMPs tend to act on more generalized and multiple targets, making development of resistance more difficult. Conventional approaches toward AMP identification include bioactivity-guided fractionation and genome mining. Complementary methods leveraging bioactivity-guided fractionation, cysteine motif-guided in silico AMP prediction, and mass spectrometric approaches can be combined to expand botanical AMP discovery. Herein, we present an integrated workflow which serves to streamline implementation toward a robust botanical AMP discovery pipeline.

1. Introduction

Antimicrobial resistance is a public health issue that results in at least 2.8 million infections and more than 35,000 deaths each year (Centers for Disease Control and Prevention, 2020). Antimicrobial peptides (AMPs) are explored as a way to combat ever increasing multi-drug resistant bacteria (Browne et al., 2020; Magana et al., 2020; Mahlapuu, Björn, & Ekblom, 2020; Sierra, Fusté, Rabanal, Vinuesa, & Viñas, 2017). AMPs are ubiquitously expressed across all life, and are generally amphipathic, highly basic, and cysteine-rich, containing disulfide bonds which influence tertiary structure folding (Tam, Wang, Wong, & Tan, 2015). Generally, their basic nature facilitates interaction with bacterial membranes and promotes specificity toward these bacterial membranes rather than zwitterionic mammalian membranes (da Costa, Cova, Ferreira, & Vitorino, 2015). AMPs commonly act via a membrane-lytic mechanism of action, in which they will disrupt the cytoplasmic barrier, following interaction with the negatively charged surface and insertion into the hydrophobic membrane (da Costa et al., 2015; Moretta et al., 2021). As AMPs tend to act on these more generalized targets, resistance toward them is not as prevalent as with antibiotics (Assoni et al., 2020). Among sources of AMPs, plant AMPs remain relatively underexplored compared to their mammalian counterparts (Wang, Li, & Wang, 2015) and are a potential wealth of novel bioactives. Plant AMPs are ribosomally derived and often expressed as precursor proteins containing N-terminal signal peptides, which are cleaved in the processing of the mature AMP (Tam et al., 2015). They are classified into AMP families based on characteristics such as higher order structure, cysteine motifs, and disulfide bonding patterns (Tam et al., 2015).

Methods for AMP discovery can be challenging. Most conventionally, bioactivity-guided fractionation techniques are employed in which bioactive fractions are subjected to iterative chromatographic separations and bioassays to lead to the compound of interest (Sharma & Gupta, 2015). This can be time consuming and often biased toward abundant and highly active compounds, potentially leading to identification of previously known compounds (Yang et al., 2013). Furthermore, bioactivity-guided fractionation may fail to identify AMPs if they are simply at too low of a concentration to exert any activity or if they are specific toward different pathogens that are not screened against. Isolation of compounds of interest and subsequent confirmation of the bioactive molecule can often be difficult given the complexities of the sample matrices, especially in plant extracts (Agarwal, D'Souza, Johnson, Dethe, & Chandrasekaran, 2014). To address these challenges, contributors to bioactivity within more complex samples/fractions can be determined with statistical modeling, in which peptidyl abundance is correlated with observed bioactivity as is done in PepSAVI-MS (Kirkpatrick et al., 2017).

Alternatively, AMP discovery can be achieved via genome mining (Russell & Truman, 2020), where biosynthetic gene clusters that produce putative AMPs are detected within genomic data. This can be challenging *in planta*, as there are few conserved features across ribosomally synthesized and post-translationally modified peptide (RiPP) biosynthetic gene clusters (Russell & Truman, 2020). Furthermore, this method gives no information about antimicrobial activity.

Herein, we detail a multi-faceted approach to provide efficient AMP identification from complex botanical extracts. In silico cysteine-motif guided AMP prediction is combined with mass spectrometric approaches such as bottom-up proteomics/mass shift analysis and bioactivity-guided fractionation (Fig. 1) to increase overall success for novel AMP discovery.

1.1 In silico AMP prediction

Cysteine motif-guided AMP prediction can be applied to thoroughly probe increasingly available plant genomic and/or transcriptomic data for putative AMPs. As mature AMPs are generally cysteine rich, carrying highly conserved cysteine motifs (Tam et al., 2015), these features can be leveraged to search for AMPs within protein databases. Furthermore, as AMP precursors often carry N-terminal signal peptides (Tam et al., 2015), these too can be leveraged. SignalP 5.0 (Almagro Armenteros et al., 2019), a deep neural

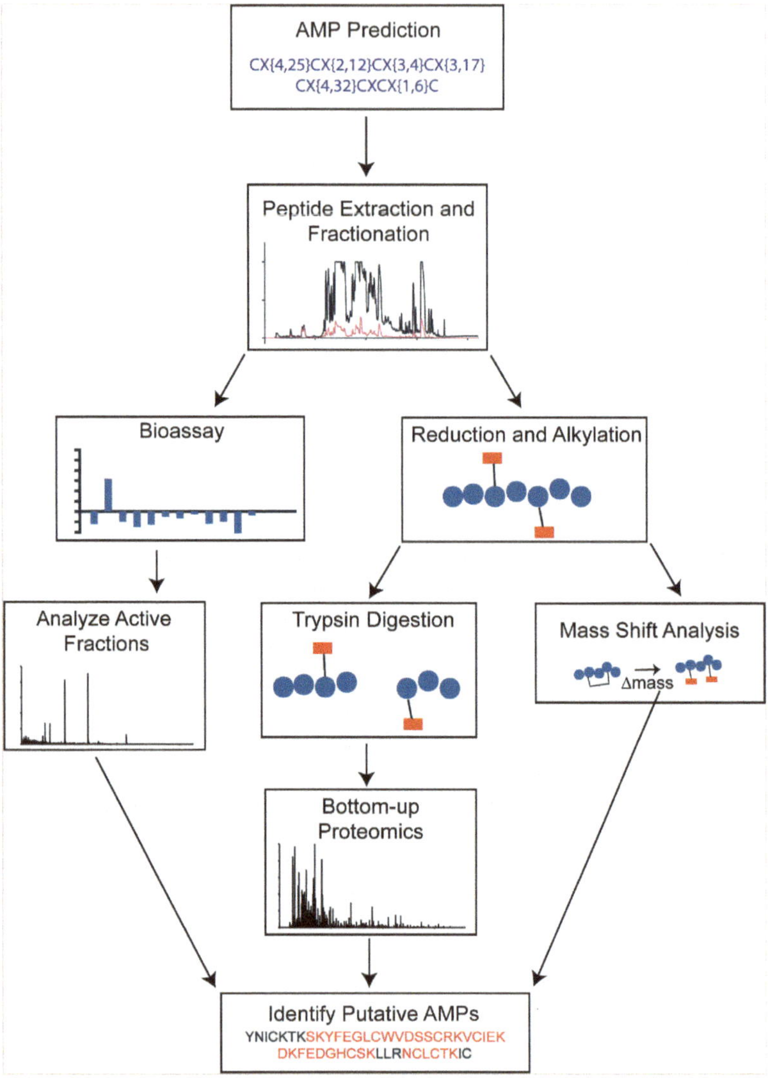

Fig. 1 Overall AMP identification workflow. Bioactivity-guided assays and in silico AMP prediction are combined with mass spectrometric approaches including bottom-up proteomics and mass shift analysis.

network that predicts and cleaves N-terminal signal peptides in silico, is used to suggest mature sequences and generate a focused set of peptides for further analysis (Fig. 2A). CysmotifSearcher (Shelenkov, Slavokhotova, & Odintsova, 2018) reveals cysteine motifs within the SignalP output and sorts results into AMP families (Fig. 2B). These methods lead to a concise list of predicted AMPs, with sequence and classification information.

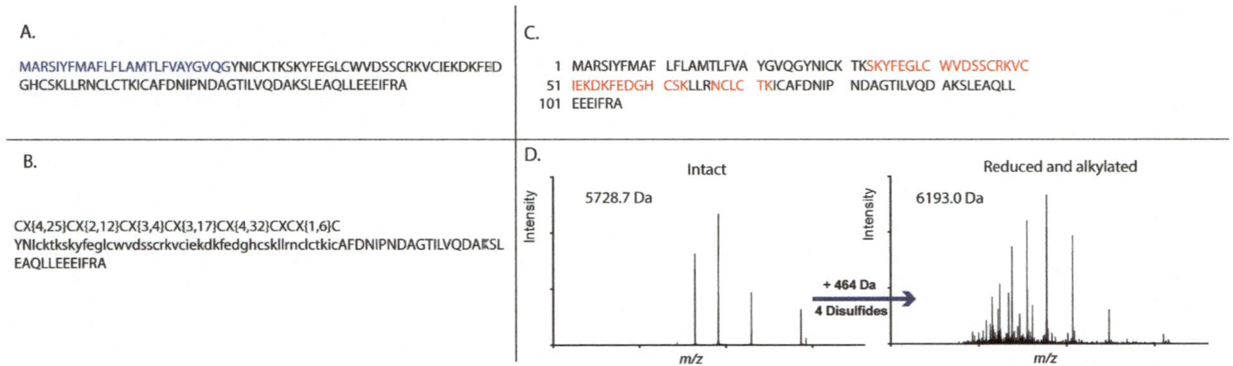

Fig. 2 Identification of a putative defensin. (A) Full sequence of the precursor protein with the signal peptide predicted by SignalP denoted in blue. (B) Defensin sequence predicted by Cysmotif Searcher, with the identified cysteine motif denoted above the sequence, and the motif within the sequence denoted by lowercase letters. The cysteine motif is represented with C corresponding to cysteine residues, X corresponding to any other amino acid residue, and bracketed numbers indicating the range of residues between cysteines (e.g., CX {4,25}C means that anywhere from 4 to 25 residues can exist between these cysteine residues). (C) Database search results, with the identified tryptic peptides denoted in red within the full protein sequence. (D) Mass shift analysis of the peptide, with the masses and mass spectra of both the intact and reduced/iodoacetamide-alkylated peptide.

A key limitation of this approach is its inability to account for additional protein cleavages, such as in the case of a C-terminal pro-domain removal. For example, consider a predicted defensin (Fig. 2B), which suggests a mature sequence that contains an additional C-terminal domain occurring after the cysteine motif. Closer inspection of this sequence shows that this domain is likely an acidic C-terminal pro-domain, which are known to exist in some defensin precursors (Lay et al., 2014). Additional challenges in accurate sequence prediction can be encountered when a mature AMP is post-translationally modified, especially given how extensive the occurrence of post-translational modifications (PTMs) are within AMPs (Wang, 2012), as Cysmotif Searcher cannot predict these. In order to elucidate the mature AMP sequences, mass spectrometric approaches building upon these in silico predictions are necessary.

1.2 Mass spectrometric characterization

While predictive approaches are useful in prioritizing botanical species for AMP screening as well as providing a targeted list for identification, mass spectrometric approaches are a necessary complement in addressing the limitations of mature sequence prediction. Furthermore, in vivo translation is not guaranteed, making direct detection necessary. Mass spectrometry-based bottom-up proteomics can be used to detect these predicted AMPs within an extract, in which the plant extract is subjected to reduction and alkylation of disulfide bonds, followed by digestion with a protease such as trypsin. These tryptic peptides are then subjected to tandem mass spectrometry (MS/MS), coupled with a separation method such as liquid chromatography (LC), and the results are searched against a protein database digested in silico to identify proteins within the sample (Dupree et al., 2020; Zhang, Fonslow, Shan, Baek, & Yates 3rd, 2013; Zhang, Wu, Stenoien, & Paša-Tolić, 2014).

Bottom-up proteomics has several strengths in that the proteolytically generated peptides analyzed will display easier separation, as well as better ionization and fragmentation, as compared to larger peptides and full proteins (Dupree et al., 2020). However, key disadvantages include limited protein coverage (i.e., the amount of the full protein that is detected), especially when tryptic peptides are too small to be identified by MS (Dupree et al., 2020). For example, the defensin that was predicted by Cysmotif Searcher (Fig. 2B) only had 46% coverage with a trypsin digestion, and the mature sequence is still unclear (Fig. 2C). In addition, highly similar protein sequences within a proteome have the potential to share tryptic peptides and, when only non-unique tryptic peptides are detected it is difficult to guarantee detection of a specific protein of interest,

Mass shift analysis is another mass spectrometric approach that can be utilized in the characterization of cysteine-rich AMPs (Narayani, Chadha, & Srivastava, 2017). This involves utilizing characteristic mass shifts that are observed during reduction and alkylation of disulfide-bound peptides in order to enumerate these disulfides in the mature structure. Reduction with dithiothreitol and subsequent alkylation with iodoacetamide results in a 58 Da mass shift per disulfide-bound cysteine residue (Fig. 3). Comparison of the mass of the mature peptide to that of the reduced and alkylated peptide indicates the total number of disulfide bonds in the AMP structure. This data can be highly complementary to the data obtained from bottom-up proteomics. For example, a mass of 5728.7 Da is observed in a sample containing intact peptides, with no reduction or alkylation. Upon reduction and alkylation, a mass of 6193.0 Da is observed, which would correspond to eight cysteine residues involved in four disulfide bonds (Fig. 2D). Coupled with in silico predictions and bottom-up proteomics, it can be determined that this likely corresponds to the defensin that was identified (Fig. 2B), in which there is an additional cleavage after the final cysteine residue in the cysteine motif, yielding a mature AMP structure with a mass of 5728.7 Da and four disulfide bonds.

1.3 Incorporation of bioactivity screening

While it is possible to identify AMPs solely through predictive and mass spectrometric approaches, incorporating bioactivity-guided fractionation in tandem with these methods can help to prioritize characterization as well as reveal novel AMPs that may not be predicted through Cysmotif Searcher.

In addition to screening for in silico predicted AMPs via mass spectrometric methods, peptide fractions can also be directly subjected to bioassays against pathogens of interest. In the case of observed bioactivity, those

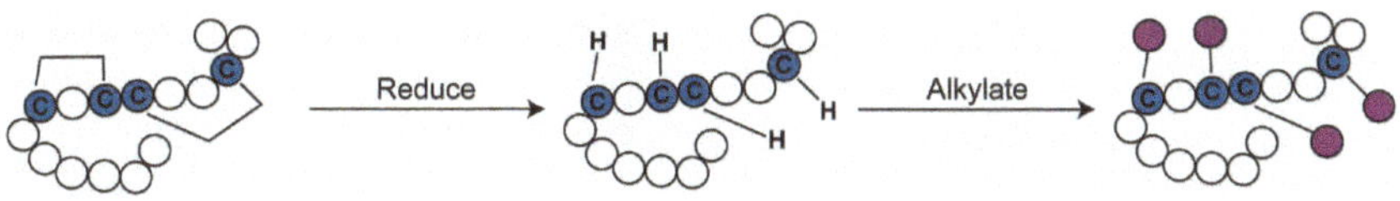

Fig. 3 Mass shift analysis workflow for determining the number of disulfide bonds in a peptide. The intact, disulfide-bound peptide is reduced with dithiothreitol, followed by alkylation with iodoacetamide. A carbamidomethyl group is added to each reduced cysteine, shown as purple circles, resulting in a shift of +58 Da for each.

fractions may be prioritized for characterization. This was seen with the recent discovery of two botanical AMPs from ghost pepper, CC-AMP1 and CC-AMP2, in which peptides were fractionated via reversed-phase LC and several of these fractions displayed antibacterial activity and were further examined (Culver, Allen, Shaw, & Hicks, 2021). While CC-AMP1 was a variation of a predicted AMP sequence, CC-AMP2 contained an entirely novel cysteine motif and was not predicted by CysmotifSearcher, only being identified via these antibacterial assays (Culver et al., 2021). This example further illustrates the complementary nature of using these different approaches in tandem.

Through incorporation of each of these complementary methods of in silico prediction, mass spectrometric bottom-up proteomics, mass shift analysis, mass spectrometric analyses of both the intact and reduced/alkylated peptides for sequence confirmation, and bioactivity-guided fractionation incorporated with PepSAVI-MS for analyses of complex mixtures of peptides, a comprehensive protocol for botanical AMP identification is achieved.

2. Materials

2.1 Databases

(1) Protein database in FASTA format, such as one downloaded from UniProt (https://www.uniprot.org/). Alternatively, transcriptomic data can be downloaded from sources such as 1000 plants (https://sites.google.com/a/ualberta.ca/onekp/), or genomic data can be downloaded from sources such as Phytozome (https://phytozome-next.jgi.doe.gov/)

(2) Database of common laboratory contaminants (https://www.thegpm.org/crap/)

(3) EMBOSS (http://emboss.sourceforge.net/download/), if translating transcriptomic or genomic assemblies

2.2 Reduction and alkylation

(1) 100 mM ammonium bicarbonate, pH 7.8

(2) Reduction buffer: 500 mM dithiothreitol in 100 mM ammonium bicarbonate, pH 7.8

(3) Alkylation buffer: 100 mM iodoacetamide in 100 mM ammonium bicarbonate, pH 7.8

2.3 Trypsin digestion

(1) 0.5 µg/µL Promega (Madison, WI, USA) trypsin gold, mass spectrometry grade, in 50 mM acetic acid
(2) 5% formic acid (FA, LC-MS grade) in water (LC-MS grade)

2.4 Desalting

(1) Millipore (Burlington, MA, USA) C18 ZipTips
(2) 0.1% trifluoroacetic acid (TFA, LC-MS grade)
(3) 5% TFA (LC-MS grade)
(4) 1% TFA (LC-MS grade), 2% acetonitrile (ACN, LC-MS grade)
(5) ACN (LC-MS grade)
(6) 70% ACN (LC-MS grade), 0.1% TFA (LC-MS grade)

2.5 LC-MS/MS analysis

(1) LC-MS total recovery vials (Waters)
(2) 5% ACN (LC-MS grade), 0.1% TFA (LC-MS grade)
(3) Symmetry C18 trap column (100 Å, 5 µm, 180 µm × 20 mm; Waters)
(4) HSS T3 C18 column (100 Å, 1.8 µm, 75 µm × 250 mm; Waters)
(5) Mobile phase A: 0.1% FA (LC-MS grade) in water (LC-MS grade)
(6) Mobile phase B: 0.1% FA (LC-MS grade) in ACN (LC-MS grade)
(7) Acquity UPLC M-Class system (Waters)
(8) Q Exactive HF X Hybrid Quadrupole Orbitrap mass spectrometer (ThermoFisher, Waltham, MA, USA)

2.6 Data analysis

(1) SignalP 5.0b for Linux or Darwin. Download link available at DTU Health Tech (https://services.healthtech.dtu.dk/software.php)
(2) Perl 5.8 or later (https://www.perl.org/get.html)
(3) Cysmotif Searcher for cysteine motif mining (https://github.com/fallandar/cysmotifsearcher)
(4) MSConvert (ProteoWizard, http://proteowizard.sourceforge.net/download.html)
(5) Mascot Daemon v3.5.1 (Matrix Science, Boston, MA, USA)
(6) Progenesis QI for Proteomics v2.0 (Nonlinear Dynamics, Durham, NC, USA)
(7) Python script for mass shift analysis (https://github.com/hickslab/MassShiftAnalysis)

2.7 Disulfide bond connectivity determination via partial acid hydrolysis

(1) 3 M hydrochloric acid (HCl, ACS Plus grade)
(2) Glass vial

3. Protocol

3.1 Cysteine motif-guided AMP prediction

(1) If translating transcriptomic or genomic assemblies, run getorf using the EMBOSS package with the input as the FASTA file containing the nucleotide sequences. This program will find and extract open reading frames, as well as translate them into protein sequences. Use the default options.
(2) Upload the FASTA file of the protein database of interest to the system that will run SignalP.
(3) Run SignalP on the protein database FASTA file using the following options:
  a. Set the output format to short
  b. Set organism to eukaryote
  c. Generate a mature sequence FASTA file
(4) With the resultant mature FASTA file from SignalP, run Cysmotif Searcher with the option of skipping translation of input sequences.
(5) Unzip the Cysmotif Searcher results file and obtain the *_motifs_pre.FASTA file, which will contain the full set of predicted AMPs.

Tip: When parsing through Cysmotif Searcher results, it is important to note that CYSRICH classifications arise when a sequence is found to have a known cysteine motif, but to also contain additional cysteine residues not belonging to that motif. While this can potentially result in the discovery of novel cysteine motifs, CYSRICH classifications will often be observed in larger precursor proteins that have additional cleavages outside of just the signal peptide cleavage. If these pro-domains contain any cysteine residues, the sequence is classified as CYSRICH. This is also observed in precursors that produce multiple mature AMP sequences, as in some α-Hairpinins (Slavokhotova & Rogozhin, 2020), in which the cysteine motif is observed several times throughout the sequence. In analyzing these results, it can often be useful to search CYSRICH sequences on UniProt BLAST

(Pundir et al., 2016), which may reveal regions of homology with known AMP families and provide more insight into the precursor protein and possible mature AMP regions.

3.2 Peptide extraction and fractionation

For a detailed protocol on plant peptide extraction and subsequent chromatographic fractionation, refer to Brechbill, Moyer, Parsley, and Hicks (2021). Briefly, the steps are as follows:

(1) Grind the plant material into a powder under liquid nitrogen and extract with the extraction buffer.
(2) Filter out solid material, then filter out large proteins from the extract.
(3) Dialyze the extract, then subject it to strong cation exchange (SCX) chromatography to remove the majority of small molecules.
(4) Using C18 solid phase extraction (SPE), desalt the SCX elution.
(5) Fractionate the desalted sample using reversed-phase chromatography.
(6) Resuspend the fractions in LC-MS water for downstream analyses.

3.3 Bacterial bioassay

For a detailed bioassay protocol, refer to Brechbill et al. (2021). Briefly, the steps are as follows:

(1) Streak the bacterial pathogen on an appropriate agar plate and incubate at 37 °C overnight.
(2) Add an isolated bacterial colony to MHB and incubate overnight.
(3) Dilute the culture and incubate for an additional hour, reaching mid log phase.
(4) Combine bacterial culture with peptide samples in a 96 well plate, using water as a negative control and an appropriate antibiotic as a positive control.
(5) Incubate for 4 h, then measure optical density at 600 nm (OD_{600}).
(6) Add resazurin to the wells, incubate for an additional hour, then measure fluorescence with an excitation wavelength of 544 nm and emission wavelength of 590 nm.
(7) Calculate percent activity against the positive and negative controls.

Tip: Peptide fractions that display bioactivity can be prioritized for further analyses. If an active fraction is an isolated peptide, then this is likely the active component. If a complex mixture of peptides is observed, then a statistical modeling approach, such as PepSAVI-MS

(Kirkpatrick et al., 2017) (https://cran.r-project.org/web/packages/PepSAVIms/index.html) can be utilized to determine the active component(s).

3.4 Reduction and alkylation

(1) From each peptide fraction, aliquot three samples at 10 μL each (approximately 1 μg/μL). One for analysis of mature peptides, one for reduction and alkylation, and one for reduction, alkylation, and trypsin digestion. The mature peptides do not need to be desalted and are ready for LC-MS/MS analysis (Section 3.7).

(2) To the 10 μL of both the reduction/alkylation and the reduction/alkylation/trypsin digestion samples, add 40 μL of 100 mM ammonium bicarbonate, pH 7.8. Reduce by adding 1 μL of 500 mM dithiothreitol (~10 mM final concentration) and incubating in a thermomixer at 45 °C and 850 rpm for 30 min in the dark.

(3) Alkylate by adding 10 μL of 100 mM iodoacetamide (~16 mM final concentration) and incubating at 25 °C at 850 rpm for 15 min in the dark.

(4) For samples that will not be trypsin digested, quench with 10 μL of 5% FA (LC-MS grade) and dry in a vacuum centrifuge.

Tip: The iodoacetamide and dithiothreitol stocks should be made fresh and light exposure should be avoided.

3.5 Trypsin digestion

(1) To the reduced and alkylated samples that will be digested, add 0.5 μL of 0.5 μg/μL Promega (Madison, WI, USA) trypsin gold, mass spectrometry grade, in 50 mM acetic acid for an enzyme:protein ratio of approximately 1:50 (w/w).

(2) Incubate overnight, approximately 16–17 h, at 37 °C and 850 rpm.

(3) Add 10 μL of 5% FA (LC-MS grade) and dry in a vacuum centrifuge.

3.6 Desalting

(1) Resuspend samples in 20 μL of 1% TFA (LC-MS grade), 2% ACN (LC-MS grade).

(2) Confirm with a pH test strip that the pH of the sample is <3. If pH is not <3, spike in additional small aliquots of 5% TFA (LC-MS grade) until the pH is <3.

(3) In advance, prepare an Eppendorf tube for elution containing 10 μL of 70% ACN (LC-MS grade), 0.1% TFA (LC-MS grade).
(4) Using a Millipore (Burlington, MA, USA) C18 ZipTip attached to a 10 μL pipette, set pipette to 10 μL and draw up 100% ACN (LC-MS grade) 3 times, expelling the waste liquid each time down to the resin, without letting the resin go dry.
(5) Draw up 0.1% TFA (LC-MS grade) 3 times.
(6) Draw up sample 10 times, expelling back into the sample tube each time.
(7) Draw up 0.1% TFA (LC-MS grade) 6 times.
(8) From the prepared elution tube, draw up 70% ACN (LC-MS grade), 0.1% TFA (LC-MS grade) 10 times, expelling back into the same tube each time.
(9) Dry elutions using a vacuum centrifuge.

3.7 LC-MS/MS analysis

(1) Resuspend mature, reduced/alkylated, and reduced/alkylated/trypsin digested peptide samples in 5% ACN (LC-MS grade), 0.1% TFA (LC-MS grade) and transfer to LC-MS total recovery vials (Waters).
(2) Inject samples and perform LC-MS/MS analysis using an Acquity UPLC M-Class system (Waters) coupled to a Q Exactive HF-X Hybrid Quadrupole Orbitrap mass spectrometer (ThermoFisher) via a Nanospray Flex Ion Source (ThermoFisher). Inject the peptide mixture to a Symmetry C18 trap column (100 Å, 5 μm, 180 μm × 20 mm; Waters) with a flow rate of 5 μL/min for 3 min using 99% A and 1% B, then separate on a HSS T3 C18 column (100 Å, 1.8 μm, 75 μm × 250 mm; Waters) using a gradient of increasing mobile phase B at a flow rate of 300 nL/min. Hold mobile phase B at 5% for 1 min before increasing from 5% to 50% in 30 min and ramping to 85% in 2 min. Hold at 85% for 3 min before returning to 5% in 1 min and re-equilibrating for 23 min.
(3) Use the following MS acquisition parameters: Use a tune file set with positive polarity, 2.1 kV spray voltage, 325 °C capillary temperature, and 40 S-lens RF level. In the instrument method, include lock masses best of 371.10124 and 445.12003 background polysiloxane ions. Select full MS/DD-MS2 scan type and set method duration to 60 min and default charge state to 2. Perform MS survey scan in profile mode across 350–2000 *m/z* at 120,000 resolution until 50 ms maximum IT or 3×10^6 AGC target is reached. Select the top 20 features above 5000

counts excluding ions with unassigned, +1, or >+8 charge state. Collect MS2 scans at 30,000 resolution with NCE at 28 until 100 ms maximum IT or 1×10^5 AGC target. Set the dynamic exclusion window for precursor *m/z* to 10 s and an isolation window of 1.5 *m/z*.

3.8 Data analysis

3.8.1 Bottom-up proteomics via Mascot database search

(1) Upload LC-MS/MS results files (.raw) into MSConvert and convert to Mascot Generic Format (.mgf) files, selecting "Write index," "Use zlib compression," and "TPP compatibility." Use the following filter options:

a. peakPicking: vendor msLevel = 1–2
b. zeroSamples: removeExtra 1-
c. activation: HCD
d. threshold: count 5000 most intense

(2) Using Mascot Daemon, search against the protein database of interest, as well as common laboratory contaminants (https://www.thegpm.org/crap/). Select MS/MS ion search with a tolerance of 0.1 Da and decoy database. Set peptide charge to +1, +2, and +3 and set the peptide tolerance to 15 ppm. Select trypsin with max missed cleavages of 3. Select a variable modification of methionine oxidation and a fixed modification of cysteine carbamidomethyl. Run the search and after it is complete, adjust the false discovery rate of the significant peptide identifications to be less than 1% using the embedded Percolator algorithm. Export results as a (.csv) file.

(3) Parse through Mascot results to identify any Cysmotif Searcher predicted AMPs detected in the peptide samples. One way to do this is by finding matching protein accessions between the two lists and then discarding Mascot identifications of other proteins.

Tip: Other database search engines can be used to analyze the data. Examples include SEQUEST (Eng, McCormack, & Yates, 1994), ProteinPilot (Shilov et al., 2007), and Andromeda (Cox et al., 2011).

Tip: After percolation, peptide scores of 13 represent a *p*-value of 0.05 and peptide matches with scores <13 should generally be discarded. Coverage on accessions of interest can be verified by manually checking the quality of the matched spectra.

3.8.2 Mass shift analysis

(1) Upload spectral result file (.raw) from LC-MS/MS analyses of both mature peptide samples and reduced/alkylated samples individually into Progenesis QI for Proteomics (Nonlinear Dynamics).

(2) Use default peak picking parameters, except for applying a minimum peak width of 0.05 min.

(3) From the "Identify Peptides" section, export peptide ion data with m/z, retention time, mass, charge, and raw abundance selected.

(4) With the peptide ion data of both mature and reduced/alkylated sample, sort each by mass. Copy and paste the mass columns of each into a new, single .csv file, labeling column A as the intact peptides and column B as the reduced/alkylated peptides.

(5) Using the .csv file, run the mass shift analysis python script (https://github.com/hickslab/MassShiftAnalysis), identifying mass pairs that correspond to the specified number of disulfide bonds.

Tip: Mass pairs identified by the mass shift analysis python script should be manually confirmed by checking that the reduced and alkylated mass is not present in the mature sample and that the mature mass is not present in the reduced and alkylated sample. Furthermore, Progenesis may misidentify monoisotopic peaks, which should also be checked by manually inspecting the MS spectra. This is more likely to occur in the case of low abundance peptides.

Tip: When the mass of a putative AMP is determined and the number of disulfides is known, the reduced and alkylated peptide can be subjected to MS/MS experiments to fully determine and/or confirm the sequence, in addition to coverage obtained from a trypsin digestion. This can be used to reveal any amino acid substitutions or PTMs in the sequence. The theoretical mass of the determined sequence should match that of the observed mass in the intact peptide sample. When a peptide's sequence is confirmed and the peptide is relatively isolated, the specific disulfide bond connectivity may be determined via Section 3.9.

3.9 Disulfide bond connectivity determination via partial acid hydrolysis

(1) Add 20 μL of peptide sample (approximately 1 μg/μL) and 40 μL of 3 M HCl to a glass vial.

(2) Heat at 85 °C for a total of 80 min, removing 15 μL aliquots every 20 min.

(3) Dry down samples under nitrogen gas and resuspend in 5% ACN (LC-MS grade), 0.1% TFA (LC-MS grade).

(4) Analyze each sample via LC-MS/MS following Section 3.7.

(5) The full peptide will be cleaved into a complex mixture of smaller, disulfide-linked peptides via the partial acid hydrolysis. Parse through precursor masses that could correspond to disulfide-linked acid hydrolysis dipeptides and confirm the residues and disulfide linkages via the corresponding MS/MS spectra.

Tip: Acid hydrolysis will convert asparagine and glutamine residues to aspartic acid and glutamic acid residues, respectively. This should be accounted for when analyzing the data.

Tip: Depending on the distribution of lysine and arginine residues within the peptide sequence, an alternative approach to partial acid hydrolysis is to perform a trypsin digestion without any prior reduction or alkylation. This will cleave the peptide after arginine and lysine residues, while leaving disulfide bonds intact.

4. Summary

The workflow detailed herein outlines a robust approach toward botanical AMP identification. Increasingly available genomic and/or transcriptomic data from sources such as the 1000 plants project (Leebens-Mack et al., 2019), Phytozome (Goodstein et al., 2012), and UniProt (Pundir et al., 2016) provide a foundation for discovery via the use of AMP prediction. Protein sequences can be mined for signal peptides using SignalP (Almagro Armenteros et al., 2019), obtaining sequences with these signal peptides cleaved off. This output can then be mined for conserved cysteine motifs using CysmotifSearcher (Shelenkov et al., 2018), providing a focused list of putative AMPs.

Bioactivity-guided fractionation approaches can be used in tandem with mass spectrometric approaches such as bottom-up proteomics and mass shift analysis. Bioassays can provide direct assessments of AMP bioactivity, as well as help prioritize characterization. Bottom-up proteomics allows detection of predicted AMPs within the botanical extract, regardless of observed bioactivity. Finally, as cysteine-rich AMPs are expected to have multiple disulfide bonds, mass shift analysis provides a means to identify mass species that contain disulfide bonds, as well as enumerate these disulfide bonds. In addition, MS/MS analysis of the reduced and alkylated peptide can confirm the full sequence, revealing any amino acid substitutions or unexpected PTMs.

Taken together, these approaches encompass a robust AMP discovery workflow, which helps to address the disadvantages of any individual approach.

Acknowledgment

This work was supported by NIH-NIGMS under award R01 GM125814 to L.M.H.

References

Agarwal, A., D'Souza, P., Johnson, T. S., Dethe, S. M., & Chandrasekaran, C. V. (2014). Use of in vitro bioassays for assessing botanicals. *Current Opinion in Biotechnology*, *25*, 39–44. https://doi.org/10.1016/j.copbio.2013.08.010.

Almagro Armenteros, J. J., Tsirigos, K. D., Sønderby, C. K., Petersen, T. N., Winther, O., Brunak, S., et al. (2019). SignalP 5.0 improves signal peptide predictions using deep neural networks. *Nature Biotechnology*, *37*(4), 420–423. https://doi.org/10.1038/s41587-019-0036-z.

Assoni, L., Milani, B., Carvalho, M. R., Nepomuceno, L. N., Waz, N. T., Guerra, M. E. S., et al. (2020). Resistance mechanisms to antimicrobial peptides in gram-positive bacteria. *Frontiers in Microbiology*, *11*, 593215. https://doi.org/10.3389/fmicb.2020.593215.

Brechbill, A. M., Moyer, T. B., Parsley, N. C., & Hicks, L. M. (2021). Creating optimized peptide libraries for AMP discovery via PepSAVI-MS. *Methods in Enzymology*, *663*.

Browne, K., Chakraborty, S., Chen, R., Willcox, M. D., Black, D. S., Walsh, W. R., et al. (2020). A new era of antibiotics: The clinical potential of antimicrobial peptides. *International Journal of Molecular Sciences*, *21*(19), 7047. https://doi.org/10.3390/ijms21197047.

Centers for Disease Control and Prevention. (2020). *Antibiotic/antimicrobial resistance (AR/AMR)*. https://www.cdc.gov/drugresistance/index.html.

Cox, J., Neuhauser, N., Michalski, A., Scheltema, R. A., Olsen, J. V., & Mann, M. (2011). Andromeda: A peptide search engine integrated into the MaxQuant environment. *Journal of Proteome Research*, *10*(4), 1794–1805. https://doi.org/10.1021/pr101065j.

Culver, K. D., Allen, J. L., Shaw, L. N., & Hicks, L. M. (2021). Too hot to handle: Antibacterial peptides identified in ghost pepper. *Journal of Natural Products*. https://doi.org/10.1021/acs.jnatprod.1c00281.

da Costa, J. P., Cova, M., Ferreira, R., & Vitorino, R. (2015). Antimicrobial peptides: An alternative for innovative medicines? *Applied Microbiology and Biotechnology*, *99*(5), 2023–2040. https://doi.org/10.1007/s00253-015-6375-x.

Dupree, E. J., Jayathirtha, M., Yorkey, H., Mihasan, M., Petre, B. A., & Darie, C. C. (2020). A critical review of bottom-up proteomics: The good, the bad, and the future of this field. *Proteomes*, *8*(3), 14. https://doi.org/10.3390/proteomes8030014.

Eng, J. K., McCormack, A. L., & Yates, J. R. (1994). An approach to correlate tandem mass spectral data of peptides with amino acid sequences in a protein database. *Journal of the American Society for Mass Spectrometry*, *5*(11), 976–989. https://doi.org/10.1016/1044-0305(94)80016-2.

Goodstein, D. M., Shu, S., Howson, R., Neupane, R., Hayes, R. D., Fazo, J., et al. (2012). Phytozome: A comparative platform for green plant genomics. *Nucleic Acids Research*, *40*(Database issue), D1178–D1186. https://doi.org/10.1093/nar/gkr944.

Kirkpatrick, C. L., Broberg, C. A., McCool, E. N., Lee, W. J., Chao, A., McConnell, E. W., et al. (2017). The "PepSAVI-MS" pipeline for natural product bioactive peptide discovery. *Analytical Chemistry*, *89*(2), 1194–1201. https://doi.org/10.1021/acs.analchem.6b03625.

Lay, F. T., Poon, S., McKenna, J. A., Connelly, A. A., Barbeta, B. L., McGinness, B. S., et al. (2014). The C-terminal propeptide of a plant defensin confers cytoprotective and subcellular targeting functions. *BMC Plant Biology*, *14*(1), 41. https://doi.org/10.1186/1471-2229-14-41.

Leebens-Mack, J. H., Barker, M. S., Carpenter, E. J., Deyholos, M. K., Gitzendanner, M. A., Graham, S. W., et al. (2019). One thousand plant transcriptomes and the phylogenomics of green plants. *Nature*, *574*(7780), 679–685. https://doi.org/10.1038/s41586-019-1693-2.

Magana, M., Pushpanathan, M., Santos, A. L., Leanse, L., Fernandez, M., Ioannidis, A., et al. (2020). The value of antimicrobial peptides in the age of resistance. *The Lancet Infectious Diseases*, *20*(9), e216–e230. https://doi.org/10.1016/S1473-3099(20)30327-3.

Mahlapuu, M., Björn, C., & Ekblom, J. (2020). Antimicrobial peptides as therapeutic agents: Opportunities and challenges. *Critical Reviews in Biotechnology*, *40*(7), 978–992. https://doi.org/10.1080/07388551.2020.1796576.

Moretta, A., Scieuzo, C., Petrone, A. M., Salvia, R., Manniello, M. D., Franco, A., et al. (2021). Antimicrobial peptides: A new Hope in biomedical and pharmaceutical fields. *Frontiers in Cellular and Infection Microbiology*, *11*, 668632. https://doi.org/10.3389/fcimb.2021.668632.

Narayani, M., Chadha, A., & Srivastava, S. (2017). Cyclotides from the Indian medicinal plant Viola odorata (Banafsha): Identification and characterization. *Journal of Natural Products*, *80*(7), 1972–1980. https://doi.org/10.1021/acs.jnatprod.6b01004.

Pundir, S., Martin, M. J., O'Donovan, C., & UniProt Consortium. (2016). UniProt tools. *Current Protocols in Bioinformatics*, *53*, 1.29.1–1.29.15. https://doi.org/10.1002/0471250953.bi0129s53.

Russell, A. H., & Truman, A. W. (2020). Genome mining strategies for ribosomally synthesised and post-translationally modified peptides. *Computational and Structural Biotechnology Journal*, *18*, 1838–1851. https://doi.org/10.1016/j.csbj.2020.06.032.

Sharma, S. B., & Gupta, R. (2015). Drug development from natural resource: A systematic approach. *Mini Reviews in Medicinal Chemistry*, *15*(1), 52–57. https://doi.org/10.2174/138955751501150224160518.

Shelenkov, A. A., Slavokhotova, A. A., & Odintsova, T. I. (2018). Cysmotif searcher pipeline for antimicrobial peptide identification in plant transcriptomes. *Biochemistry (Moscow)*, *83*(11), 1424–1432. https://doi.org/10.1134/S0006297918110135.

Shilov, I. V., Seymour, S. L., Patel, A. A., Loboda, A., Tang, W. H., Keating, S. P., et al. (2007). The paragon algorithm, a next generation search engine that uses sequence temperature values and feature probabilities to identify peptides from tandem mass spectra. *Molecular & Cellular Proteomics*, *6*(9), 1638–1655. https://doi.org/10.1074/mcp.T600050-MCP200.

Sierra, J. M., Fusté, E., Rabanal, F., Vinuesa, T., & Viñas, M. (2017). An overview of antimicrobial peptides and the latest advances in their development. *Expert Opinion on Biological Therapy*, *17*(6), 663–676. https://doi.org/10.1080/14712598.2017.1315402.

Slavokhotova, A. A., & Rogozhin, E. A. (2020). Defense peptides from the α-Hairpinin family are components of plant innate immunity. *Frontiers in Plant Science*, *11*, 465. https://doi.org/10.3389/fpls.2020.00465.

Tam, J., Wang, S., Wong, K., & Tan, W. (2015). Antimicrobial peptides from plants. *Pharmaceuticals*, *8*(4), 711–757. https://doi.org/10.3390/ph8040711.

Wang, G. (2012). Post-translational modifications of natural antimicrobial peptides and strategies for peptide engineering. *Current Biotechnology*, *1*(1), 72–79. https://doi.org/10.2174/2211550111201010072.

Wang, G., Li, X., & Wang, Z. (2015). APD3: The antimicrobial peptide database as a tool for research and education. *Nucleic Acids Research*, *44*(D1), D1087–D1093. https://doi.org/10.1093/nar/gkv1278.

Yang, J. Y., Sanchez, L. M., Rath, C. M., Liu, X., Boudreau, P. D., Bruns, N., et al. (2013). Molecular networking as a dereplication strategy. *Journal of Natural Products*, *76*(9), 1686–1699. https://doi.org/10.1021/np400413s.

Zhang, Y., Fonslow, B. R., Shan, B., Baek, M.-C., & Yates, J. R., 3rd. (2013). Protein analysis by shotgun/bottom-up proteomics. *Chemical Reviews*, *113*(4), 2343–2394. https://doi.org/10.1021/cr3003533.

Zhang, Z., Wu, S., Stenoien, D. L., & Paša-Tolić, L. (2014). High-throughput proteomics. *Annual Review of Analytical Chemistry*, 7(1), 427–454. https://doi.org/10.1146/annurev-anchem-071213-020216.

CHAPTER EIGHT

Combined thermal and carboxypeptidase Y stability assays for probing the threaded fold of lasso peptides

Julian D. Hegemann*
Helmholtz Institute for Pharmaceutical Research Saarland (HIPS), Helmholtz Centre for Infection Research (HZI), Saarland University Campus, Saarbrücken, Germany
*Corresponding author: e-mail address: jdhegemann@googlemail.com

Contents

Abstract

Lasso peptides are natural products belonging to the superfamily of ribosomally synthesized and post-translationally modified peptides (RiPPs). The defining characteristic of lasso peptides is their threaded structure, which is reminiscent of a lariat knot. When working with lasso peptides, it is therefore of major importance to understand and evidence their threaded folds. While the full elucidation of their three-dimensional structures via NMR spectroscopy or crystallization remains the gold standard, these methods are time-consuming, require large quantities of highly pure lasso peptides, and therefore might not always be applicable. Instead, the unique properties of lasso peptides in

Methods in Enzymology, Volume 663
ISSN 0076-6879
https://doi.org/10.1016/bs.mie.2021.09.011

context of their behavior at elevated temperatures and toward carboxypeptidase Y treatment can be leveraged as a tool to investigate and evidence the threaded lasso fold using only minute amounts of compound that does not need to be purified first. This chapter will provide insights into the thermal stability properties of lasso peptides and their behavior when treated with carboxypeptidase Y in comparison to a branched-cyclic peptide with the same amino acid sequence. Furthermore, it will be described in detail how to set up a combined thermal and carboxypeptidase Y stability assay and how to analyze its outcomes.

1. Introduction

Lasso peptides are intriguing bacterial natural products featuring a unique threaded fold that is reminiscent of a lariat knot and provides the basis of their name (Fig. 1A) (Arnison et al., 2013; Cheng & Hua, 2020; Hegemann, 2020; Hegemann, Dit-Foque, & Xie, 2020; Hegemann, Zimmermann, Xie, & Marahiel, 2015; Montalbán-López et al., 2021). They represent a subfamily of ribosomally synthesized and post-translationally modified peptides (RiPPs) and therefore their biosynthesis begins with the ribosomal assembly

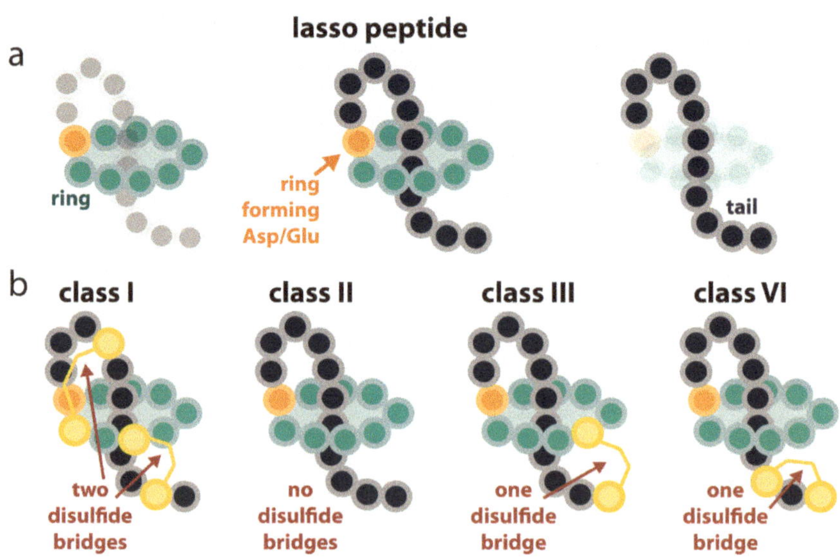

Fig. 1 Schematic showing the general features of lasso peptide structures. (A) The typical lasso fold is shown in the middle. On the left, the residues comprising the N-terminal macrolactam ring are highlighted, while on the right, the residues forming the C-terminal tail are emphasized. (B) The four different lasso peptide classes are depicted and their class defining features are highlighted.

of a precursor peptide that is then converted into the mature lasso peptide by a number of processing enzymes. In general, the lasso peptide precursor can be subdivided into an N-terminal leader region, needed for enzymatic recognition, and a C-terminal core peptide, which is converted into the mature lasso peptide. Genome mining efforts have revealed that lasso peptides are widely distributed throughout the bacterial kingdom (Bratovanov et al., 2020; Cheung-Lee & Link, 2019; Hegemann, Zimmermann, Zhu, Klug, & Marahiel, 2013; Maksimov, Pelczer, & Link, 2012; Sabino et al., 2020; Tietz et al., 2017) and many interesting biological properties have been observed for them (Braffman et al., 2019; Cheng & Hua, 2020; Guerrero-Garzón et al., 2020; Hegemann et al., 2020, 2015; Pati et al., 2018). These properties most commonly relate to antimicrobial activities, but also antiviral and anticancer activities have been reported.

The lasso fold comprises a macrolactam ring, formed between the N-terminal α-amine and the carboxylic acid sidechain of an Asp or Glu residue at position 7, 8, or 9, that is threaded by the linear C-terminal tail (Arnison et al., 2013; Hegemann et al., 2020, 2015; Montalbán-López et al., 2021). Such lasso folds are primarily stabilized through the positioning of plug amino acids in the tail region, which are amino acids with large sidechains that provide a steric bulk and trap the tail inside the ring. In addition, the lasso fold can be stabilized by disulfide bridges, whose presence and positioning allows further subclassification (Fig. 1B) (Cheung-Lee, Cao, & Link, 2019; Guerrero-Garzón et al., 2020; Hegemann et al., 2020; Knappe, Linne, Xie, & Marahiel, 2010; Tietz et al., 2017). Whereas class II lasso peptides only rely on steric hindrance for fold maintenance, class I lasso peptides contain two disulfide bridges and class III lasso peptides one disulfide bond connecting the ring and tail regions. The most recently discovered class IV lasso peptides also contain a single disulfide bridge that is however not connecting ring and tail. Instead, the disulfide links two cysteines in the tail of a class IV lasso peptide and thus generates a bulky ring whose steric demand vastly surpasses even the biggest sidechains of the canonical amino acids and thereby provides a much enhanced stabilization of the lasso fold.

For lasso peptide biosynthesis, three distinct protein functionalities are needed (Burkhart, Hudson, Dunbar, & Mitchell, 2015; DiCaprio, Firouzbakht, Hudson, & Mitchell, 2019; Hegemann, Schwalen, Mitchell, & van der Donk, 2018; Koos & Link, 2019; Sumida, Dubiley, Wilcox, Severinov, & Tagami, 2019; Yan et al., 2012; Zhu, Fage, et al., 2016). First, a RiPP recognition element (RRE) is recognizing and binding to a specific recognition sequence in the N-terminal leader region of the precursor peptide. The RRE then recruits a leader peptidase that cleaves the precursor and thereby releases the C-terminal core peptide. In turn, a macrocyclase takes the core peptide, activates the ring forming Asp/Glu carboxylic acid sidechain in an ATP-dependent manner, and finally mediates the nucleophilic attack of the N-terminal primary amine on the activated AMP ester to form the macrolactam through a simple condensation reaction. To accomplish the threaded fold, a prefolding step would be necessary, which would facilitate that the macrolactam ring can form around the right position of the C-terminal tail. The prefolding step is however still only a hypothetical notion, as it could not yet be proven experimentally.

As the lasso fold is the defining feature of the members of this RiPP subfamily and is usually a prerequisite for their biological activity, it is of major importance to make sure that a newly isolated compound has the correct threaded fold. While a full elucidation of the three-dimensional structure of a lasso peptide via NMR spectroscopy or crystallization is commendable, such structure elucidations are time-consuming procedures that require large quantities of pure compound (Braffman et al., 2019; Fage et al., 2016; Hegemann et al., 2020, 2016, 2014; Jeanne Dit Fouque, Lavanant, et al., 2018; Mathavan et al., 2014; Nar, Schmid, Puder, & Potterat, 2010; Xie & Marahiel, 2012). Obtaining large amounts of pure compound is not always possible and might not be feasible when utilizing high-throughput genome mining approaches or when screening variant libraries generated through mutation of the precursor encoding gene in a heterologous production system.

Thus, several strategies have been developed that can rapidly confirm the threaded structure of a lasso peptide with minute amounts of compounds. One of the simplest and most efficient methods for this goal will be discussed in this chapter.

2. Concerning the thermal stability of lasso peptides

Although (class II) lasso peptides are merely stabilized by the steric interactions of the plug amino acids with the macrolactam ring, they typically exhibit a very high stability against proteolysis and denaturation through chaotropic reagents (Fage et al., 2016; Hegemann, 2020; Hegemann et al., 2020, 2015, 2016, 2014; Hegemann, Zimmermann, Xie, & Marahiel, 2013; Knappe, Linne, Robbel, & Marahiel, 2009; Maksimov, Koos, Zong, Lisko, & Link, 2015; Maksimov & Link, 2013). In addition, thermal stability is often associated with lasso peptides as well

(Hegemann, 2020). The first time the thermal stability of a lasso peptide was noted, was in 1992, when Salomon and Farias described how the lasso peptide microcin J25 retained its antimicrobial activity (and thereby its fold) even after autoclaving at 120 °C (Salomon & Farias, 1992). As a matter of fact, it was even believed for a long time that the thermal stability would be an intrinsic property inherent to all lasso peptides. This changed in 2013, when it was observed that caulosegnins I and III were the first examples of heat-sensitive lasso peptides (Hegemann, Zimmermann, Xie, et al., 2013; Hegemann et al., 2016).

But what exactly does heat sensitivity mean in the context of lasso peptides? In simple terms, albeit generally very stable, a heat-sensitive lasso peptide can adopt certain conformations where ring and tail align in a way that enables the tail to slip out of the ring, effectively leading to the unthreading of a threaded lasso into a simple branched-cyclic peptide (Fig. 2).

For the heat-sensitive lasso peptides so far described in the literature (Hegemann, 2020), these unthreading events happen rarely, if at all, at the temperatures corresponding to the native habitats of the producing strains. However, when increasing the temperature of a lasso peptide solution, the molecular movements also increase, and it becomes more likely that at one point a conformation is adopted that facilitates the unthreading to occur (Allen et al., 2016; Hegemann et al., 2016).

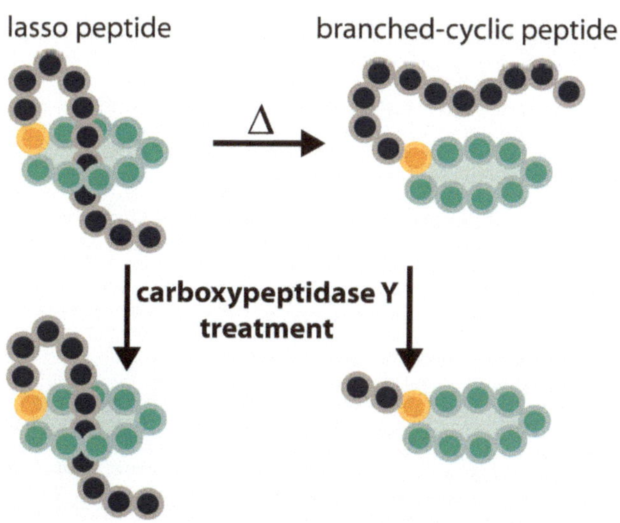

Fig. 2 Schematic showing the heat-induced unthreading of a heat-sensitive lasso peptide and how carboxypeptidase Y treatment can distinguish between a lasso and a branched-cyclic peptide.

Experiments have shown that the thermal stability of a lasso peptide often correlates to both the size of the macrolactam ring and the plug amino acids present (Allen et al., 2016; Fage et al., 2016; Guerrero-Garzón et al., 2020; Hegemann, 2020; Hegemann, Zimmermann, Xie, et al., 2013; Hegemann et al., 2016, 2014; Knappe et al., 2009; Long et al., 2021; Metelev et al., 2017; Zhu, Hegemann, et al., 2016; Zimmermann, Hegemann, Xie, & Marahiel, 2013, 2014; Zong, Wu, Qin, & Link, 2017). The smaller the ring, the more easily a heat-stable lasso fold can be accomplished. Of course, the use of more bulky sidechains also increases the overall stability of these threaded [1]rotaxane structures. In several cases, a heat-sensitive lasso peptide could even be converted into a heat-stable one by mutating the native plug amino acid to a more bulky residue like tryptophan (Allen et al., 2016; Hegemann, 2020; Hegemann et al., 2016; Long et al., 2021; Zimmermann et al., 2013, 2014).

In itself, the unthreading of a lasso peptide can usually be observed rather easily. By unthreading, the residues otherwise shielded by the surrounding macrolactam ring suddenly become solvent exposed. Hence, the interactions of a lasso peptide with the liquid and mobile phases during reversed-phase high-performance liquid chromatography (HPLC) are typically distinctly different when compared to the interactions a branched-cyclic peptide with the same primary structure exhibits. These differences are usually observable through a heat-induced shift in retention time between the two topoisomers. Hence, partial unthreading becomes apparent by appearance of a new peak in the chromatogram that has a different retention time, but still the same mass as the lasso peptide, and prolonged incubation at elevated temperatures eventually leads to the complete conversion of the lasso peptide into its branched-cyclic analog.

Yet, due to the isobaric nature of the two topoisomers, it is usually necessary to validate the temperature-induced conversion into a branched-cyclic peptide by further experiments. Here, the treatment with carboxypeptidase Y has proven to be a most useful and efficient method (Allen et al., 2016; Cheung-Lee, Cao, et al., 2019; Cheung-Lee, Parry, Jaramillo Cartagena, Darst, & Link, 2019; Cheung-Lee et al., 2020; DiCaprio et al., 2019; Fage et al., 2016; Guerrero-Garzón et al., 2020; Hegemann, 2020; Hegemann et al., 2020; Hegemann, Zimmermann, Xie, et al., 2013; Hegemann, Zimmermann, Zhu, et al., 2013; Hegemann et al., 2016, 2014; Li et al., 2015; Martin-Gomez, Linne, Albericio, Tulla-Puche, & Hegemann, 2018; Metelev et al., 2017; Tietz et al., 2017; Zhu, Hegemann, et al., 2016; Zimmermann et al., 2013, 2014; Zong et al., 2017; Zyubko et al., 2019). Carboxypeptidases are by definition peptidases that sequentially remove the most C-terminal residues of a peptide until either the peptide is fully degraded, or a point is reached that inhibits further degradation. In case of a branched-cyclic peptide, the isopeptide bond connecting the carboxylic acid sidechain and the N-terminal primary amine is a structural feature that stops carboxypeptidases from further degrading the peptide chain. Thereby, a branched-cyclic peptide can be clearly distinguished from a lasso peptide. Whereas the C-terminal tail of a branched-cyclic peptide can be completely or almost completely removed through carboxypeptidases, the threading of a lasso peptide effectively shields the C-terminus from carboxypeptidase access and thereby prevents proteolysis. From the commercially available carboxypeptidases, carboxypeptidase Y is the one most suited for distinguishing lasso and branched-cyclic peptides, as it has a very relaxed substrate specificity and hence can effectively degrade C-terminal peptides (almost) independently of their sequences.

3. Combined heat and carboxypeptidase Y stability assays

The lasso fold is the key feature defining the RiPP subfamily of lasso peptides. The macrolactam ring can effectively shield the threading linear C-terminal peptide segment from access by enzymes like carboxypeptidase Y. Hence, resistance against carboxypeptidase Y can clearly differentiate a lasso peptide from a simple branched-cyclic peptide with the same primary structure. Further evidence of the threaded topology of a lasso peptide can be obtained by assaying the compound in question before and after prolonged incubation at elevated temperatures. As heating can induce the conversion of a heat-sensitive lasso peptide into its branched-cyclic counterpart, a clear difference would be observed after carboxypeptidase Y treatment and often also through a shift in retention time on a C18 HPLC column. However, some lasso peptides are heat-stable and will not unthread when heated. In this case, a comparison to a synthetically produced branched-cyclic analog can be helpful. In either case, the assessment of the thermal stability properties of a lasso peptide has become a routine procedure for the characterization of novel representatives of this RiPP subfamily and as such is an important assay to perform. Further evidence for the threaded structure of a lasso fold can subsequently be obtained from more sophisticated tandem MS and ion mobility MS experiments (Dit Fouque et al., 2018, 2021; Fouque et al., 2017; Hegemann, Jeanne Dit Fouque, Santos-Fernandez, & Fernandez-Lima, 2021; Jeanne Dit Fouque, Afonso, et al., 2015; Jeanne Dit Fouque, Bisram, et al., 2019; Jeanne Dit Fouque, Hegemann, et al., 2019;

Jeanne Dit Fouque, Lavanant, et al., 2018, 2015; Jeanne Dit Fouque, Moreno, et al., 2018; Martin-Gomez et al., 2018; Perot-Taillandier et al., 2012; Zhu, Hegemann, et al., 2016; Zirah et al., 2011).

3.1 Equipment

1. Microreaction tubes.
2. Micropipettes with suitable volume ranges.
3. Microcentrifuge able to spin at $>10,000 \times g$.
4. A thermocycler with a heated lid (preferred) or a simple heat block.
5. An HPLC-MS setup with a UV detector. High-resolution MS instruments are preferred as they are more sensitive and yield more accurate data. When using an instrument with lower resolution, it is recommended to confirm the identity of the observed lasso peptide and the masses of its possible degradation products by tandem MS analysis. An analytical HPLC column packed with C18 chromatography material should be used for separation. The author prefers analytical columns with 125×2 mm dimensions, but similar sizes will also work. [*MS instruments using direct injection or MALDI-TOF-MS instruments can be used if no HPLC-MS instruments are available. However, with such an experimental setup it can only be assessed if carboxypeptidase Y is able to remove any residues from the C-terminus, but the information about a possible shift in retention time induced by heating is lost.*]

   (optional: 6. A lyophilisator or a speed vacuum concentrator for microreaction tubes in case organic solvents need to be removed prior to the carboxypeptidase Y assays or in case the lasso peptide sample needs to be more concentrated)

3.2 Chemicals

1. 2-(N-morpholino)ethanesulfonic acid (MES, CAS number 4432-31-9).
2. Calcium chloride ($CaCl_2$).
3. MES buffer solution (50 mM MES, 1 mM $CaCl_2$, pH 6.75).
4. Lyophilized carboxypeptidase Y (CAS number 9046-67-7) [*Depending on the amino acid composition of the C-terminal tail, carboxypeptidase Y can sometimes be substituted by carboxypeptidase A (in conjunction with carboxypeptidase B, if the sequence also contains any Lys or Arg residues). However, due its broad substrate specificity, carboxypeptidase Y is vastly superior for the described assays and should be the preferred choice whenever available.*]

5. A sufficiently concentrated solution of the lasso peptide in question in MES buffer (organic solvents like MeOH and MeCN need to be completely removed from the sample as they will inhibit carboxypeptidase Y). To ensure that the amounts of peptide present are suitable for the planned experiments, a reference containing the final concentration of lasso peptide and the final volume used in the assay should be injected in the HPLC-MS instrument. The MS signal corresponding to the lasso peptide should be easily detectable and ideally also give a clear UV peak at a 220 nm absorption wavelength. The solution can either be of a pure lasso peptide of a defined concentration (preferred), or a lasso peptide-containing extract (see Section 4.3 for more details). As a rule of thumb, 5–10 μg of an average lasso peptide per sample will give a very good signal in the HPLC-MS analysis.
6. Methanol (MeOH).
7. Suitable HPLC solvent systems, e.g., ddH_2O containing 0.1% formic acid (FA) as solvent A, and HPLC-grade acetonitrile (MeCN) or MeOH containing 0.1% FA as solvent B.

   (optional: 6. Micropipette tips with C18-beds for desalting and concentration of samples in case MS measurements can only been done via direct injection or MALDI-TOF-MS)

   (optional: 7. Tris-(2-carboxyethyl)-phosphine hydrochloride (TCEP-HCl, CAS number 51805-45-9) for reduction of disulfides when working with disulfide bond-containing lasso peptides (see Section 4.4))

4. Protocols

A flowchart summarizing the key steps of the described protocols is shown in Fig. 3.

4.1 Thermal stability assay

4.1.1. Prepare 250 μL of an organic-solvent-free solution of lasso peptide in MES buffer (50 mM MES, 1 mM $CaCl_2$, pH 6.75) (e.g., by diluting a stock solution of lasso peptide in H_2O into MES buffer or dissolving a dried peptide sample directly in MES buffer). The final concentration of the lasso peptide after dilution should be around 0.1–0.2 mg/mL (smaller amounts can work too, but should be tested in advance for suitability with the HPLC-MS instrument used). Split the sample into two 125 μL aliquots.

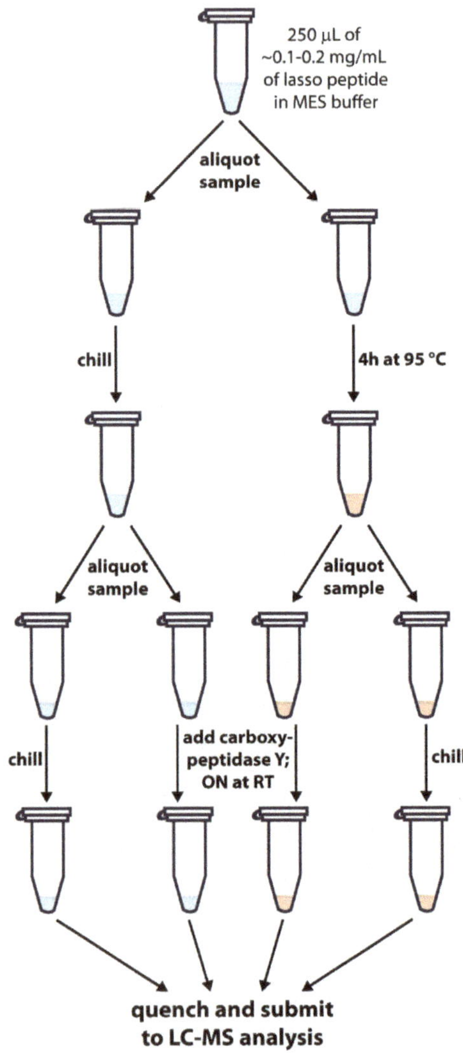

Fig. 3 Flowchart depicting a short summary of the combined thermal and carboxypeptidase Y stability assays as described in Section 4.

4.1.2. Freeze one aliquot for later use. Incubate the second aliquot for 4 h at 95 °C in a thermocycler with a heated lid to avoid the formation of condensation on the lid of the tube. If using a normal heat block, regularly spin down the condensation formed and quickly mix the sample before returning it into the heat block. Afterward, freeze the sample for later use as well.

4.2 Carboxypeptidase Y assays

4.2.1. Thaw the reference and heat-treated samples. Remove 50 μL of each sample for the carboxypeptidase Y assays and freeze the rest to measure as reference later.

4.2.2. Prepare a stock solution of carboxypeptidase Y by dissolving the lyophilized protein in ice-cold MES buffer to obtain a final concentration of 1 mg/mL of carboxypeptidase Y. Keep on ice. [*Leftover carboxypeptidase Y solution should be stored in a fridge or a cold room at 4 °C. There, it should remain active for at least a few days up to a few weeks. Make sure to run a positive control to ensure that the carboxypeptidase Y is still active, when using the protease after prolonged storage of the stock solution.*]

4.2.3. Mix each of the 50 μL of the samples with 10 μL of the carboxypeptidase Y stock solution. After mixing, incubate overnight at room temperature (RT). [*Most of the times, 4 h at RT should be sufficient for the assay to run to completion. Running it overnight makes certain that full degradation is accomplished.*]

4.2.4. On the next day, quench the assays by adding 5 μL of pure MeOH and mixing.

4.2.5. Thaw the two reference samples. For comparability, take 50 μL of each sample and mix with 10 μL of MES buffer and 5 μL of pure MeOH.

4.2.6. Either directly use all samples for HPLC-MS analysis or freeze the samples when planning to do the measurements at another time.

4.3 Modified protocol when using lasso peptide-containing extracts

4.3.1. When performing the described assays with lasso peptide-containing extracts from the cell pellet or the culture supernatant, take five times the amount of sample giving a good signal in an HPLC-MS measurement, freeze- or vacuum-dry, and then dissolve the dried extract in 250 μL of MES buffer. Centrifuge for 10–20 min at >10,000 × *g* to ensure the removal of any potential precipitate present and then transfer two 125 μL aliquots into fresh microreaction tubes. Then, use these samples to proceed as described from step 4.1.2 onward.

4.4 Modified protocol when working with a class I, III, or IV lasso peptide

The disulfide bridges present in class I, III, and IV lasso peptides will effectively inhibit a thermally-induced unthreading of the lasso fold as well as carboxypeptidase Y-mediated degradation of a branched-cyclic peptide. Nonetheless, it can still be assessed if such a lasso peptide remains heat-stable in the absence of these disulfide bridges. For such assays, the disulfide bonds need to be broken by reduction with TCEP-HCl (Cheung-Lee, Cao, et al., 2019; Guerrero-Garzón et al., 2020; Jeanne Dit Fouque, Bisram, et al., 2019; Li et al., 2015).

4.4.1. Prepare a 100 mM stock solution of TCEP-HCl in MES buffer.

4.4.2. To the lasso peptide solution from step 1.1 of Section 4.1 add 2.5 μL of the 100 mM TCEP-HCl stock solution. Mix and incubate at 50 °C for 15 min for reduction of the disulfide bonds and then proceed as described from step 4.1.2 onward. [*If the reduction at 50 °C already causes significant unthreading of the lasso peptide to occur, it can instead be tried to incubate the sample at RT and regularly check for reduction by MS (+2 Da per reduced disulfide bond). Extend the incubation period if necessary. This should work in most cases, however, depending on the present structures, some disulfide bonds can be more resistant to reduction at RT than others.*]

4.5 HPLC-MS analysis

4.5.1. Prior to running the samples, a suitable analytical HPLC gradient for the lasso peptide investigated should be established. Each run should start with an isocratic step at a low solvent concentration that lasts a few minutes (depending on the dead volume of the whole system) and is diverted into the waste to wash off all the salt from the assay buffers and not inject it into the MS instrument. The subsequent gradient for separation should not be too steep and ideally the lasso peptide should elute somewhere in the middle of the gradient. This way, potential shifts in retention time caused by heat-induced unthreading will be easy to observe. Consider that the shift in retention time can occur both to higher and lower solvent concentrations as the change in interactions with the stationary phase depend on which residues are shielded by the macrolactam ring in the threaded fold. In some cases, there can also be no observable shift in retention time, despite the conversion into a branched-cyclic peptide taking place.

[*As example for how such LC-MS analyses could be carried out, consider these exemplary gradients taken from the literature, which were employed for HPLCs run at flow rates of 0.2–0.3 mL/min and utilizing reversed-phase*

C18 columns with a diameter of 2 mm and a length of 125 mm. The solvents used were ddH_2O containing 0.1% FA as solvent A and HPLC-grade MeCN containing 0.1% FA as solvent B. *For the analysis of the stability assays of sphingopyxin I, the HPLC gradient started with holding 20% B for 5 min, followed by a linear increase from 20% to 50% B in 10 min, then ramping from 50% to 95% B in 1 min, and holding 95% B for another 1 min (*Fage et al., *2016). In a genome mining study (*Hegemann, Zimmermann, Zhu, et al., 2013*), the thermal and carboxypeptidase Y stability assays of a variety of newly identified lasso peptides were analyzed using an HPLC gradient starting with the holding of 2% B for 2 min, followed by a linear increase from 2% B to 20% or 30% B in 18 min, then a linear increase to 95% B in 15 min and holding 95% B for another 2 min.]*

4.5.2. After establishing a suitable HPLC-MS program, analyze the samples prepared. For each measurement, inject ~90% of your sample.

4.5.3. After the measurements are completed, analyze the data obtained (see also Section 5). Pay attention to peaks shifting between samples, new peaks appearing and/or reference peaks disappearing, and the masses associated with each peak. If a change in mass occurs, sequentially delete residues from the C-terminus of your lasso peptide sequence until the degradation product with a matching mass is identified. [*When performing the assays with lasso peptide-containing extracts and not pure samples, consider following these additional points: (i) Use extracted ion currents selecting for masses corresponding to every intermediate between the full length lasso peptide and the mere macrolactam ring to check for the presence of additional minor side products. (ii) When using a high-resolution MS instrument, make sure that the masses observed are close enough to the ones predicted and that the charge state of each mass is correct. (iii) When using a low-resolution MS instrument, repeat the measurements and fragment all masses that potentially represent unthreaded and/or degraded lasso peptide derivatives to further confirm their primary structures.*]

5. Analysis and data interpretation

5.1 General guidelines

First and foremost, the careful comparison of LC traces and MS signals before and after heating and/or carboxypeptidase Y treatment is key for the analysis of these assays and for determining the fold present. There are several basic scenarios that could occur, which will be discussed briefly (for special cases deviating from the norm see Section 5.2):

(i) The compound is completely stable against both thermally-induced unthreading (= no shift in retention time) and carboxypeptidase Y (= no amino acids are removed by carboxypeptidase Y neither before nor after heat treatment). In this case (Fig. 4A), it is likely that a heat-stable lasso peptide is present. Tandem MS analysis should be employed to confirm the primary structure of the compound and its peptidic nature to ensure that the mass observed does not belong to an unrelated non-peptidic compound, an uncyclized peptide, or a peptide carrying a different type of cyclization (e.g., head-to-tail cyclization).

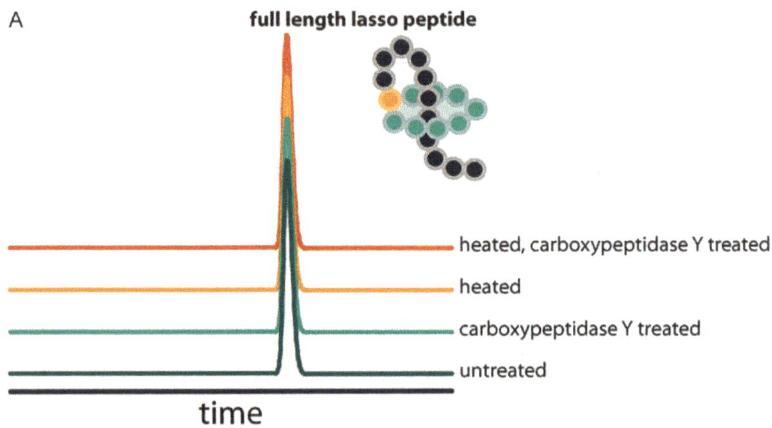

Fig. 4A Schematic depicting how the HPLC chromatogram of a fictitious heat-stable lasso peptide completely resistant toward carboxypeptidase Y might look like.

(ii) The compound is completely stable against thermally-induced unthreading (= no shift in retention time), however loses a limited number of amino acids when treated with carboxypeptidase Y independent if it was priorly incubated at elevated temperatures or not; i.e., the exact same results are obtained for the unheated and the heated samples after carboxypeptidase Y treatment (Fig. 4B). In this scenario, the compound is likely a heat-stable lasso peptide with a long C-terminal tail that extends way beyond the lower plug amino acid. The explanation in this case is that while the macrolactam ring effectively shields the threading portion of the tail from degradation, a long linear peptide tail can extend so far beyond the threading portion that it becomes accessible to proteolytic removal by carboxypeptidase Y. However, once the peptidase has removed a limited number of residues, it cannot proceed any further as then finally the steric bulk of the ring does not allow the peptidase to access the C-terminus any longer. Such behavior was observed for a number of heat-stable lasso peptides and many lasso peptides with long tails are isolated after heterologous production exclusively as truncated compounds lacking a certain number of C-terminal residues (Cheung-Lee, Cao, et al., 2019; Fage et al., 2016; Hegemann, Zimmermann, Zhu, et al., 2013; Hegemann et al., 2014; Maksimov et al., 2015; Maksimov & Link, 2013; Zimmermann et al., 2013; Zong et al., 2017). Thus, the accessibility of the C-terminus in these cases already allowed proteolytic removal to occur inside the producer cells after the lasso peptide was formed.

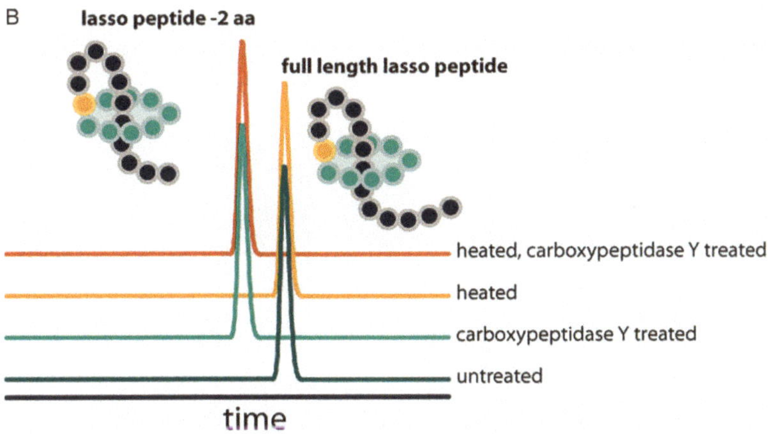

Fig. 4B Schematic depicting how the HPLC chromatogram of a fictitious heat-stable lasso peptide whose long lower tail region allows limited proteolysis through carboxypeptidase Y might look like.

(iii) The compound shows a distinct shift in the LC chromatogram in response to incubation at 95 °C; i.e., a new peak with a different retention time but an identical mass appears when heating (Fig. 4C). Ultimately, prolonged heating will cause the first peak to disappear completely and trigger its full conversion into what is likely a branched-cyclic peptide obtained through thermally-induced unthreading. As such, carboxypeptidase Y will remove no or only a select few residues from the C-terminus of the original compound, but will degrade much more of the compound post-heating (in some cases until only the macrolactam ring is left). Therefore, such an observation is a strong indication of a heat-sensitive lasso peptide being investigated.

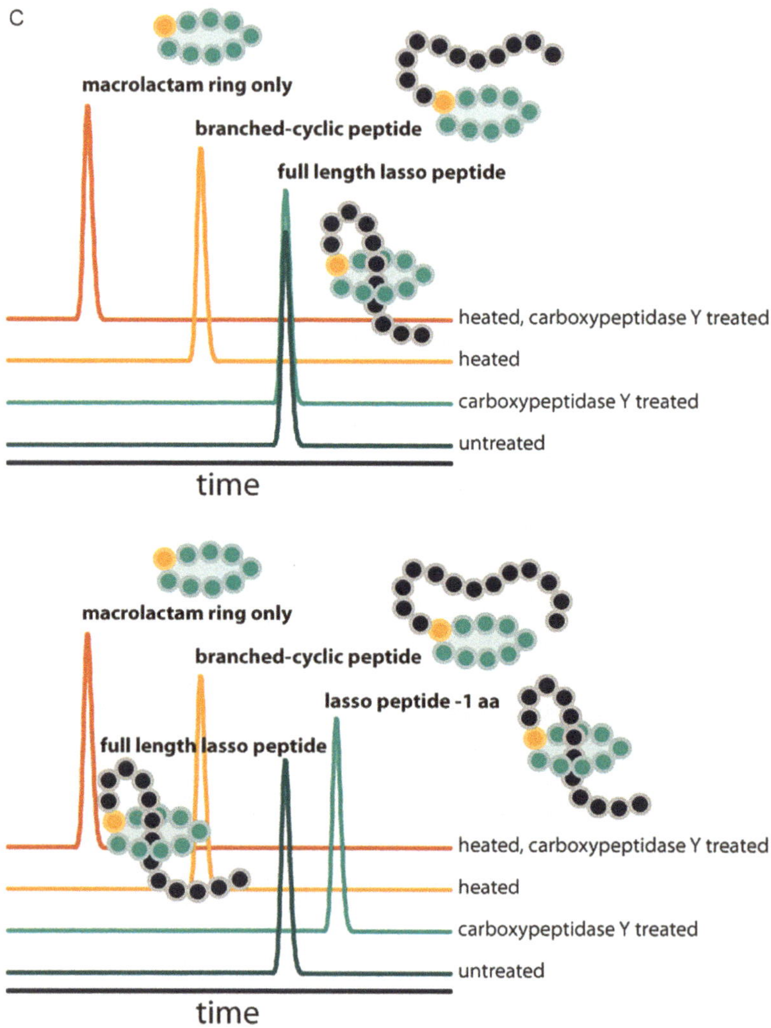

Fig. 4C Schematic depicting how the HPLC chromatogram of fictitious heat-sensitive lasso peptides could look like. In the example above, the lasso peptide is completely resistant toward carboxypeptidase Y before heating, while in the example below, the long tail of the lasso peptide allows limited degradation through carboxypeptidase Y to occur prior to heating. In both cases, the lasso peptides completely unthread when heated and in their unthreaded form, carboxypeptidase Y is able to completely degrade their C-terminal tails until only the macrolactam ring remains.

(iv) Like in (iii), but without a visible shift in retention time (Fig. 4D). In this case, it is also likely that a heat-sensitive lasso peptide is being investigated, where coincidentally both the lasso peptide and branched-cyclic analog have a (nigh) identical retention time.

Normally, the behavior prior and subsequent to carboxypeptidase Y incubation still allows a clear distinction between both species and thus the lack of a shift in retention time is of no consequence. Yet, if desired, it can be tried to optimize the gradient or alter the solvents used in attempts to make the unthreading event already observable through a shift in retention time.

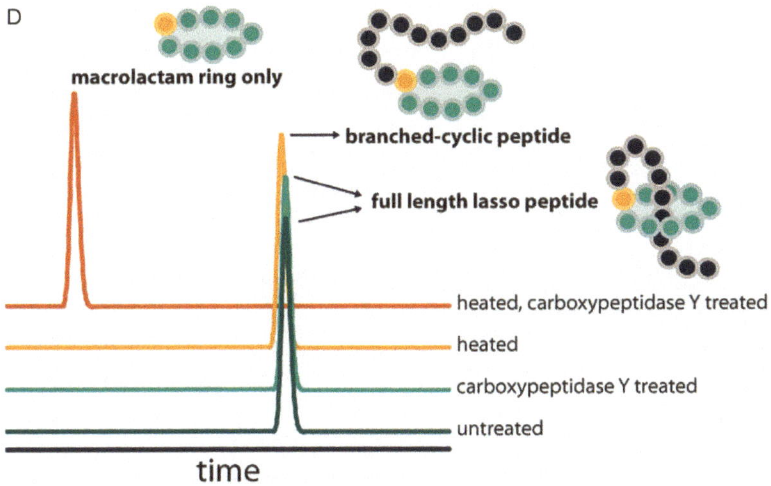

Fig. 4D Schematic depicting how the HPLC chromatogram of a fictitious heat-sensitive lasso peptide could look like. In this example, the lasso peptide is completely resistant toward carboxypeptidase Y before heating. Whereas the lasso peptide completely unthreads upon heating, the branched-cyclic peptide retains a nigh identical retention time. However, comparison of the results obtained after carboxypeptidase Y treatment of the compound before and after heating allows the clear identification of the compound being a heat-sensitive lasso peptide.

(v) If the retention time does not change upon heating, yet the peptide is highly sensitive toward carboxypeptidase Y independent of being priorly heated or not (Fig. 4E), the compound in question might already be a branched-cyclic peptide. In this case, there might be several possible explanations like, for example, the compound was never a lasso peptide to begin with or the lasso peptide is very prone to unthreading and unthreading was triggered already before isolation or through the isolation protocol administered.

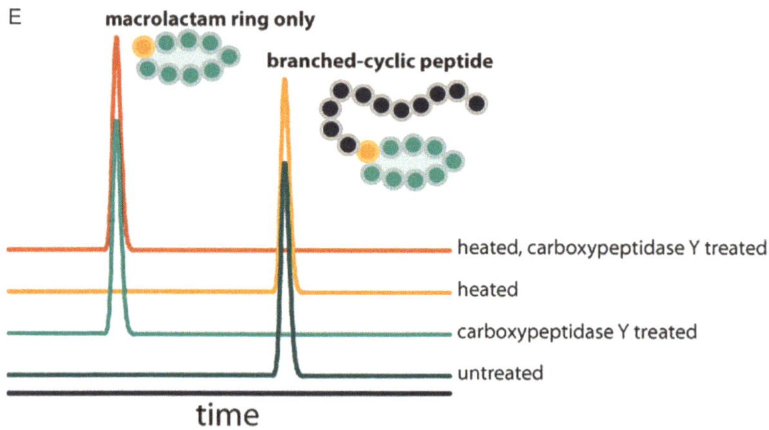

Fig. 4E Schematic depicting how the HPLC chromatogram of a fictitious branched-cyclic peptide could look like. As the conformation of the compound does not change upon heating, no shift in retention time occurs. The high sensitivity of the compound toward carboxypeptidase Y even before the heat treatment strongly points toward the investigated compound already having a branched-cyclic conformation in the very beginning.

In general, the results of cases like (iii) and (iv) are clearer indications of the isolated compounds having a proper threaded lasso fold. Yet, if the primary structure has been confirmed by tandem MS analysis or other means, the (nigh) complete resistance against carboxypeptidase Y described in (i) and (ii) is also already a strong proof for the lasso conformation, especially if the original compound was isolated from a host strain after heterologous expression of a corresponding lasso peptide biosynthetic gene cluster. These results can be further corroborated by comparison with a reference synthetic branched-cyclic peptide of the same primary structure, but the synthesis of such peptides is usually time-consuming and/or expensive. For other means to provide further evidence of the lasso fold also see Section 5.3.

In addition to the standard assays described in Section 4, it is also possible to determine more details about the thermal stability parameters of a heat-sensitive lasso peptide (Allen et al., 2016; Hegemann, 2020; Hegemann, Zimmermann, Xie, et al., 2013; Hegemann et al., 2016; Jeanne Dit Fouque, Bisram, et al., 2019; Zhu, Hegemann, et al., 2016; Zimmermann et al., 2013; Zong et al., 2017). Here, the assay conditions can be altered to follow the thermal unthreading process at different temperatures and/or durations. While it is usually not necessary to perform such elaborate assays for simply evidencing the presence of a threaded lasso peptide, it might be interesting in certain cases, especially when looking at the physicochemical properties determining the thermal stability of a lasso peptide. For example, it might be helpful to alter the temperatures and durations of the assays when investigating a series of lasso peptide variants with exchanges of the lower plug residue to assess which amino acid sidechain still provides enough bulk for stabilizing the lasso fold, and which amino acid sidechain will do so in a heat-stable manner.

5.2 Special cases

There can also be cases where the data obtained deviates from what is described in Section 5.1, although such cases have so far only been observed in a select few instances. This section will explain what these special cases are, when and why they occur, how such special cases affect the assay outcome, and how they can possibly be further analyzed for a lasso topology.

(i) Some (lasso) peptides will break apart upon heating (Fig. 4F). This is highly sequence specific and occurs mostly in peptides with an Asp-Pro sequence, a sequence known to readily self-hydrolyze via the Asp-dependent autocatalytic formation of a cyclic imide intermediate (Cheung-Lee, Cao, et al., 2019; Hegemann, Zimmermann, Xie, et al., 2013; Skribanek, Mezo, Mak, & Hudecz, 2002; Zhu, Hegemann, et al., 2016). In such cases, the HPLC MS analysis of a lasso peptide post-heating will contain distinct signals witnessing the breakage of the Asp-Pro bond. In some cases, even a complete decomposition of the lasso peptide can be induced through the assay conditions described in Section 4. This is of course problematic, can however be circumvented by reducing the duration of the 95 °C incubation to 30–60 min and/or reducing the temperature of the heat treatment. Usually, the assay conditions can be optimized in this way to yield conditions were full length lasso peptide and, if heat sensitivity is given, branched-cyclic analog resulting from thermally-induced unthreading can be observed. While a full conversion of a heat-sensitive lasso into a branched-cyclic peptide might not be accomplished anymore under these conditions, the comparison to references and the carboxypeptidase Y-only assays should still allow the identification of the unthreading event that occurred.

F

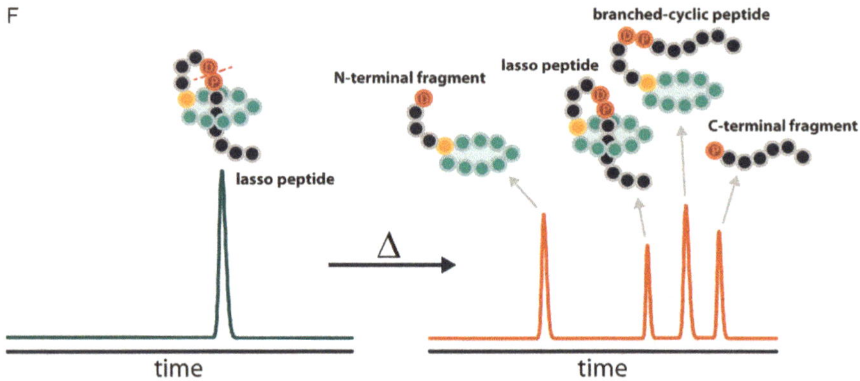

Fig. 4F Schematic depicting the thermally-induced partial decomposition of a fictitious heat-sensitive lasso peptide containing an Asp-Pro sequence. When heated, the lasso peptide not only slowly undergoes an unthreading process, but also self-hydrolysis occurs, which becomes evident through the detection of distinct peptide fragment relating to the self-cleavage of the peptide backbone between the Asp and Pro residues.

(ii) Benenodin-1 is a special case as it comprises the first example of a lasso peptide that changes back and forth between two distinct lasso peptide conformations when heated without ever completely unthreading into a branched-cyclic peptide (Fig. 4G) (Zong et al., 2017). Prolonged incubation at elevated temperatures causes an equilibrium to occur between the two lasso peptide conformations and the ratio of both species in equilibrium is temperature-dependent (Zong et al., 2017). While such special cases are rarely observed, the explanation how such behavior is caused is simple enough. Basically, the naturally used lower plug residue in the isolated lasso peptide is too small to confer a heat-stable lasso fold. However, shortly subsequent to the naturally occurring lower plug residue, an amino acid with an even bulkier sidechain is present. This sidechain has to be too large to pass through the macrolactam ring and thus stops the further unthreading into a branched-cyclic peptide by effectively acting as a kind of secondary plug in a second lasso conformation. Whereas such a lasso peptide can therefore never fully unthread, the ring can jump back and forth over the original lower plug residue and thus alter between the original and the alternative lasso conformation. In context of the assay described here, the obtained results might first seem confusing. Heat treatment will cause a new, isobaric peak to appear in the chromatogram. However, neither will the full conversion of the original lasso peptide peak ever be observed even after prolonging the incubation time at 95 °C, nor will either of the two peaks exhibit sensitivity toward carboxypeptidase Y treatment. To confirm that indeed a lasso peptide able to thermally switch between two conformations is present, two additional experiments can be carried out. After heat treatment, the newly appearing HPLC peak can be collected, heated again, and then reapplied to HPLC. If a thermally switchable lasso peptide is present, this should cause both the original peak to reappear as well as the ratio between both conformations to adopt the same ratio as observed before (at least if the heat treatment was sufficiently long to allow the reaching of a state of equilibrium). In addition, the heating step can be carried out at different temperatures, which will exhibit different equilibrium ratios. These additional assays combined with the observed carboxypeptidase Y stability of the lasso peptide species relating to the two different peaks in the chromatogram are strong indicators that indeed a lasso peptide was identified that is able to switch between two distinct lasso folds in a heat-dependent manner.

G

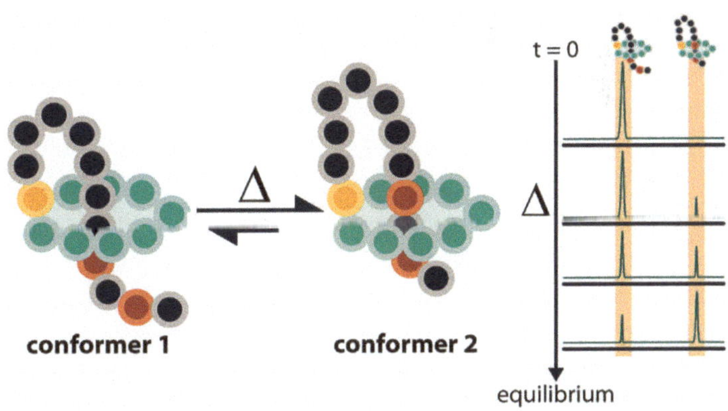

Fig. 4G Schematic depicting the heat-dependent interconversion between two distinct fictitious lasso peptide conformations. The primary and secondary lower plug residues are highlighted in red. This behavior becomes possible when the naturally used, primary lower plug amino acid is small enough to allow the ring to skip over it at elevated temperatures. In addition, the secondary plug residue needs to be large enough to suppress any further unthreading to occur. Hence, when heated, the ring will jump back and forth over the primary plug residue and thus skip between the two distinct lasso conformations. This process can be monitored via HPLC-MS, where two isobaric peaks will be observed when heated that will eventually reach a state of equilibrium and will be resistant against carboxypeptidase Y-mediated proteolysis. In contrast, a normal heat-sensitive lasso peptide would completely unthread eventually and only the peak relating to the obtained branched-cyclic peptide would remain.

5.3 Brief overview of additional analytics to complement stability assays

Following the in-depth discussion about how to perform thermal and carboxypeptidase Y stability assays and how to analyze the data obtained, it should also be considered how these assays can be complemented by additional experiments.

First and foremost, a full structure elucidation via NMR spectroscopy or crystallization would facilitate data interpretation and yield a much more complete understanding of the lasso fold present (Braffman et al., 2019; Fage et al., 2016; Hegemann et al., 2020, 2016, 2014; Jeanne Dit Fouque, Lavanant, et al., 2018; Mathavan et al., 2014; Nar et al., 2010; Xie & Marahiel, 2012). However, due to time-constraints or the amount of lasso peptide available, this might not always be an option. Therefore, a number of methods utilizing different tandem MS techniques as well as ion mobility—mass spectrometry (IMS) have been developed (Dit Fouque et al., 2018, 2021; Fouque et al., 2017; Hegemann et al., 2021; Jeanne Dit Fouque, Afonso, et al., 2015; Jeanne Dit Fouque, Bisram, et al., 2019; Jeanne Dit Fouque, Hegemann, et al., 2019; Jeanne Dit Fouque, Lavanant, et al., 2018, 2015; Jeanne Dit Fouque, Moreno, et al., 2018; Martin-Gomez et al., 2018; Perot-Taillandier et al., 2012; Zhu, Hegemann, et al., 2016; Zirah et al., 2011).

IMS allows the separation of isobaric ions based on their size, or more accurately, their collision cross sections. Separation in IMS is accomplished by having the selected ions traverse a drift tube along an electrical field. Collisions with neutral buffer gas molecules in the drift tube will slow down the ions, hence the larger the ion, the longer its drift time will be. At low charge states the differences in the collision cross sections of a lasso peptide and its branched-cyclic topoisomer are small and can only be detected when employing IMS instruments with sufficiently high resolving power (Dit Fouque et al., 2018, 2021; Hegemann, 2020; Hegemann et al., 2020; Jeanne Dit Fouque, Bisram, et al., 2019; Jeanne Dit Fouque, Hegemann, et al., 2019; Jeanne Dit Fouque, Moreno, et al., 2018). However, higher charge states, which can be accomplished through the use of supercharging agents like sulfolane, cause stronger electrostatic repulsion, which forces the branched-cyclic peptides to adopt more spacious gas phase conformations, while the rigid lasso fold suppresses this effect and thus, the separation of the topoisomers based on their drift times becomes more pronounced (Fouque et al., 2017; Hegemann, 2020; Hegemann et al., 2020; Jeanne Dit Fouque,

Afonso, et al., 2015; Jeanne Dit Fouque, Lavanant, et al., 2015; Martin-Gomez et al., 2018; Zhu, Hegemann, et al., 2016). Therefore, analyzing a heat-sensitive lasso peptide before and after heating via IMS allows further validation of the thermally-induced conversion of a lasso into a branched-cyclic conformation. In addition, IMS also allows direct comparison of the collision cross sections of a lasso peptide with a synthetic branched-cyclic analog, and hence can be used to evidence that an isolated heat-stable lasso peptide does not adopt a mere branched-cyclic conformation.

Gas phase fragmentation via collision-induced dissociation (CID) does not only allow validation of the primary structure and the location of the macrolactam ring of a lasso peptide, but can, for lasso peptides fulfilling certain criteria, also evidence their threaded topology. It was demonstrated that for lasso peptides with loop regions spanning more than four residues, CID spectra often contain mechanically interlocked fragment ions (Hegemann et al., 2021; Jeanne Dit Fouque, Bisram, et al., 2019; Jeanne Dit Fouque, Lavanant, et al., 2018; Zirah et al., 2011), i.e., fragments of the b-series containing the macrolactam ring that still entrap a y-series fragment of a part of the C-terminal tail containing the plug amino acids. This factual conversion of a [1]rotaxane into a [2]rotaxane in the gas phase is only possible when a threaded lasso but not a branched-cyclic peptide is submitted to CID.

When fragmentation is induced via electron capture dissociation (ECD) or electron transfer dissociation (ETD) instead of CID, hydrogen migration patterns between c and z fragments (−1 Da shifts of c fragment ions by loss, +1 Da shifts of z fragment ions by capture of hydrogen radicals) can also evidence a lasso fold. Such hydrogen migration events are facilitated through the close proximity of the residues in the loop of a lasso peptide, which causes them to occur much more frequently than when investigating a branched-cyclic peptide of the same sequence (Hegemann et al., 2021; Jeanne Dit Fouque, Bisram, et al., 2019; Jeanne Dit Fouque, Lavanant, et al., 2018; Zirah et al., 2011). Unlike for the detection of interlocked fragments in CID-based fragmentation, there is no lower loop size limit for observing such distinct hydrogen migration events in ECD/ETD spectra (Jeanne Dit Fouque, Lavanant, et al., 2018) and hence this technique can be applied to all kinds of lasso peptides.

Taken together, IMS, CID, and ECD/ETD are MS techniques that can provide further evidence for the presence or the absence of a threaded fold and thereby nicely complement the discussed combined thermal and carboxypeptidase Y stability assays. Due to the high sensitivity of MS instruments, these assays can all be performed quickly and with minute amounts of lasso peptides, which makes them particularly interesting when the production/isolation of large amounts of a given lasso peptide is not feasible.

6. Summary

Characterization of a novel lasso peptide demands proof of its threaded topology. Besides the time-consuming full structure elucidation via NMR spectroscopy or crystallization that requires large amounts of pure material, combined thermal and carboxypeptidase Y assays can be rapidly performed and can already be done using lasso peptide-containing crude culture extracts (Fage et al., 2016; Hegemann et al., 2021; Hegemann, Zimmermann, Zhu, et al., 2013). In case of a heat-sensitive lasso peptide, the observations of both the heat-induced unthreading and the change in behavior toward carboxypeptidase Y provide strong evidence for the initially threaded fold. However, while the outcomes of these assays when studying a heat-stable lasso peptide are not providing quite as much evidence, the high resistance against carboxypeptidase Y-mediated degradation is still a benchmark for the lasso fold, particularly when the primary structure of the compound has been confirmed via tandem MS analysis.

In conclusion, the assays described in this chapter are highly useful when working with new representatives of this RiPP subfamily, especially as the assessment of the thermal stability properties has become a routine procedure for the characterization of newly discovered lasso peptides.

References

Allen, C. D., Chen, M. Y., Trick, A. Y., Le, D. T., Ferguson, A. L., & Link, A. J. (2016). Thermal unthreading of the lasso peptides Astexin-2 and Astexin-3. *ACS Chemical Biology*, *11*(11), 3043–3051.

Arnison, P. G., Bibb, M. J., Bierbaum, G., Bowers, A. A., Bugni, T. S., Bulaj, G., et al. (2013). Ribosomally synthesized and post-translationally modified peptide natural products: Overview and recommendations for a universal nomenclature. *Natural Product Reports*, *30*(1), 108–160.

Braffman, N. R., Piscotta, F. J., Hauver, J., Campbell, E. A., Link, A. J., & Darst, S. A. (2019). Structural mechanism of transcription inhibition by lasso peptides microcin J25 and capistruin. *Proceedings of the National Academy of Sciences of the United States of America*, *116*(4), 1273–1278.

Bratovanov, E. V., Ishida, K., Heinze, B., Pidot, S. A.-O., Stinear, T. P., Hegemann, J. A.-O., et al. (2020). Genome mining and heterologous expression reveal two distinct families of lasso peptides highly conserved in endofungal bacteria. *ACS Chemical Biology*, *15*(5), 1169–1176.

Burkhart, B. J., Hudson, G. A., Dunbar, K. L., & Mitchell, D. A. (2015). A prevalent peptide-binding domain guides ribosomal natural product biosynthesis. *Nature Chemical Biology*, *11*(8), 564–570.

Cheng, C., & Hua, Z. C. (2020). Lasso peptides: Heterologous production and potential medical application. *Frontiers in Bioengineering and Biotechnology*, *8*, 571165.

Cheung-Lee, W. L., Cao, L., & Link, A. J. (2019). Pandonodin: A proteobacterial lasso peptide with an exceptionally long C-terminal tail. *ACS Chemical Biology*, *14*(12), 2783–2792.

Cheung-Lee, W. L., & Link, A. J. (2019). Genome mining for lasso peptides: Past, present, and future. *Journal of Industrial Microbiology & Biotechnology*, *46*(9–10), 1371–1379.

Cheung-Lee, W. L., Parry, M. E., Jaramillo Cartagena, A., Darst, S. A., & Link, A. J. (2019). Discovery and structure of the antimicrobial lasso peptide citrocin. *The Journal of Biological Chemistry*, *294*(17), 6822–6830.

Cheung-Lee, W. L., Parry, M. E., Zong, C., Cartagena, A. J., Darst, S. A., Connell, N. D., et al. (2020). Discovery of ubonodin, an antimicrobial lasso peptide active against members of the Burkholderia cepacia complex. *Chembiochem*, *21*(9), 1335–1340.

DiCaprio, A. J., Firouzbakht, A., Hudson, G. A., & Mitchell, D. A. (2019). Enzymatic reconstitution and biosynthetic investigation of the lasso peptide fusilassin. *Journal of the American Chemical Society*, *141*(1), 290–297.

Dit Fouque, K. J., Moreno, J., Hegemann, J. D., Zirah, S., Rebuffat, S., & Fernandez-Lima, F. (2018). Identification of lasso peptide topologies using native nanoelectrospray ionization-trapped ion mobility spectrometry-mass spectrometry. *Analytical Chemistry*, *90*(8), 5139–5146.

Dit Fouque, K. J., Scutelnic, V., Hegemann, J. D., Rebuffat, S., Maître, P., Rizzo, T. R., et al. (2021). Structural insights from tandem mass spectrometry, ion mobility-mass spectrometry, and infrared/ultraviolet spectroscopy on Sphingonodin I: Lasso vs branched-cyclic Topoisomers. *Journal of the American Society for Mass Spectrometry*, *32*(4), 1096–1104.

Fage, C. D., Hegemann, J. D., Nebel, A. J., Steinbach, R. M., Zhu, S., Linne, U., et al. (2016). Structure and mechanism of the Sphingopyxin I lasso peptide isopeptidase. *Angewandte Chemie, International Edition*, *55*(41), 12717–12721.

Fouque, K. J. D., Lavanant, H., Zirah, S., Hegemann, J. D., Zimmermann, M., Marahiel, M. A., et al. (2017). Signatures of mechanically interlocked topology of lasso peptides by ion mobility-mass spectrometry: Lessons from a collection of representatives. *Journal of the American Society for Mass Spectrometry*, *28*(2), 315–322.

Guerrero-Garzón, J. F., Madland, E., Zehl, M., Singh, M., Rezaei, S., Aachmann, F. L., et al. (2020). Class IV lasso peptides synergistically induce proliferation of cancer cells and sensitize them to doxorubicin. *iScience*, *23*(12), 101785.

Hegemann, J. D. (2020). Factors governing the thermal stability of lasso peptides. *Chembiochem*, *21*(1–2), 7–18.

Hegemann, J. D., Dit-Foque, K. J., & Xie, X. (2020). 2.10—X lasso peptides: Structure, biosynthesis, activities, and beyond. In H.-W. Liu, & T. P. Begley (Eds.), *Comprehensive natural products III* (pp. 206–228). Oxford: Elsevier.

Hegemann, J. D., Fage, C. D., Zhu, S., Harms, K., Di Leva, F. S., Novellino, E., et al. (2016). The ring residue proline 8 is crucial for the thermal stability of the lasso peptide caulosegnin II. *Molecular BioSystems*, *12*(4), 1106–1109.

Hegemann, J. D., Jeanne Dit Fouque, K., Santos-Fernandez, M., & Fernandez-Lima, F. (2021). A bifunctional leader peptidase/ABC transporter protein is involved in the maturation of the lasso peptide Cochonodin I from Streptococcus Suis. *Journal of Natural Products*. https://doi.org/10.1021/acs.jnatprod.1c00514.

Hegemann, J. D., Schwalen, C. J., Mitchell, D. A., & van der Donk, W. A. (2018). Elucidation of the roles of conserved residues in the biosynthesis of the lasso peptide paeninodin. *Chemical Communications*, *54*(65), 9007–9010.

Hegemann, J. D., Zimmermann, M., Xie, X., & Marahiel, M. A. (2013). Caulosegnins I-III: A highly diverse group of lasso peptides derived from a single biosynthetic gene cluster. *Journal of the American Chemical Society, 135*(1), 210–222.

Hegemann, J. D., Zimmermann, M., Xie, X., & Marahiel, M. A. (2015). Lasso peptides: An intriguing class of bacterial natural products. *Accounts of Chemical Research, 48*(7), 1909–1919.

Hegemann, J. D., Zimmermann, M., Zhu, S., Klug, D., & Marahiel, M. A. (2013). Lasso peptides from proteobacteria: Genome mining employing heterologous expression and mass spectrometry. *Peptide Science, 100*(5), 527–542.

Hegemann, J. D., Zimmermann, M., Zhu, S., Steuber, H., Harms, K., Xie, X., et al. (2014). Xanthomonins I–III: A new class of lasso peptides with a seven-residue macrolactam ring. *Angewandte Chemie, International Edition, 53*(8), 2230–2234.

Jeanne Dit Fouque, K., Afonso, C., Zirah, S., Hegemann, J. D., Zimmermann, M., Marahiel, M. A., et al. (2015). Ion mobility-mass spectrometry of lasso peptides: Signature of a rotaxane topology. *Analytical Chemistry, 87*(2), 1166–1172.

Jeanne Dit Fouque, K., Bisram, V., Hegemann, J. D., Zirah, S., Rebuffat, S., & Fernandez-Lima, F. (2019). Structural signatures of the class III lasso peptide BI-32169 and the branched-cyclic topoisomers using trapped ion mobility spectrometry—Mass spectrometry and tandem mass spectrometry. *Analytical and Bioanalytical Chemistry, 411*(24), 6287–6296.

Jeanne Dit Fouque, K., Hegemann, J. D., Zirah, S., Rebuffat, S., Lescop, E., & Fernandez-Lima, F. (2019). Evidence of cis/trans-isomerization at Pro7/Pro16 in the lasso peptide Microcin J25. *Journal of the American Society for Mass Spectrometry, 30*(6), 1038–1045.

Jeanne Dit Fouque, K., Lavanant, H., Zirah, S., Hegemann, J. D., Fage, C. D., Marahiel, M. A., et al. (2018). General rules of fragmentation evidencing lasso structures in CID and ETD. *Analyst, 143*(5), 1157–1170.

Jeanne Dit Fouque, K., Lavanant, H., Zirah, S., Lemoine, J., Rebuffat, S., Tabet, J. C., et al. (2015). Gas-phase conformations of capistruin—Comparison of lasso, branched-cyclic and linear topologies. *Rapid Communications in Mass Spectrometry, 29*(15), 1411–1419.

Jeanne Dit Fouque, K., Moreno, J., Hegemann, J. D., Zirah, S., Rebuffat, S., & Fernandez-Lima, F. (2018). Metal ions induced secondary structure rearrangements: Mechanically interlocked lasso vs. unthreaded branched-cyclic topoisomers. *Analyst, 143*(10), 2323–2333.

Knappe, T. A., Linne, U., Robbel, L., & Marahiel, M. A. (2009). Insights into the biosynthesis and stability of the lasso peptide capistruin. *Chemistry & Biology, 16*(12), 1290–1298.

Knappe, T. A., Linne, U., Xie, X., & Marahiel, M. A. (2010). The glucagon receptor antagonist BI-32169 constitutes a new class of lasso peptides. *FEBS Letters, 584*(4), 785–789.

Koos, J. D., & Link, A. J. (2019). Heterologous and in vitro reconstitution of Fuscanodin, a lasso peptide from Thermobifida fusca. *Journal of the American Chemical Society, 141*(2), 928–935.

Li, Y., Ducasse, R., Zirah, S., Blond, A., Goulard, C., Lescop, E., et al. (2015). Characterization of Sviceucin from Streptomyces provides insight into enzyme exchangeability and disulfide bond formation in lasso peptides. *ACS Chemical Biology, 10*(11), 2641–2649.

Long, T., Liu, L., Tao, Y., Zhang, W., Quan, J., Zheng, J., et al. (2021). Light-controlled tyrosine nitration of proteins. *Angewandte Chemie, International Edition, 60*(24), 13414–13422.

Maksimov, M. O., Koos, J. D., Zong, C., Lisko, B., & Link, A. J. (2015). Elucidating the specificity determinants of the AtxE2 lasso peptide isopeptidase. *The Journal of Biological Chemistry, 290*(52), 30806–30812.

Maksimov, M. O., & Link, A. J. (2013). Discovery and characterization of an isopeptidase that linearizes lasso peptides. *Journal of the American Chemical Society, 135*(32), 12038–12047.

Maksimov, M. O., Pelczer, I., & Link, A. J. (2012). Precursor-centric genome-mining approach for lasso peptide discovery. *Proceedings of the National Academy of Sciences of the United States of America, 109*(38), 15223–15228.

Martin-Gomez, H., Linne, U., Albericio, F., Tulla-Puche, J., & Hegemann, J. D. (2018). Investigation of the biosynthesis of the lasso peptide Chaxapeptin using an E. coli-based production system. *Journal of Natural Products, 81*(9), 2050–2056.

Mathavan, I., Zirah, S., Mehmood, S., Choudhury, H. G., Goulard, C., Li, Y., et al. (2014). Structural basis for hijacking siderophore receptors by antimicrobial lasso peptides. *Nature Chemical Biology, 10*(5), 340–342.

Metelev, M., Arseniev, A., Bushin, L. B., Kuznedelov, K., Artamonova, T. O., Kondratenko, R., et al. (2017). Acinetodin and Klebsidin, RNA polymerase targeting lasso peptides produced by human isolates of Acinetobacter gyllenbergii and Klebsiella pneumoniae. *ACS Chemical Biology, 12*(3), 814–824.

Montalbán-López, M., Scott, T. A., Ramesh, S., Rahman, I. R., van Heel, A. J., Viel, J. H., et al. (2021). New developments in RiPP discovery, enzymology and engineering. *Natural Product Reports, 38*(1), 130–239.

Nar, H., Schmid, A., Puder, C., & Potterat, O. (2010). High-resolution crystal structure of a lasso peptide. *ChemMedChem, 5*(10), 1689–1692.

Pati, N. B., Doijad, S. P., Schultze, T., Mannala, G. K., Yao, Y., Jaiswal, S., et al. (2018). Enterobacter bugandensis: A novel enterobacterial species associated with severe clinical infection. *Scientific Reports, 8*(1), 5392.

Perot-Taillandier, M., Zirah, S., Rebuffat, S., Linne, U., Marahiel, M. A., Cole, R. B., et al. (2012). Determination of peptide topology through time-resolved double-resonance under electron capture dissociation conditions. *Analytical Chemistry, 84*(11), 4957–4964.

Sabino, Y. N. V., de Araújo, K. C., de Assis, F., Moreira, S. M., Lopes, T. D. S., Mendes, T. A. O., et al. (2020). In silico screening unveil the great potential of ruminal bacteria synthesizing lasso peptides. *Frontiers in Microbiology, 11*, 576738.

Salomon, R. A., & Farias, R. N. (1992). Microcin 25, a novel antimicrobial peptide produced by Escherichia coli. *Journal of Bacteriology, 174*(22), 7428–7435.

Skribanek, Z., Mezo, G., Mak, M., & Hudecz, F. (2002). Mass spectrometric and chemical stability of the Asp-Pro bond in herpes simplex virus epitope peptides compared with X-Pro bonds of related sequences. *Journal of Peptide Science, 8*(8), 398–406.

Sumida, T., Dubiley, S., Wilcox, B., Severinov, K., & Tagami, S. (2019). Structural basis of leader peptide recognition in lasso peptide biosynthesis pathway. *ACS Chemical Biology, 14*(7), 1619–1627.

Tietz, J. I., Schwalen, C. J., Patel, P. S., Maxson, T., Blair, P. M., Tai, H. C., et al. (2017). A new genome-mining tool redefines the lasso peptide biosynthetic landscape. *Nature Chemical Biology, 13*(5), 470–478.

Xie, X., & Marahiel, M. A. (2012). NMR as an effective tool for the structure determination of lasso peptides. *Chembiochem, 13*(5), 621–625.

Yan, K. P., Li, Y., Zirah, S., Goulard, C., Knappe, T. A., Marahiel, M. A., et al. (2012). Dissecting the maturation steps of the lasso peptide microcin J25 in vitro. *Chembiochem, 13*(7), 1046–1052.

Zhu, S., Fage, C. D., Hegemann, J. D., Mielcarek, A., Yan, D., Linne, U., et al. (2016). The B1 protein guides the biosynthesis of a lasso peptide. *Scientific Reports, 6*, 35604.

Zhu, S., Hegemann, J. D., Fage, C. D., Zimmermann, M., Xie, X., Linne, U., et al. (2016). Insights into the unique phosphorylation of the lasso peptide paeninodin. *The Journal of Biological Chemistry, 291*(26), 13662–13678.

Zimmermann, M., Hegemann, J. D., Xie, X., & Marahiel, M. A. (2013). The astexin-1 lasso peptides: Biosynthesis, stability, and structural studies. *Chemistry & Biology*, *20*(4), 558–569.

Zimmermann, M., Hegemann, J. D., Xie, X., & Marahiel, M. A. (2014). Characterization of caulonodin lasso peptides revealed unprecedented N-terminal residues and a precursor motif essential for peptide maturation. *Chemical Science*, *5*(10), 4032–4043.

Zirah, S., Afonso, C., Linne, U., Knappe, T. A., Marahiel, M. A., Rebuffat, S., et al. (2011). Topoisomer differentiation of molecular knots by FTICR MS: Lessons from class II lasso peptides. *Journal of the American Society for Mass Spectrometry*, *22*(3), 467–479.

Zong, C., Wu, M. J., Qin, J. Z., & Link, A. J. (2017). Lasso peptide Benenodin-1 is a thermally actuated [1]Rotaxane switch. *Journal of the American Chemical Society*, *139*(30), 10403–10409.

Zyubko, T., Serebryakova, M., Andreeva, J., Metelev, M., Lippens, G., Dubiley, S., et al. (2019). Efficient in vivo synthesis of lasso peptide pseudomycoidin proceeds in the absence of both the leader and the leader peptidase. *Chemical Science*, *10*(42), 9699–9707.

CHAPTER NINE

Evaluation of endogenous peptide stereochemistry using liquid chromatography-mass spectrometry-based spiking experiments

Baba M. Yussif[a] and James W. Checco[a,b,*]

[a]Department of Chemistry, University of Nebraska-Lincoln, Lincoln, NE, United States
[b]The Nebraska Center for Integrated Biomolecular Communication (NCIBC), University of Nebraska-Lincoln, Lincoln, NE, United States
*Corresponding author: e-mail address: checco@unl.edu

Contents

Abstract

Functional D-amino acid-containing peptides are endogenously found throughout nature, generated through non-ribosomal peptide synthesis or through post-translational modification of ribosome-derived peptides. Despite the high functional importance of peptide stereochemistry in biomolecular recognition, identification of amino acid residue isomerization can be challenging using most common peptide characterization workflows. Here, we describe a relatively simple approach to test hypotheses regarding the stereochemistry of endogenous peptides via liquid chromatography-mass spectrometry (LC-MS) spiking experiments with synthetic isotope-labeled peptide standards. Our protocol details the synthesis of ^{13}C-labeled synthetic peptide diastereomers via solid-phase peptide synthesis, their purification and characterization, and LC-MS experiments to evaluate putative stereochemical assignments. This approach does not require demanding purification of the endogenous

Methods in Enzymology, Volume 663
ISSN 0076-6879
https://doi.org/10.1016/bs.mie.2021.10.009

peptide and is compatible with small quantities of endogenous extracts. Overall, this procedure complements current endogenous peptide characterization methods, granting the ability to relatively quickly verify the sequence and stereochemistry in endogenous peptide diastereomers directly in complex biological extracts.

1. Introduction

Although the ribosome generates proteins and peptides using exclusively L-amino acids, polypeptides containing D-amino acid residues are found throughout nature. Perhaps the most well-known D-amino acid-containing peptides are those in the bacterial peptidoglycan (Lovering, Safadi, & Strynadka, 2012). However, endogenous antimicrobial peptides, peptide toxins, neuropeptides, and peptide hormones can also contain D-amino acid residues (Checco et al., 2018; Cotter et al., 2005; Erspamer et al., 1989; Heck et al., 1994, 1996; Kessler, Schuhmann, Morneweg, Linne, & Marahiel, 2004; Knerr & van der Donk, 2013; Kreil et al., 1989; Ling et al., 2015; Livnat et al., 2016; Mast, Checco, & Sweedler, 2019; Süssmuth & Mainz, 2017). In these cases, the D-residue is incorporated either through non-ribosomal peptide synthesis or as an enzymatic post-translational modification following synthesis by the ribosome. Additionally, peptide isomerization has also been found in long-lived proteins, usually generated through spontaneous isomerization of aspartic acid or asparagine residues (Lambeth et al., 2019; Maeda et al., 2015; Tao & Julian, 2014; Tao, Quebbemann, & Julian, 2012). In most known D-amino acid-containing peptides, the D-residue has critical functional consequences, oftentimes greatly increased biological function relative to their all-L-residue analogues (Mast, Checco, & Sweedler, 2021; Ollivaux, Soyez, & Toullec, 2014).

Despite the functional importance of D-residues in most D-amino acid-containing peptides, peptide stereochemistry is often difficult to detect. This is because most modern peptide discovery approaches rely primarily on mass spectrometry (MS)-based peptide identifications, and diastereomeric peptides have identical masses. This makes isomerization "invisible" to most common peptide characterization workflows. Significant effort has been devoted to developing methods to reliably identify peptide diastereomers, including chiral amino acid analysis methods (Bhushan & Brückner, 2004; Fujii, Ikai, Mayumi, et al., 1997; Fujii, Ikai, Oka, Suzuki, &

Harada, 1997), novel mass spectrometry methods (Adams & Zubarev, 2005; Bai et al., 2013; Jia, Lietz, Yu, & Li, 2014; Koehbach, Gruber, Becker, Kreil, & Jilek, 2016; Tao et al., 2012; Tao & Julian, 2014), and enzymatic screening approaches (Ewing, Wang, Sheeley, & Sweedler, 2008; Livnat et al., 2016; Maeda et al., 2015). However, many methods of stereochemical confirmation require arduous purification of the endogenous peptide or sophisticated MS/MS methods not readily available to most laboratories. Here, we outline a relatively simple method of confirming peptide stereochemistry through spiking experiments with synthetic isotope-labeled peptide standards (Checco et al., 2018; Mast et al., 2019). This method does not require rigorous purification of the endogenous peptide, making this protocol ideal for relatively quickly assessing putative stereochemical assignments for endogenous extracts with limited sample quantity. This protocol has been written to aid laboratories with little or no prior experience in peptide synthesis, and outlines relatively simple MS and MS/MS analysis that can be performed on most LC-MS instruments readily available in core instrumentation facilities.

2. Overview of the protocol

The method described in this chapter is designed to test hypotheses generated about endogenous peptide stereochemistry (Fig. 1). To accomplish these goals, Section 4.1 of this protocol outlines detailed methods for the synthesis, purification, and analysis of ^{13}C labeled synthetic peptide analogues. Following synthesis, Section 4.2 outlines spiking these isotope-labeled peptides into endogenous peptide extracts, followed by analysis using reversed-phase high-performance liquid chromatography (HPLC) coupled to MS. This method relies on the fact that most peptide diastereomers elute at different retention times during reversed-phase HPLC (Bai et al., 2013; Checco et al., 2018; Fujisawa et al., 1992; Heck et al., 1994; Livnat et al., 2016; Mast et al., 2019, 2021; Morishita et al., 1997). As a result, co-elution of a synthetic standard with the endogenous peptide of interest when spiked into the endogenous peptide extracts provides strong evidence that the correct stereochemical assignment was made (provided the chromatographic method used is able to separate peptide diastereomers under consideration). The use of isotope-labeled peptides is important to unambiguously differentiate synthetic peptides from endogenous peptides, preventing any possibility of unintentional cross-contamination which

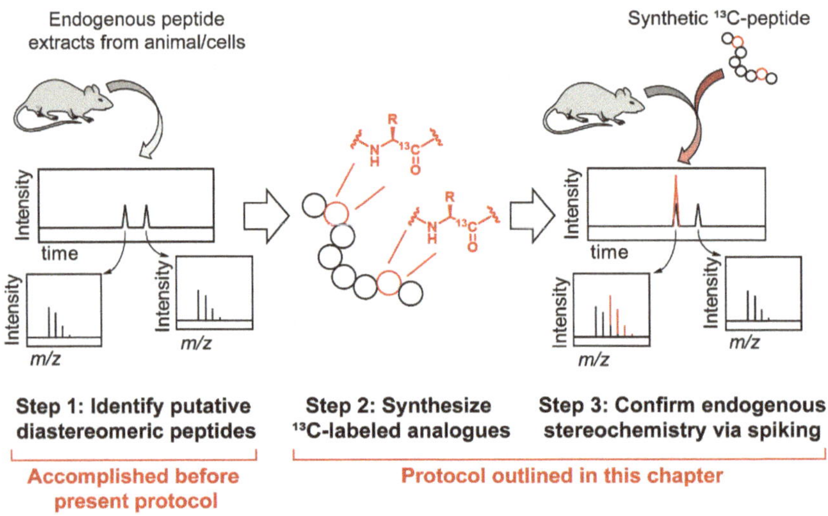

Fig. 1 Overview of the method. In Step 1, peptides suspected to contain D-residues are identified from animal tissues through a variety of methods. The cartoon depicts an extracted ion chromatogram (EIC) displaying two distinct peaks with identical MS1 spectra, a common characteristic of endogenous diastereomers. (Diastereomeric peaks will often also have very similar MS2 spectra.) In Step 2, ^{13}C-labeled synthetic peptides corresponding to the putative diastereomers are synthesized by solid-phase peptide synthesis and purified by reverse-phase HPLC. In Step 3, ^{13}C-labeled synthetic peptide is spiked into endogenous peptide extracts. Co-elution of endogenous and synthetic peptide adds high confidence to the stereochemical assignment of the endogenous peptide.

would lead to false-positive results. For the purpose of this chapter, we show data for naturally occurring diastereomers of the *Aplysia californica* allatotropin-related peptide (ATRP) (Checco et al., 2018). Two endogenous forms of ATRP exist in the *Aplysia* central nervous system: the all-L-residue form (L-ATRP), and its diastereomer bearing a D-Phe residue at position 2 (D2-ATRP).

3. Before you begin—Determining putative positions of isomerization

Before carrying out spiking experiments, the researcher must have already identified candidate DAACPs and putative positions of isomerization. Several methods are available for untargeted identification of candidate DAACPs, which are covered in detail elsewhere (Jansson, 2018; Livnat et al., 2016; Mast et al., 2019, 2021). Putative positions of isomerization must also be posited because the researcher must synthesize the hypothesized diastereomers for the spiking experiments outlined here. Although it is possible to synthesize and evaluate all possible diastereomers of short peptides (e.g., peptides of 3 or 4 residues), this strategy dramatically loses feasibility as peptide length increases. As a result, preliminary studies must be performed to narrow down the possible isomerized position(s).

There are several ways that one may make putative stereochemical assignments for evaluation without needing to purify large amounts of material for chiral amino acid analysis (Mast et al., 2021). At the simplest level, homology to a previously discovered DAACP can often point the researcher in productive directions. Developing and testing hypotheses based on sequence homology has proven successful for many discovered DAACPs, including *Aplysia* ATRP (Checco et al., 2018) and other bioactive neuropeptides from mollusks (Bai et al., 2013; Ohta et al., 1991; Yasuda-Kamatani et al., 1997), DAACPs from frog skin secretions (Kreil et al., 1989), peptide toxins (Buczek et al., 2005), and diastereomers of crustacean hyperglycemic hormone and vitellogenesis inhibiting hormone (Aguilar et al., 1995; Ollivaux, Gallois, Amiche, Boscameric, & Soyez, 2009).

In the absence of sequence homology, controlled protease digestion of the endogenous peptide of interest can be an effective technique for narrowing down the position of isomerization. In practice, there are at least two fundamentally different types of enzymatic digestion experiments that have been successful to localize isomerization. The first strategy is to utilize exopeptidases to identify regions of the peptide that are stable to degradation. Because proteases hydrolyze amide bonds near D-residues significantly slower than those near L-residues, exopeptidases dramatically slow their digestion as they approach an isomerized residue. In the presence of an exopeptidase such as aminopeptidase M, endogenous peptides stable to degradation likely indicate the presence of a D-residue near the N-terminus (Checco et al., 2018; Livnat et al., 2016; Mast et al., 2019; Torres et al., 2002). Similarly, putative internal D-residues may be identified by carefully examining the peptide fragments stable to degradation after terminal L-residues are removed through LC-MS analysis (Mast et al., 2019). A second distinct approach is to digest both the endogenous peptide and synthetic all-L-residue diastereomer with an endopeptidase and compare the chromatographic properties of resulting peptide fragments (Buczek, Yoshikami, Bulaj, Jimenez, & Olivera, 2005; Soyez et al., 1994). In this approach, peptide fragments that contain the isomerized residue(s) will differ in retention time between the synthetic all-L-version and the endogenous peptide, while peptide fragments that contain only L-residues will be identical between the two. From careful analysis of these digests, regions of the peptide likely to differ between the diastereomers can be identified.

In the ATRP example highlighted in this chapter, the putative location of the D-residue (Phe2) was determined using a combination of homology and exopeptidase digestion (Checco et al., 2018). It is also important to note that endogenous ATRP also displayed LC-MS and LC-MS/MS behavior characteristic of peptides endogenously present as multiple diastereomers: two chromatographically resolved peaks with identical MS parent mass and nearly identical MS/MS fragment spectra (Fig. 2). Peptides showing this behavior could be considered as possible DAACP candidates.

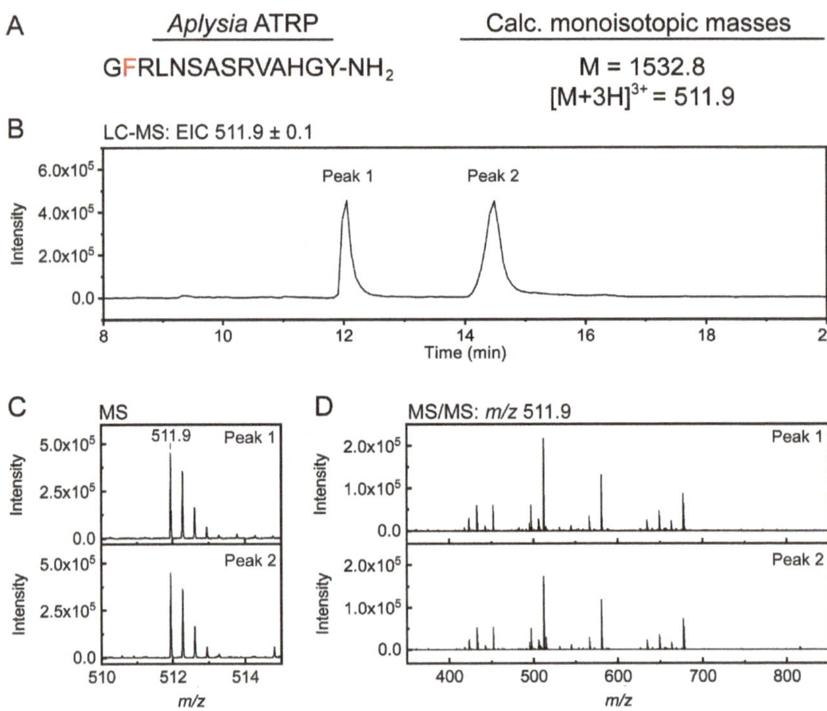

Fig. 2 LC-MS and LC-MS/MS analysis of putative peptide diastereomers. Shown here are experimental results examining two distinct peaks from *Aplysia californica* neuropeptide extracts. (A) Sequence of *Aplysia* ATRP and relevant calculated monoisotopic masses. Suspected isomerized residue is highlighted in red. (B) Extracted ion chromatogram (EIC) for the ATRP parent mass ($m/z = 511.9 \pm 0.1$, $z = +3$). (C) Representative MS spectra for Peak 1 and Peak 2. (D) Representative MS/MS fragmentation spectra (CID) for Peak 1 and Peak 2. *Figure adapted from Checco, J. W., Zhang, G., Yuan, W. D., Le, Z. W., Jing, J., & Sweedler, J. V. (2018).* Aplysia *allatotropin-related peptide and its newly identified D-amino acid-containing epimer both activate a receptor and a neuronal target.* Journal of Biological Chemistry, 293, *16862–16873.*

4. Step-by-step methods

4.1 Chemical synthesis of isotope-labeled peptide analogs

Once putative stereochemical assignments of peptide diastereomers have been generated, one should synthesize the all-L-residue isomer in addition to any suspected D-amino acid-containing peptides, all containing one or more ^{13}C-isotope labels. (More ^{13}C-isotope labels provide a greater mass difference between endogenous and synthetic peptides, and reduce overlap of isotopic envelopes. For the example highlighted in this chapter, we utilize synthetic peptides containing two ^{13}C atoms.) Note that the protocol described in this chapter differs slightly from the protocol previously reported to generate ^{13}C-labeled L- and D2-ATRP (Checco et al., 2018). Both protocols are likely to be successful for most peptides, but we include the protocol described in this chapter because it uses fewer reagents and thus may be more amenable for labs aiming to begin peptide synthesis with little experience. Fig. 3 shows an overview of the chemical reactions necessary to assemble peptides via solid-phase peptide synthesis.

4.1.1 Materials and reagents

- Protected amino acid monomers suitable for Fmoc solid-phase peptide synthesis. Monomers should be protected on their N-terminus with an Fmoc group, and sidechain protected with appropriate base-stable, acid-labile protecting group. We obtain these monomer building blocks from Novabiochem, Chem Impex, or Anaspec. The appropriate monomers for standard synthesis are the following:
  - Alanine: Fmoc-Ala-OH
  - Cysteine: Fmoc-Cys(Trt)-OH
  - Aspartic acid: Fmoc-Asp(OtBu)-OH
  - Glutamic acid: Fmoc-Glu(OtBu)-OH
  - Phenylalanine: Fmoc-Phe-OH
  - Glycine: Fmoc-Gly-OH
  - Histidine: Fmoc-His(Trt)-OH
  - Isoleucine: Fmoc-Ile-OH
  - Lysine: Fmoc-Lys(Boc)-OH
  - Leucine: Fmoc-Leu-OH
  - Methionine: Fmoc-Met-OH
  - Asparagine: Fmoc-Asn(Trt)-OH
  - Proline: Fmoc-Pro-OH

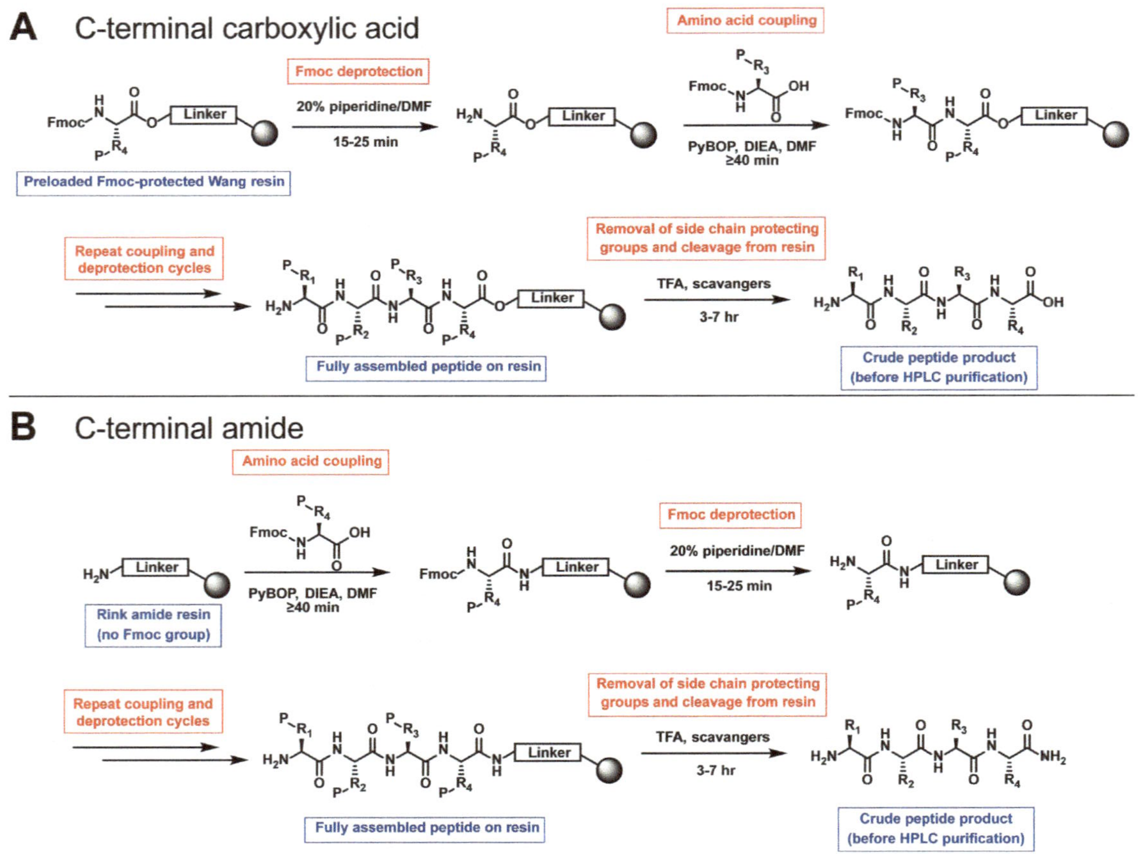

Fig. 3 Schematic outline of solid-phase peptide synthesis for (A) C-terminal carboxylic acid using pre-loaded Wang resin, or (B) C-terminal amide using Rink Amide resin. Grey circle indicates resin beads. "P" indicates a side-chain protecting group. DMF washing steps, which take place between each reaction, are omitted in this schematic.

  - Glutamine: Fmoc-Gln(Trt)-OH
  - Arginine: Fmoc-Arg(Pbf)-OH
  - Serine: Fmoc-Ser(tBu)-OH
  - Threonine: Fmoc-Thr(tBu)-OH
  - Valine: Fmoc-Val-OH
  - Tryptophan: Fmoc-Trp(Boc)-OH
  - Tyrosine: Fmoc-Tyr(tBu)-OH
  - Desired Fmoc-protected D-amino acids with appropriate side chain protecting groups
- Resin for solid phase peptide synthesis. The choice of resin primarily depends on the functionality on the C-terminus of the desired peptide.
  - For C-terminal carboxylic acid, we recommend buying Wang resin preloaded with the C-terminal amino acid. For example, for the sequence GFFD-OH, we would purchase Fmoc-Asp(OtBu)-Wang resin (Novabiochem, 856005).
  - For C-terminal amide, we recommend Rink Amide resin, such as NovaPEG Rink Amide resin (Novabiochem, 855047).
- Appropriate ^{13}C-labeled Fmoc-protected amino acid monomers. For the example outlined in this chapter, we used Fmoc-^{13}C-Gly-OH (Cambridge Isotope Laboratories, CLM-3639-1 or Anaspec, AS-61572-1), which contains a single ^{13}C atom at the carbonyl carbon. However, if the peptide(s) of interest do not contain a glycine residue for replacement, another labeled amino acid will have to be chosen (e.g., Fmoc-^{13}C-Ala-OH, Cambridge Isotope Laboratories, CLM-818-PK).
- Benzotriazole-1-yl-oxytris-pyrrolidino-phosphonium hexafluorophosphate (PyBOP, Novabiochem, 8510090100).
- Dimethylformamide (DMF), peptide synthesis or sequencing grade.
- Dichloromethane (DCM).
- Piperidine.
- *N,N*-diisopropylethylamine (DIEA).
- Trifluoroacetic acid (TFA).
- Triisopropylsilane (TIS).
- 1,2-Ethanedithiol (EDT) (required only if peptide contains one or more cysteine residues).
- Ultrapure water.
- Diethyl ether.
- 5 mL peptide synthesis reaction vessels (Torviq, SF-0500).
- Pressure caps for reaction vessels (Torviq, PC-SF).

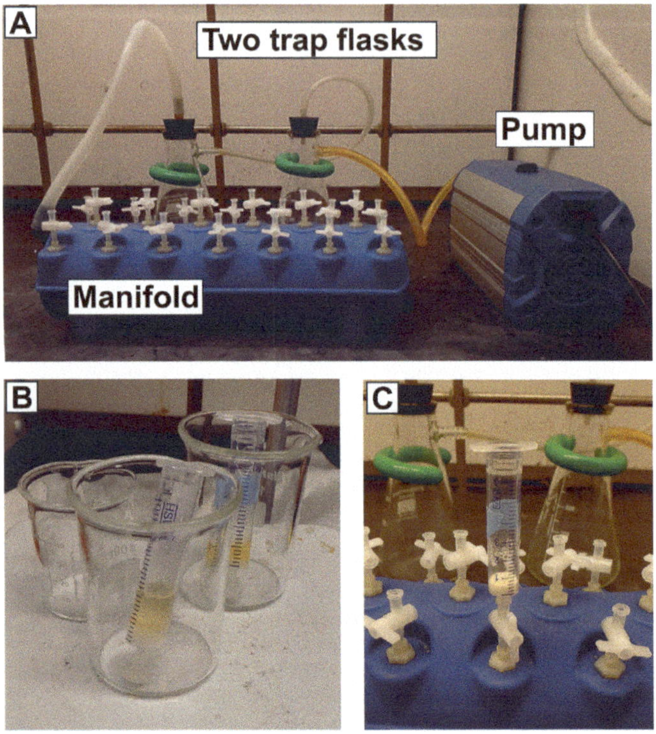

Fig. 4 Vacuum manifold system for manual solid-phase peptide synthesis. (A) Set-up for manifold system, including manifold, two trap flasks, and a small vacuum pump. (Pump may be replaced with house vacuum line or water aspirator.) (B) Two reaction vessels on stir plate during coupling or deprotection reaction. (C) Reaction vessel being drained on vacuum manifold during washing steps.

- System for vacuum aspiration of solvent from peptide synthesis reaction vessels. There are many possible options for this. Our setup uses a vacuum manifold designed for DNA and RNA purification, with modification of chemically-resistant stopcocks. Set-up of apparatus is shown in Fig. 4.
  - Vac-Man Vacuum Manifold (Promega, A7231).
  - Chemically resistant stopcocks for vacuum manifold (e.g., polypropylene). (Note that the polycarbonate stopcocks sold with the Vac-Man Vacuum Manifold are not resistant to the solvents used in solid-phase peptide synthesis.)
- Micro stir bars, ~7 mm length (Fisher Scientific, 1451363SIX).
- HPLC-grade H_2O with 0.1% TFA.
- HPLC-grade acetonitrile with 0.1% TFA.

- "Preparatory"-scale HPLC instrument, C18 column, vials, and collection tubes capable of reverse-phase separations and fraction collection.
  - We use C18 columns with internal diameter range of ~9–21.2mm for "preparatory"-scale purifications, usually running flow rates between 5 and 15mL/min.
- "Analytical"-scale HPLC instrument, or alternative method for evaluating purity of peptide after large-scale HPLC Purification.
  - We use C18 column with internal diameter of 4.2mm running a flow rate of 1mL/min.
- 10mL syringe.
- 0.22μm syringe filters (we use 33mm diameter units with PVDF membrane).
- Method for removing solvents after HPLC purification. We prefer lyophilization, but other methods such as centrifuge-based vacuum concentrators and rotary evaporation can also work.
- Micropipettes and tips capable of accurately dispensing 1–1000μL.
- Microcentrifuge tubes (~1.5mL).
- Conical centrifuge tubes (50mL).
- Temperature-controlled centrifuge capable of centrifuging 50mL conical tubes at 3000×*g*, 4°C.
- Mass spectrometer capable of measuring the mass of synthesized peptide. This instrument must have sufficient resolving power and mass accuracy to confirm incorporation of the isotope label. We prefer to use matrix-assisted laser desorption ionization (MALDI) MS for this step, although methods using electrospray ionization (ESI) will also produce comparable results.
- UV-Visible spectrophotometer with cuvette capable of measuring absorbance at 280nm (if peptide contains tyrosine or tryptophan residues) or 214nm (if the peptide does not contain tyrosine or tryptophan residues). ("QS"-grade quartz cuvettes work well for wavelengths examined here. Plastic cuvettes are not usable because they absorb strongly in the wavelength range of interest.)

4.1.2 Before you begin

- Choose a location for incorporation of ^{13}C-labeled amino acid residue. If peptide sequence has glycine or alanine residues, we recommend replacing these residues using Fmoc-^{13}C-Gly-OH or Fmoc-^{13}C-Ala-OH as the building blocks, respectively, due to relatively low cost.

However, other isotope-labeled amino acids can be obtained and used for the same effect.

- Although peptide sequences are always written from N-terminus to C-terminus, chemical synthesis proceeds from C-terminus to N-terminus. Researchers new to peptide synthesis should take special care to ensure that they are synthesizing the peptide in the correct order.
- Choose a scale for your synthesis. We recommend 10–25 μmol scale for most peptides.
- Based on the peptide's sequence, choose a resin as follows:
  - For C-terminal acid, Wang resin preloaded with the C-terminal amino acid.
  - For C-terminal amide, Rink Amide resin.
- Prepare a solution of 20% piperidine in DMF for deprotection of the Fmoc protecting group. This solution is generally usable for up to 3 days.
- For all washing steps, fill reaction vessel with ≥3 mL of appropriate solvent and then filter using aspiration apparatus such as a vacuum manifold (see Fig. 4).

4.1.3 Protocol: Solid-phase peptide synthesis

1. Based on the chosen scale and the resin's loading capacity, weigh the appropriate amount of resin into reaction vessel with micro stir bar. (For example, for a 25 μmol scale synthesis and resin with 0.70 mmol/g loading capacity, weigh 35.7 mg of resin into the reaction vessel.)
2. Wash resin 3–5 times with DCM, then let resin swell in DCM for at least 30 min.
3. Wash resin 3–5 times with DMF.
4. If resin is protected with an Fmoc group, deprotect Fmoc group by adding 1–3 mL of 20% piperidine in DMF and stirring for 15–25 min at room temperature. (If resin is not Fmoc protected, move directly to coupling the first amino acid, as in Step 6.)

Note: Extended deprotection reaction times (≥40 min) should be avoided as they may lead to unwanted side reactions.

5. Wash resin 3–5 times with DMF.
6. For coupling amino acid monomer onto the chain, dissolve 4 equivalents (equiv.) Fmoc-protected amino acid, 4 equiv. PyBOP, and 8 equiv. DIEA in 1–3 mL DMF. Vortex to ensure completely dissolved, then add to resin in reaction vessel. Stir reaction for ≥40 min at room temperature.

Note: If using resin that is not preloaded with the C-terminal residue, the first coupling will be the C-terminal residue in the sequence. If using a resin preloaded with the C-terminal residue, the first coupling will be the penultimate residue in the sequence.

Note: Equivalents (equiv.) are based on scale of synthesis, as determined by resin loading. As an example, for a 25 μmol synthesis, 4 equiv. is 100 μmol.

Note: Duration of coupling reactions can vary significantly, and are flexible. We typically allow coupling reactions to proceed for 40–60 min, although longer reactions times (up to 12–18 h) are permissible without major loss in product quality. Note that depending on the specific stir bar used, overnight reactions may cause grinding of the resin into small particles, making resin filtration difficult. If performing overnight coupling reactions, we recommend replacing stirring with agitation/rocking to avoid this.

Note: ^{13}C-labeled amino acids and D-amino acids are incorporated in solid-phase peptide synthesis using the same conditions as standard L-amino acids.

7. Wash resin 3–5 times with DMF.

8. Deprotect Fmoc group by adding 1–3 mL of 20% piperidine in DMF. Stir for 15–25 min.

Note: Extended deprotection reaction times (≥40 min) should be avoided as they may lead to unwanted side reactions.

9. Wash resin 3–5 times with DMF.

10. Repeat Steps 6–9 for each amino acid in the sequence, moving from C-terminus to N-terminus.

11. After final Fmoc deprotection and wash with DMF, wash resin 3–5 times with DCM. Allow resin to air dry.

Note: Remember to remove the Fmoc group on the N-terminal residue with 20% piperidine in DMF solution at the end of the synthesis. This protecting group is not fully removed during TFA cleavage reaction.

Pause point: Assembly of the peptide can be paused after any step by washing the resin with DMF, followed by DCM and air drying. Peptide on dried resin is relatively stable, and can be stored at 4 °C for weeks–months without significant issue. When resuming synthesis, first swell resin in DCM, wash resin with DMF, then continue with the next step.

12. Cleave the peptide from the resin and remove sidechain protecting groups using 2–4 mL of cleavage cocktail:

(a) For peptides without cysteine residues, use: 95% TFA, 2.5% ultrapure water, 2.5% TIS.

(b) For peptides with cysteine residues, use: 94% TFA, 2.5% ultrapure water, 2.5% EDT, 1% TIS.

Note: TFA is extremely corrosive, and is a significant hazard. Be sure to only handle only if properly trained and while using the appropriate PPE in a chemical fume hood.

13. Allow cleavage reaction to proceed for 3–7 h with gentle agitation. Agitation can be provided via stirring, or plunger and cap can be placed on reaction vessel and allowed to mix on rocker.

14. When cleavage reaction is completed, use plunger of reaction vessel to transfer the solution to clean 50 mL conical centrifuge tube (leaving the resin in the reaction vessel). The cleaved peptide is now in this solution.

15. Using a gentle stream of air or nitrogen in a fume hood, carefully evaporate the cleavage solution until ~1 mL remains in the 50 mL conical tube.

16. Add ~40 mL cold diethyl ether (pre-cooled on dry ice) to precipitate peptide.

17. Centrifuge $\geq 3000 \times g$, 4 °C for 10 min. A visible white/off-white pellet should collect at the bottom of the tube.

Note: Some peptides, particularly those that are very short (<6 residues), may not effectively precipitate during this step. If no visible pellet is seen after centrifugation, we recommend evaporating the ether/TFA mixture via rotary evaporation and continuing directly to HPLC purification.

18. Decant supernatant. Dry the remaining pellet with a gentle stream of air or nitrogen.

Pause point: Dried, crude peptide can be stored at -20 °C for extended periods of time (e.g., months) prior to purification.

4.1.4 Protocol: HPLC purification

Synthetic peptides should be purified to remove undesired side products and other unwanted material via reverse-phase HPLC equipped with a UV detector and fraction collector. There are many acceptable ways to purify peptides via HPLC other than the approach outlined in this chapter. Specific mobile phase solvents, columns, gradients, injection volumes, and others can be altered to meet the individual researcher's needs or laboratory preferences. Here, we walk through our lab's standard protocol, which we find works well to obtain peptides of high purity (>95%) for analysis in spiking experiments. We generally use columns with C18 stationary phase and internal diameter from 9 to 21.2 mm. An example of HPLC purification for ^{13}C-D2-ATRP is shown in Fig. 5.

One important consideration for HPLC purification and drying is to avoid any possibilities of cross contamination of biological samples with the synthetic peptide. Thus, it is recommended to purify the synthetic peptide using instruments and columns that will not come in contact with biological extracts.

1. Dissolve crude peptide in 2–5 mL of 50% water/acetonitrile +0.1% TFA.

Note: It is critical that the product is fully dissolved in solution at this step. If product is found to be insoluble in the above solution, the researcher may wish to try other ratios of water:acetonitrile, or alternative HPLC-compatible solvents such as dimethyl sulfoxide (DMSO).

2. Filter solution through 0.22 μm syringe filter to remove insoluble particulates.
3. To determine the optimal gradient for peptide purification, inject a small portion of sample (e.g., 100 μL) onto HPLC instrument using the following method:
   - **(a)** Solvent A = water + 0.1% TFA; Solvent B = acetonitrile + 0.1% TFA
   - **(b)** Flow rate = 12–15 mL/min on C18 column (21.2 mm internal diameter)
   - **(c)** Starting composition = 5% B. Gradient = 0–50 min, 5–55% B; 50–50.1 min, 55–90% B; 50.1–55 min, 90% B; 55–55.1 min, 90–5% B; 55.1–60 min, 5% B.

Note: If peptide looks to be of relatively high purity at this stage (i.e., one major peak without closely eluting products), one may choose to purify using exclusively the above gradient (1% B/min), performing several larger

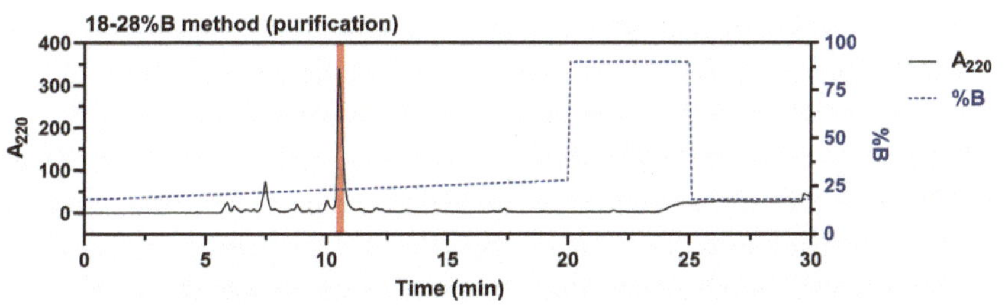

Fig. 5 HPLC purification of ^{13}C- D2-ATRP (Sequence = [G*][dF]RLNSASRVAH[G*]Y-NH$_2$, where [dF] = D-Phe, and [G*] = ^{13}C-Gly containing one ^{13}C atom). Purification gradient was 18–28% B over 20 min for purification of major peptide peak (21.2 mm ID column, 12 mL/min flow rate). Peptide was purified over multiple injections similar to the one shown. Black solid line indicates absorbance at 220 nm, while the blue dashed line indicates solvent gradient. Red bar indicates region collected via fraction collector for purification.

volume injections and collecting the major peak via fraction collector. However, if chromatogram shows multiple peaks or side products closely eluting to major peak, it is recommended to purify on a shallower gradient (0.5% B/min), as described below.

Note: It is often helpful to collect all major peaks via fraction collector at this stage and analyze by mass spectrometry to determine which peak corresponds to the desired product. However, if easy access to a mass spectrometer is not available, one can proceed by collecting all major peaks at this stage and determining which peak is the desired product during subsequent analysis.

4. Based on the results from injection on the 5–55% B gradient described above, determine the approximate % B in which the product peptide elutes. Purify peptide using a 0.5% B/min gradient centered around the % B where the peptide elutes.

Note: In the example in Fig. 5, ^{13}C-D2-ATRP eluted at ~23% B during the initial 5–55% B gradient injection, so it was purified using an 18–28% B gradient with the following method: Solvent A = water + 0.1% TFA, Solvent B = acetonitrile + 0.1% TFA. Flow rate = 12 mL/min on C18 column (21.2 mm internal diameter). Starting composition = 18% B. Gradient = 0–20 min, 18–28% B; 20–20.1 min, 28–90% B; 20.1–25 min, 90% B; 25–25.1 min, 90–18% B; 25.1–30 min, 18% B.

Note: It is recommended to purify over several injections of 300–1000 μL, depending on the yield and current volume of the crude peptide solution. One wants to be sure to inject a small enough volume to ensure adequate peak separation and avoid column overloading, but inject enough each run to ensure purification of crude peptide in a reasonable amount of time.

Note: When calculating the % B at which the peptide elutes, take into account the system's dwell/dead times, which delay elution from the column relative to when pumps run at a given composition.

5. Combine collected fractions corresponding to the peak of interest and remove solvent via lyophilization, vacuum concentrator, or alternative method.

Pause point: Most dried purified peptides can be stored at -20 °C for extended periods of time (years) without significant degradation. Peptides containing cysteine or methionine residues may oxidize over time, so it is recommended to confirm purity of these compounds immediately before use via HPLC and repurify if necessary.

4.1.5 Analysis of synthetic peptide

After synthesis and purification of a synthetic, isotope-labeled peptide, it is recommended to perform three analytical tests: (1) UV absorbance analysis to quantify peptide yield (2) mass spectrometry analysis to confirm peptide identify, and (3) analytical HPLC analysis to assess peptide purity. These steps can be performed in any order. Following these tests, the synthetic peptide may be aliquoted and dried for long-term storage, or diluted and used immediately for spiking experiments.

(1) Dissolve purified peptide in 1–3 mL of HPLC-grade water and determine concentration by UV absorbance on a UV-Visible spectrophotometer.

(a) If the sequence contains a tyrosine or tryptophan residue, it is recommended to calculate concentration using absorbance at 280 nm (Conibear, Daly, & Craik, 2012; Gill & von Hippel, 1989). At 280 nm, the extinction coefficient of the purified peptide can be calculated as:

$$\varepsilon_{280\text{nm}} = (\text{no. of Trp} \times 5500) + (\text{no. of Tyr} \times 1490) + (\text{no. of disulfide bridges} \times 125)$$

(b) If the peptide sequence contains no tyrosine or tryptophan residues, you may also calculate the concentration based on the absorbance at 214 nm (Buck, Olah, Weitzmann, & Cooperman, 1989; Conibear et al., 2012). At 214 nm, the extinction coefficient of the purified peptide can then be calculated as:

$$\varepsilon_{214\text{nm}} = ([\text{no. of Amino acids} - 1 + \text{no. of Asn} + \text{no. of Gln}] \times 2846) + (\text{no. of Phe} \times 7200) + (\text{no. of His} \times 6309) + (\text{no. of Trp} \times 22735) + (\text{no. of Tyr} \times 5755)$$

Note: The first term in the above equation ([*no. of Amino acids* – *1* + *no. of Asn* + *no. of Gln*]) is a calculation of the number of amide bonds in the peptide, and is applicable for peptides bearing C-terminal carboxylic acids. For peptides bearing C-terminal amides, the first term should be modified to [*no. of Amino acids* + *no. of Asn* + *no. of Gln*].

Note: We recommend that researchers calculate concentration based on A_{214} only when the peptide contains no Trp or Tyr residues, as we have found peptide concentrations based on A_{280}

to be more accurate than those based on A_{214}. If peptide sequence does not have any Trp or Tyr residues, measuring peptide mass using accurate analytical balance is also acceptable.

Note: If purified peptide is found to be insoluble in pure water, it may be helpful to try dissolving in water +0.1% TFA (or formic acid), or in a solution of 25–50% acetonitrile/water.

Note: It is recommended to prepare several dilutions of peptide (e.g., 1:5–1:100) and consider only those that give absorbance values between 0.1 and 1 at the wavelength being monitored. Absorbance values significantly outside of this range may not lead to accurate estimations of peptide concentration.

(c) Calculate the concentration of peptide stock using a rearranged form of the Beer-Lambert law: $c = A/(\varepsilon \times l)$, where c is concentration (in M), A is measured absorbance at the predetermined wavelength, ε is calculated extinction coefficient at the predetermined wavelength (in $M^{-1}\,cm^{-1}$), and l is path length (in cm).

(d) This peptide solution with determined concentration can then be used directly for follow-up experiments, or peptide can be aliquoted into known amounts and dried via lyophilization or vacuum concentrator.

(2) Analyze purified peptide by analytical-scale HPLC to evaluate relative purity (Fig. 6A). Use integrated peak areas to determine relative purity. If below 90% purity, we recommend repurifying the peptide via HPLC. We generally evaluate purity using the method outlined below:

(a) Solvent A = water + 0.1% TFA, Solvent B = acetonitrile + 0.1% TFA

(b) Flow rate = 1 mL/min on C18 column (4.6 mm internal diameter)

(c) Starting composition = 5% B. Gradient = 0–50 min, 5–55% B; 50–50.1 min, 55–90% B; 50.1–55 min, 90% B; 55–55.1 min, 90–5% B; 55.1–60 min, 5% B.

(3) Analyze purified peptide by mass spectrometry to determine exact mass of peptide. Be sure to pay careful attention to isotopic distribution to confirm the desired incorporation of ^{13}C. Analyze synthetic peptide by LC-MS and LC-MS/MS using collision-induced dissociation (CID) in data-dependent acquisition (DDA) mode to determine MS2 fragmentation pattern on LC-MS system. See Figs. 6 and 7 for examples.

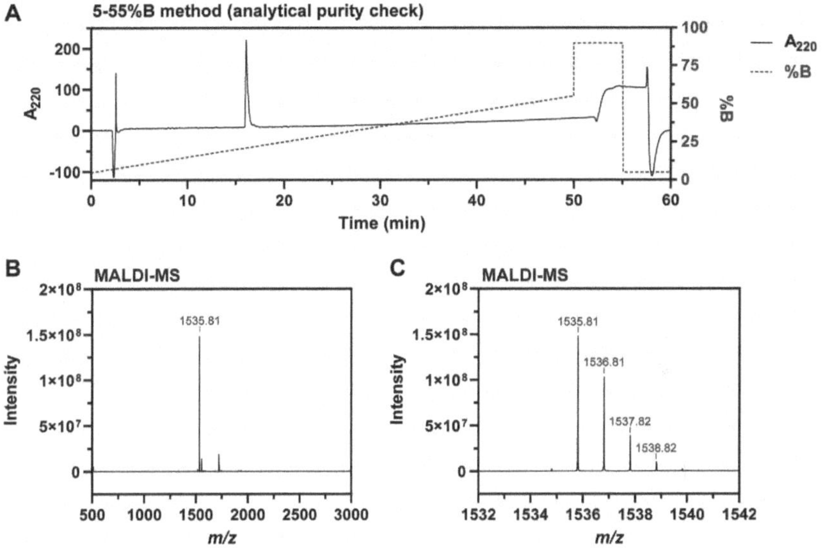

Fig. 6 Analytical characterization of synthesized and purified ^{13}C- D2-ATRP (Sequence = [G*][dF]RLNSASRVAH[G*]Y-NH_2, where [dF] = D-Phe, and [G*] = ^{13}C-Gly containing one ^{13}C atom). (A) Analytical-scale purity check (4.6 mm ID column, 1 mL/min flow rate). Black solid line indicates absorbance at 220 nm, while the blue dashed line indicates solvent gradient. (B) MALDI-MS analysis of purified peptide. Only the monoisotopic peak of the product is labeled. (C) Zoom of Panel B to examine isotopic distribution of synthetic peptide. Calculated monoisotopic $[M+H]^+ = 1535.81$ ($z = +1$) for peptide with two ^{13}C-labeled glycine residues. Calculated monoisotopic $[M+H]^+ = 1533.80$ ($z = +1$) for peptide with no ^{13}C label.

Note: Our workflow includes a mass check by MALDI-MS as part of the synthetic process to verify correct product produced (Fig. 6), followed by more detailed MS and MS/MS analysis on LC-MS instrument (Figs. 7 and 8). Our choice for a first check by MALDI-MS is due to the relative speed and ease of MALDI-MS analysis relative to LC-ESI-MS, which requires longer sample preparation and run times. However, analysis by MALDI-MS is not necessary if thorough LC-MS and LC-MS/MS analysis is performed on synthetic peptide.

4.2 LC-MS-based spiking experiments

With the isotope-labeled peptides in hand, LC-MS-based spiking experiments can provide strong evidence to support stereochemical assignments of endogenous peptides extracted from biological tissues. Individually

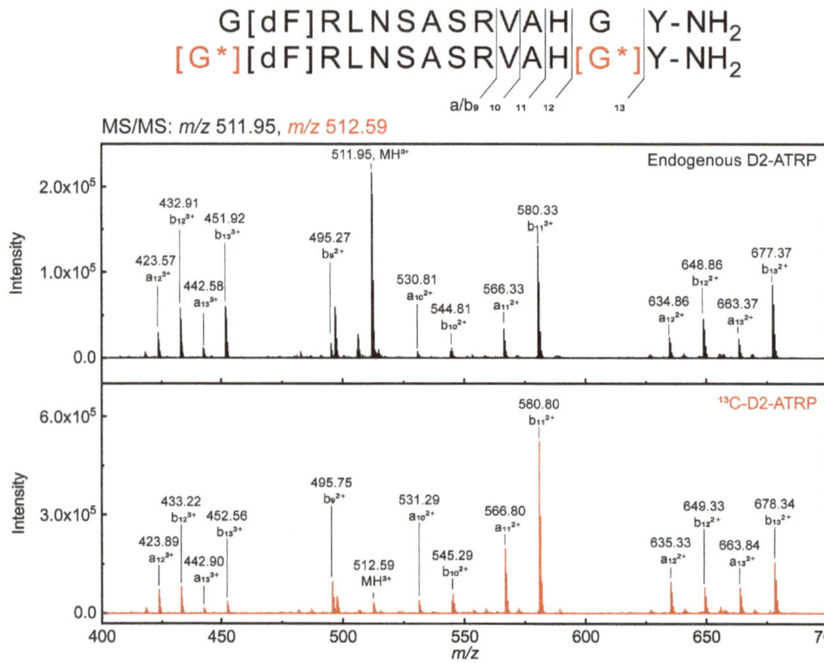

Fig. 7 Sequences and MS/MS (CID) spectra for endogenous D2-ATRP (top spectrum, black) and synthetic ^{13}C-D2-ATRP (bottom spectrum, red). Red [G*] indicates ^{13}C-labeled Gly, and [dF] indicates D-Phe residue. The spectrum for endogenous D2-ATRP is the same as that shown in Fig. 2D, replotted here for comparison. *Figure adapted from Checco, J. W., Zhang, G., Yuan, W. D., Le, Z. W., Jing, J., & Sweedler, J. V. (2018).* Aplysia *allatotropin-related peptide and its newly identified D-amino acid-containing epimer both activate a receptor and a neuronal target.* Journal of Biological Chemistry, 293, *16862-16873.*

spiking each suspected isomer into tissue extracts allows for rigorous comparisons of LC retention times, which generally differ for peptide diastereomers. Isotope labeling of synthetic standards ensure that these peptides will not be misidentified as endogenous peptides, avoiding the possibility of unintentional cross contamination leading to false positive results.

4.2.1 Materials and reagents

- Peptide extracts from tissue of interest. There are a number of different methods to extract peptides from animal tissues (or cells). The researcher should explore several extraction methods to find what works best for peptide(s) of interest. Two possible methods are highlighted below:
  - Three-stage extraction consisting of consecutive homogenizations in boiling water, acidified acetone (40:6:1 acetone:water:

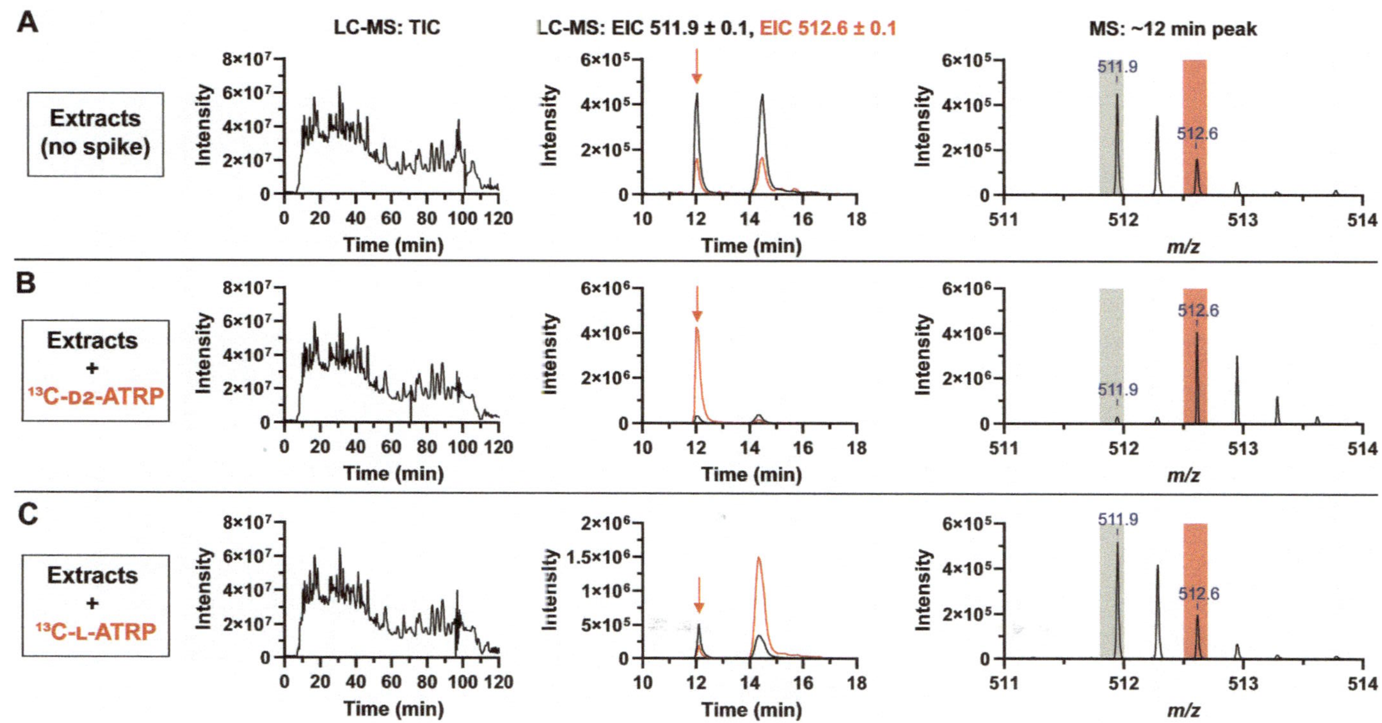

Fig. 8 See figure legend on next page.

concentrated HCl, v/v/v), and acidified water (0.25% acetic acid in water) (Lee et al., 2010; Yang, Anapindi, Romanova, Rubakhin, & Sweedler, 2017). At each stage, homogenized tissue is centrifuged and supernatant removed and dried by vacuum concentrator. Extracts from each stage can be combined or analyzed separately.
  - Homogenization in acidified acetone (40:6:1 acetone:water:concentrated HCl, v/v/v) (Bai et al., 2013; Checco et al., 2018; Livnat et al., 2016), followed by high-speed centrifugation, removal of supernatant, and drying of supernatant by vacuum concentrator.
- Isotope-labeled peptides (from Section 4.1)
- Micro BCA kit (ThermoFisher, 23235), along with UV-Visible spectrophotometer or plate reader capable of measuring absorbance at 562 nm.
- Method for performing C18 solid-phase extraction (SPE) on peptide extracts. We recommend one of the following:
  - Pierce 8 mg C18 spin columns (ThermoFisher, 89870) for <30 μg total peptide extracts.
  - HyperSep C18 Cartridges for ≥30 μg total peptide extracts: 50 mg (ThermoFisher, 60108-390), 100 mg (ThermoFisher, 60108-302), or 500 mg (ThermoFisher, 60108-304).
- HPLC-grade H_2O.
- HPLC-grade acetonitrile.
- HPLC-grade methanol.
- HPLC-grade formic acid.
- Micropipettes and tips capable of accurately dispensing 1–1000 μL.

Fig. 8 Example spiking experiment to determine endogenous stereochemistry of allatotropin-related peptide (ATRP) diastereomers in *Aplysia* cerebral ganglia peptide extracts. (A) "No spike" control showing endogenous L- and D2-ATRP diastereomers separable by LC-MS. Data is the same as that displayed in Fig. 2, replotted here for comparison. (B) Endogenous extracts are spiked with ^{13}C-labeled D2-ATRP. (C) Endogenous extracts are spiked with ^{13}C-labeled L-ATRP. The left plot in each panel shows the Total Ion Current (TIC) chromatogram for each sample. The center plot in each panel shows two overlaid extracted ion chromatograms (EICs) for endogenous ATRP (black, $m/z = 511.9 \pm 0.1$, $z = +3$) and ^{13}C-labeled synthetic ATRP (red, $m/z = 512.6 \pm 0.1$, $z = +3$). Endogenous ATRP is expected to have low signal in the $m/z = 512.6$ channel due to endogenous ^{13}C isotope abundance. The right plot in each panel shows a representative MS1 spectrum for the peak at ~12 min (indicated by arrow in the center panel). Bars indicate windows used for generation of EICs in center panel for endogenous (grey) and ^{13}C-labeled synthetic (red) peptides. *Figure adapted from Checco, J. W., Zhang, G., Yuan, W. D., Le, Z. W., Jing, J., & Sweedler, J. V. (2018).* Aplysia *allatotropin-related peptide and its newly identified D-amino acid-containing epimer both activate a receptor and a neuronal target.* Journal of Biological Chemistry, 293, *16862–16873.*

- Microcentrifuge tubes (~1.5 mL).
- Temperature-controlled centrifuge capable of centrifuging microcentrifuge tubes at >10,000 × *g*, 4 °C.
- Nano-LC instrument equipped with reversed-phase column/solvents and interfaced with a high-resolution mass spectrometer (e.g., Q-TOF, Orbitrap, etc.). For the example shown in this chapter, a Thermo Dionex Ultimate 3000 RSLC with a nanoflow selector coupled to a Bruker Impact HD UHR-QqTOF mass spectrometer with a CaptiveSpray Nanosource was used. However, many alternative instruments and configurations will also work for this purpose. The LC instrument must be able to reliably separate peptide diastereomers of interest, and the MS instrument must have sufficient resolving power and mass accuracy to definitively determine if a peptide signal is arising from endogenous peptide or isotope-labeled synthetic peptide.

4.2.2 Protocol

1. Dissolve dried peptide extracts in 20–500 µL of 5% acetonitrile/H_2O + 0.1% formic acid.

   Note: A solution of 5% methanol/H_2O + 0.1% formic acid may also be used if using a methanol/water solvent system for SPE below.
2. Determine total peptide concentration using MicroBCA assay kit, according to manufacturer's instructions. If total peptide content <30 µg, proceed to SPE with 8 mg C18 spin columns. If total peptide content ≥30 µg, proceed to SPE with HyperSep C18 cartridges (≥50 mg).
3. Desalt peptide extracts by SPE, using column/cartridge choice identified in previous step, and following manufacturer's instructions. Either H_2O/acetonitrile or H_2O/methanol solvent systems generally work well for desalting of endogenous peptides. These two different solvent systems may retain somewhat different subsets of peptides, and we recommend researchers explore both solvent systems when working with particular peptide extracts for the first time to determine what is most appropriate for their system.

   Note: Solvent composition for washing and elution steps during SPE may be modified to fit needs of specific sample. For the ATRP example highlighted in this chapter, equilibrations, sample loading, and washing steps were performed in 100% H_2O + 0.1% formic acid (no organic solvent) to minimize analyte loss due to the high polarity of the target peptides.

4. Dry eluted, desalted peptide extract (e.g., using vacuum concentrator).

Pause point: Dried, desalted peptide extracts can be stored at -20 °C for extended periods of time (usually at least several months).

5. Resuspend dried peptide extracts in 10–50 μL of LC-MS solvents matching the starting composition of your LC-MS run (e.g., 3% acetonitrile/water + 0.1% formic acid).
6. Determine concentration of resuspended peptide extracts using MicroBCA assay kit, according to manufacturer's instructions. Determining the total concentration of resuspended peptide extracts is important to control the amount of total peptide injected onto the nano-LC column.
7. Dilute peptide extracts to the appropriate concentration for sample injection and split into several microcentrifuge tubes for spiking experiment. One tube of sample is needed for each peptide spike condition, along with one tube for "No Spike" control sample.
8. For each peptide spike condition, spike in purified ^{13}C-labeled peptide at final concentration which gives signal of roughly similar intensity to endogenous peptide being examined. This concentration should be predetermined for each system using pilot experiments to evaluate the endogenous peptide signal for an injection of a given sample and separate injections of varied concentrations of ^{13}C-labeled synthetic peptides. In the examples shown in Fig. 8, ^{13}C-labeled peptides are spiked into peptide extracts at a final concentration of ~75 nM, and 5 μL of this spiked solution is injected onto the LC-MS instrument.

   Note: There are many possible LC gradients that will suffice for this analysis. The most important aspect of choosing an LC gradient for this experiment is the ability to reliably separate the diastereomeric peptides of interest. The experiment shown in Fig. 8 uses the following gradient:

   (a) Solvent A = 95% water, 5% acetonitrile, 0.1% formic acid; Solvent B = 95% acetonitrile, 5% water, 0.1% formic acid; loading solvent (for pre-column trap) = 98% water, 2% acetonitrile, 0.1% formic acid, 0.01% TFA.

   (b) Flow rate = 300 nL/min.

   (c) Starting composition = 4% B. Gradient = 0–3 min, 4% B; 3–90 min, 4–50% B; 90–93 min, 50–90% B; 93–100 min, 90% B; 100–110 min, 90–4% B; 110–120 min, 4% B.

   Note: When injecting endogenous extracts and synthetic peptides separately, researchers should carefully examine the MS2 spectra for each peptide. Although the masses of the ^{13}C-labeled synthetic peptides will differ from the endogenous peptide, the general fragmentation pattern should be nearly identical for the endogenous and synthetic peptides (see Fig. 7). If MS/MS fragments differ significantly between endogenous peptide and the ^{13}C-labeled peptide, this likely indicates the endogenous peptide sequence was incorrectly determined.

9. Prepare and inject a "No Spike" control sample by diluting peptide extracts with the same volume of solvent (no peptide) used for each spike condition.

10. Data analysis is performed by creating extracted ion chromatograms (EICs) for both the endogenous peptide and the ^{13}C-labeled synthetic peptide. Following this, the MS1 and MS2 spectra for each putative diastereomer should be closely examined to determine if putative stereochemical assignments match those of the endogenous peptides. It is important to look for the following:

(a) The "No Spike" control sample should show putative peptides with predicted endogenous isotope ratio (Fig. 8A).

(b) In each spike condition, EIC of endogenous peptide and spiked ^{13}C-labeled synthetic peptide should co-elute (Fig. 8B and C). Failure of synthetic ^{13}C-labeled peptide to co-elute with endogenous peptide indicates an incorrect assignment of endogenous peptide sequence or stereochemistry.

(c) Inspection of MS1 spectra for spiked peptide peak should show a mixture of isotope patterns for endogenous peptide and ^{13}C-labeled synthetic peptide. The non-spiked diastereomer should show only the endogenous isotope pattern (Fig. 8B and C).

Note: For peptides with masses higher than ~1000 Da, the most intense m/z peak will often be an ion with $z > +1$ using electrospray ionization (ESI). We suggest generating EICs for all m/z values associated with the peptides of interest. Further analysis can then be performed on the most intense m/z ion for the peptide of interest.

4.2.3 Pre-spiking control experiment

Performing the spiking experiment described above helps to determine the stereochemistry of endogenous peptides present in an extract after sample processing. However, it is possible that isomerization may result as an artifact of sample preparation. In order to evaluate this possibility, it is recommended to also perform a "pre-spiking" experiment (Fig. 9). In this

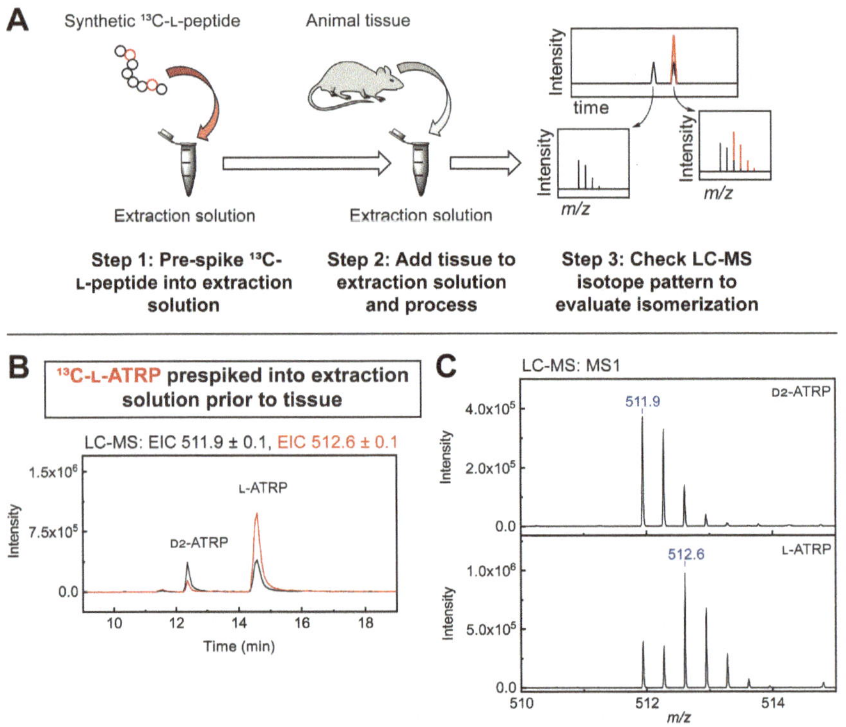

Fig. 9 Pre-spiking control experiment to evaluate possible isomerization as a result of peptide extraction and sample preparation protocols. (A) Overview of experiment. ^{13}C-labeled all-L-residue diastereomer is spiked into extraction solution before tissue is added. Tissue is then processed as usual and peptide extracts analyzed by LC-MS. Isotope label should be found in all-L-residue peptide, but absent from D-amino acid-containing diastereomer. Evidence of the isotope label in D-amino acid containing peptide indicates spontaneous isomerization during extraction and sample preparation. (B) Representative overlaid EICs for endogenous ATRP (black, $m/z = 511.9 \pm 0.1$, $z = +3$) and ^{13}C-labeled synthetic ATRP (red, $m/z = 512.6 \pm 0.1$, $z = +3$) following pre-spike with ^{13}C-L-ATRP. (C) Representative MS1 spectra for each peak shown in panel B. The isotope label is present in L-ATRP peak, but no evidence of ^{13}C-labeled synthetic peptide is found in the D2-ATRP peak. *Figure adapted from Checco, J. W., Zhang, G., Yuan, W. D., Le, Z. W., Jing, J., & Sweedler, J. V. (2018).* Aplysia *allatotropin-related peptide and its newly identified D-amino acid-containing epimer both activate a receptor and a neuronal target.* Journal of Biological Chemistry, 293, *16862–16873.*

control experiment, ^{13}C-labeled synthetic all-L-residue peptide is spiked into extraction solution *before* endogenous peptide extraction from tissue. The tissue is then processed and analyzed by LC-MS. If spontaneous isomerization occurs as part of the extraction process, the presence of isomerized ^{13}C-labeled peptide should be observable by inspecting the isotopic distributions of each peptide peak in LC-MS. (Note that the amount of synthetic peptide that is spiked for this experiment will need to be optimized based on instrument sensitivity and endogenous peptide abundance. The experiment shown in Fig. 9, 10 pmol ^{13}C- L-ATRP was spiked into 500 μL acidified acetone prior to tissue extraction.)

5. Conclusions

In order to validate the stereochemical assignment of peptide diastereomers using this protocol, three conditions must be met. First, putative diastereomeric peptides must have identical *m/z* values in MS1. Second, putative diastereomeric peptides should have nearly identical MS2 fragmentation patterns. Although it is known that changes in stereochemistry can alter the fragmentation of peptide diastereomers (especially fragment intensity) (Adams & Zubarev, 2005; Bai et al., 2013; Koehbach et al., 2016; Tao & Julian, 2014; Tao et al., 2012), the majority of fragments are usually identical between diastereomers. Observed MS2 fragmentation pattern for synthetic peptide should match that of the endogenous peptide being assigned. Significant difference in MS2 fragment patterns indicates two peptides have different sequences, even if they share the same parent mass. Third, possible peptide diastereomers must be separable during the LC-stage of the LC-MS run, and spiked synthetic standards must co-elute with peptides of interest. As demonstrated in Fig. 8 and in many other cases (Bai et al., 2013; Checco et al., 2018; Fujisawa et al., 1992; Heck et al., 1994; Livnat et al., 2016; Mast et al., 2019, 2021; Morishita et al., 1997) peptide diastereomers often elute at different retention times during reverse-phase chromatography. However, it is important to note that this may not be the case for all possible peptide isomers, and optimization of LC conditions may be required to reliably separate diastereomers. Furthermore, for most peptides it is generally not feasible to synthesize and spike every possible diastereomer. Thus, in most cases additional experiments such as homology studies, protease digestions, or others will be required prior to spiking experiments to narrow down possible positions of isomerization, as discussed in Section 2.

Acknowledgments

J.W.C. acknowledges support from the Nebraska Center for Integrated Biomolecular Communication (NIH National Institute of General Medical Sciences P20 GM113126) and from a Nebraska EPSCoR FIRST Award (OIA-1557417). Figs. 2, 7, 8, and 9 were adapted from data originally reported in (Checco et al., 2018). HPLC data shown in Figs. 5 and 6 result from a new synthesis completed at University of Nebraska-Lincoln for the purpose of this chapter. MALDI-MS data shown in Fig. 6 was generated using a 15 Tesla Bruker solariX XR Fourier transform ion cyclotron resonance mass spectrometer (Nebraska Center for Mass Spectrometry).

References

Adams, C. M., & Zubarev, R. A. (2005). Distinguishing and quantifying peptides and proteins containing D-amino acids by tandem mass spectrometry. *Analytical Chemistry*, *77*, 4571–4580.

Aguilar, M. B., Soyez, D., Falchetto, R., Arnott, D., Shabanowitz, J., Hunt, D. F., et al. (1995). Amino acid sequence of the minor isomorph of the crustacean hyperglycemic hormone (CHH-II) of the Mexican crayfish Procambarus bouvieri (Ortmann): Presence of a D-amino acid. *Peptides*, *16*, 1375–1383.

Bai, L., Livnat, I., Romanova, E. V., Alexeeva, V., Yau, P. M., Vilim, F. S., et al. (2013). Characterization of GdFFD, a D-amino acid-containing neuropeptide that functions as an extrinsic modulator of the *Aplysia* feeding circuit. *Journal of Biological Chemistry*, *288*, 32837–32851.

Bhushan, R., & Brückner, H. (2004). Marfey's reagent for chiral amino acid analysis: A review. *Amino Acids*, *27*, 231–247.

Buck, M. A., Olah, T. A., Weitzmann, C. J., & Cooperman, B. S. (1989). Protein estimation by the product of integrated peak area and flow rate. *Analytical Biochemistry*, *182*, 295–299.

Buczek, O., Yoshikami, D., Bulaj, G., Jimenez, E. C., & Olivera, B. M. (2005). Post-translational amino acid isomerization: A functionally important D-amino acid in an excitatory peptide. *Journal of Biological Chemistry*, *280*, 4247–4253.

Buczek, O., Yoshikami, D., Watkins, M., Bulaj, G., Jimenez, E. C., & Olivera, B. M. (2005). Characterization of D-amino-acid-containing excitatory conotoxins and redefinition of the I-conotoxin superfamily. *FEBS Journal*, *272*, 4178–4188.

Checco, J. W., Zhang, G., Yuan, W. D., Le, Z. W., Jing, J., & Sweedler, J. V. (2018). *Aplysia* allatotropin-related peptide and its newly identified D-amino acid-containing epimer both activate a receptor and a neuronal target. *Journal of Biological Chemistry*, *293*, 16862–16873.

Conibear, A. C., Daly, N. L., & Craik, D. J. (2012). Quantification of small cyclic disulfide-rich peptides. *Biopolymers*, *98*, 518–524.

Cotter, P. D., O'Connor, P. M., Draper, L. A., Lawton, E. M., Deegan, L. H., Hill, C., et al. (2005). Posttranslational conversion of L-serines to D-alanines is vital for optimal production and activity of the lantibiotic lacticin 3147. *Proceedings of the National Academy of Sciences of the United States of America*, *102*, 18584–18589.

Erspamer, V., Melchiorri, P., Falconieri-Erspamer, G., Negri, L., Corsi, R., Severini, C., et al. (1989). Deltorphins: a family of naturally occurring peptides with high affinity and selectivity for δ opioid binding sites. *Proceedings of the National Academy of Sciences of the United States of America*, *86*, 5188–5192.

Ewing, M. A., Wang, J., Sheeley, S. A., & Sweedler, J. V. (2008). Detecting D-amino acid-containing neuropeptides using selective enzymatic digestion. *Analytical Chemistry*, *80*, 2874–2880.

Fujii, K., Ikai, Y., Mayumi, T., Oka, H., Suzuki, M., & Harada, K.-I. (1997). A nonempirical method using LC/MS for determination of the absolute configuration of constituent amino acids in a peptide: Elucidation of limitations of Marfey's method and of its separation mechanism. *Analytical Chemistry*, *69*, 3346–3352.

Fujii, K., Ikai, Y., Oka, H., Suzuki, M., & Harada, K.-I. (1997). A nonempirical method using LC/MS for determination of the absolute configuration of constituent amino acids in a peptide: Combination of Marfey's method with mass spectrometry and its practical application. *Analytical Chemistry*, *69*, 5146–5151.

Fujisawa, Y., Ikeda, T., Nomoto, K., Yasuda-Kamatani, Y., Minakata, H., Kenny, P. T., et al. (1992). The FMRFamide-related decapeptide of *Mytilus* contains a D-amino acid residue. *Comparative Biochemistry and Physiology C: Pharmacology Toxicology and Endocrinology*, *102*, 91–95.

Gill, S. C., & von Hippel, P. H. (1989). Calculation of protein extinction coefficients from amino acid sequence data. *Analytical Biochemistry*, *182*, 319–326.

Heck, S. D., Faraci, W. S., Kelbaugh, P. R., Saccomano, N. A., Thadeio, P. F., & Volkmann, R. A. (1996). Posttranslational amino acid epimerization: Enzyme-catalyzed isomerization of amino acid residues in peptide chains. *Proceedings of the National Academy of Sciences of the United States of America*, *93*, 4036–4039.

Heck, S. D., Siok, C. J., Krapcho, K. J., Kelbaugh, P. R., Thadeio, P. F., Welch, M. J., et al. (1994). Functional consequences of posttranslational isomerization of Ser46 in a calcium channel toxin. *Science*, *266*, 1065–1068.

Jansson, E. T. (2018). Strategies for analysis of isomeric peptides. *Journal of Separation Science*, *41*, 385–397.

Jia, C., Lietz, C. B., Yu, Q., & Li, L. (2014). Site-specific characterization of D-amino acid containing peptide epimers by ion mobility spectrometry. *Analytical Chemistry*, *86*, 2972–2981.

Kessler, N., Schuhmann, H., Morneweg, S., Linne, U., & Marahiel, M. A. (2004). The linear pentadecapeptide gramicidin is assembled by four multimodular nonribosomal peptide synthetases that comprise 16 modules with 56 catalytic domains. *Journal of Biological Chemistry*, *279*, 7413–7419.

Knerr, P. J., & van der Donk, W. A. (2013). Chemical synthesis of the lantibiotic lacticin 481 reveals the importance of lanthionine stereochemistry. *Journal of the American Chemical Society*, *135*, 7094–7097.

Koehbach, J., Gruber, C. W., Becker, C., Kreil, D. P., & Jilek, A. (2016). MALDI TOF/TOF-based approach for the identification of D-amino acids in biologically active peptides and proteins. *Journal of Proteome Research*, *15*, 1487–1496.

Kreil, G., Barra, D., Simmaco, M., Erspamer, V., Erspamer, G. F., Negri, L., et al. (1989). Deltorphin, a novel amphibian skin peptide with high selectivity and affinity for δ opioid receptors. *European Journal of Pharmacology*, *162*, 123–128.

Lambeth, T. R., Riggs, D. L., Talbert, L. E., Tang, J., Coburn, E., Kang, A. S., et al. (2019). Spontaneous isomerization of long-lived proteins provides a molecular mechanism for the lysosomal failure observed in Alzheimer's disease. *ACS Central Science*, *5*, 1387–1395.

Lee, J. E., Atkins, N., Jr., Hatcher, N. G., Zamdborg, L., Gillette, M. U., Sweedler, J. V., et al. (2010). Endogenous peptide discovery of the rat circadian clock: a focused study of the suprachiasmatic nucleus by ultrahigh performance tandem mass spectrometry. *Molecular & Cellular Proteomics*, *9*, 285–297.

Ling, L. L., Schneider, T., Peoples, A. J., Spoering, A. L., Engels, I., Conlon, B. P., et al. (2015). A new antibiotic kills pathogens without detectable resistance. *Nature*, *517*, 455–459.

Livnat, I., Tai, H. C., Jansson, E. T., Bai, L., Romanova, E. V., Chen, T. T., et al. (2016). A D-amino acid-containing neuropeptide discovery funnel. *Analytical Chemistry*, *88*, 11868–11876.

Lovering, A. L., Safadi, S. S., & Strynadka, N. C. (2012). Structural perspective of peptidoglycan biosynthesis and assembly. *Annual Review of Biochemistry*, *81*, 451–478.

Maeda, H., Takata, T., Fujii, N., Sakaue, H., Nirasawa, S., Takahashi, S., et al. (2015). Rapid survey of four Asp isomers in disease-related proteins by LC-MS combined with commercial enzymes. *Analytical Chemistry*, *87*, 561–568.

Mast, D. H., Checco, J. W., & Sweedler, J. V. (2019). Differential post-translational amino acid isomerization found among neuropeptides in *Aplysia californica*. *ACS Chemical Biology*, *15*, 272–281.
Mast, D. H., Checco, J. W., & Sweedler, J. V. (2021). Advancing D-amino acid-containing peptide discovery in the metazoan. *Biochim Biophys Acta Proteins Proteom*, *1869*, 140553.
Morishita, F., Nakanishi, Y., Kaku, S., Furukawa, Y., Ohta, S., Hirata, T., et al. (1997). A novel D-amino-acid-containing peptide isolated from *Aplysia* heart. *Biochemical and Biophysical Research Communications*, *240*, 354–358.
Ohta, N., Kubota, I., Takao, T., Shimonishi, Y., Yasuda-Kamatani, Y., Minakata, H., et al. (1991). Fulicin, a novel neuropeptide containing a D-amino acid residue isolated from the ganglia of *Achatina fulica*. *Biochemical and Biophysical Research Communications*, *178*, 486–493.
Ollivaux, C., Gallois, D., Amiche, M., Boscameric, M., & Soyez, D. (2009). Molecular and cellular specificity of post-translational aminoacyl isomerization in the crustacean hyperglycaemic hormone family. *FEBS Journal*, *276*, 4790–4802.
Ollivaux, C., Soyez, D., & Toullec, J. Y. (2014). Biogenesis of D-amino acid containing peptides/proteins: Where, when and how? *Journal of Peptide Science*, *20*, 595–612.
Soyez, D., Van Herp, F., Rossier, J., Le Caer, J. P., Tensen, C. P., & Lafont, R. (1994). Evidence for a conformational polymorphism of invertebrate neurohormones. D-amino acid residue in crustacean hyperglycemic peptides. *Journal of Biological Chemistry*, *269*, 18295–18298.
Süssmuth, R. D., & Mainz, A. (2017). Nonribosomal peptide synthesis—Principles and prospects. *Angewandte Chemie International Edition*, *56*, 3770–3821.
Tao, Y., & Julian, R. R. (2014). Identification of amino acid epimerization and isomerization in crystallin proteins by tandem LC-MS. *Analytical Chemistry*, *86*, 9733–9741.
Tao, Y., Quebbemann, N. R., & Julian, R. R. (2012). Discriminating D-amino acid-containing peptide epimers by radical-directed dissociation mass spectrometry. *Analytical Chemistry*, *84*, 6814–6820.
Torres, A. M., Menz, I., Alewood, P. F., Bansal, P., Lahnstein, J., Gallagher, C. H., et al. (2002). D-Amino acid residue in the C-type natriuretic peptide from the venom of the mammal, *Ornithorhynchus anatinus*, the Australian platypus. *FEBS Letters*, *524*, 172–176.
Yang, N., Anapindi, K. D. B., Romanova, E. V., Rubakhin, S. S., & Sweedler, J. V. (2017). Improved identification and quantitation of mature endogenous peptides in the rodent hypothalamus using a rapid conductive sample heating system. *Analyst*, *142*, 4476–4485.
Yasuda-Kamatani, Y., Kobayashi, M., Yasuda, A., Fujita, T., Minakata, H., Nomoto, K., et al. (1997). A novel D-amino acid-containing peptide, fulyal, coexists with fulicin gene-related peptides in *Achatina* atria. *Peptides*, *18*, 347–354.

CHAPTER TEN

Multiplexed quantitative neuropeptidomics via DiLeu isobaric tagging

Christopher S. Sauer[a] and Lingjun Li[a,b,*]
[a]Department of Chemistry, University of Wisconsin, Madison, WI, United States
[b]School of Pharmacy, University of Wisconsin, Madison, WI, United States
*Corresponding author: e-mail address: lingjun.li@wisc.edu

Contents

Abstract

Neuropeptides are key signaling molecules in many pathways and can serve as potential biomarkers or therapeutics. Mass spectrometry has emerged as a powerful tool for studying neuropeptides with high sensitivity and accuracy. Isobaric tagging can further enhance this method by improving throughput and reducing sampling needs. In this chapter, we discuss the benefits and limitations of using isobaric tags to analyze neuropeptides. Methods for optimizing the data acquisition are also presented to enable a greater number of neuropeptides to be identified and quantified when using isobaric tags, specifically *N,N*-dimethyl leucine (DiLeu).

Methods in Enzymology, Volume 663
ISSN 0076-6879
https://doi.org/10.1016/bs.mie.2021.10.011

1. Introduction

Neuropeptides are endogenous signaling peptides synthesized in the neuroendocrine system (Christie, Stemmler, & Dickinson, 2010). These important signal modulators are formed from the cleavage of precursor proteins (pre-prohormones) containing a signal peptide and the active neuropeptide(s). After cleavage, the neuropeptide is post-translationally modified to yield the mature, bioactive neuropeptide (Hook, Lietz, Podvin, Cajka, & Fiehn, 2018). Mature neuropeptides can act in many ways, including local (paracrine) signaling, hormonal signaling, and in conjunction with other signaling molecules (e.g., neurotransmitters). They often target G protein coupled receptors (GPCRs) where they can have a wide range of functions (Romanova & Sweedler, 2015). These functions are also dependent on many factors, including concentration, location in the organism, and the presence of other neuropeptides that could have co-modulating effects (Li & Sweedler, 2008).

As they can be potent, selective signaling molecules, neuropeptides play major roles in many biochemical and physiological processes, such as behavior (Han et al., 2015; Hu et al., 2021), feeding (DeLaney et al., 2021), stress response (Buchberger, Sauer, Vu, DeLaney, & Li, 2020; Liu, Buchberger, Delaney, Li, & Li, 2019; Sauer & Li, 2021), and diseases states (Hulme et al., 2020a, 2020b; Li et al., 2019). Neuropeptide function is not independent of its environment, so profiling the entire suite of neuropeptides is critical to our understanding of their functions and related signaling pathways. Developing methods to efficiently characterize and quantify all neuropeptides in a sample is therefore essential to improving our understanding of many diseases and could even provide avenues for biomarker discovery and therapeutics.

Analyzing neuropeptides, however, is easier said than done. These signaling peptides are often present at low abundance *in vivo*, with extracellular concentrations typically in the subfemtomolar range (Maes et al., 2014), necessitating efficient extraction methods and highly sensitive instruments. The structural diversity of neuropeptides adds yet another layer of challenges to their analyses. As neuropeptide genes are transcribed into messenger RNAs, sometimes alternative splice variants can arise to yield multiple mRNA sequences that lead to variable protein precursor sequences (Hook et al., 2018). Additionally, each of these sequences can be, and often are, post-translationally modified to induce structural (and subsequently, functional) changes. These PTMs include, but are not limited to, C-terminal amidation, N-terminal acetylation, methylation, phosphorylation, and the highly variable glycosylation (Cao et al., 2017; Mast, Checco, & Sweedler, 2020). Even though neuropeptides are relatively small, typically fewer than 40 residues, multiple PTMs on a single neuropeptide can exist (Hook et al., 2018). This diversity can result in even lower abundance of a neuropeptidoform (neuropeptide analog of proteoform) of interest.

Mass spectrometry (MS) continues to solidify its place in the field as the most effective method of analyzing neuropeptides. By measuring the mass-to-charge ratio (m/z) of analytes, MS can provide users with the mass of the intact neuropeptide. Tandem MS (MS/MS or MS^2), in which precursor ions are selected for fragmentation, results in characteristic backbone fragment ions (typically b/y or c/z) that can be used to discern more detailed structural information, including amino acid sequence and the residue location of PTMs (Buchberger, Yu, & Li, 2015). Furthermore, MS can provide quantitative information in many ways, such as by evaluating the signal intensity of precursor ions, area under the curve from the chromatogram, or the intensity of diagnostic reporter ions at the MS^2 level (Sauer, Phetsanthad, Riusech, & Li, 2021). Although other methods exist for characterizing and quantifying neuropeptides, MS is particularly useful as it provides this information with higher throughput and without a priori knowledge of the sample. By rapidly analyzing all analytes in the sample, neuropeptidomic data can be generated with high throughput and without the need for expensive antibodies or the time to develop targeted assays (Sauer et al., 2021). As neuropeptide function, and the signaling pathways they are involved in, are influenced by the many neuropeptides present in a sample, MS is a powerful tool for studying neuropeptides, their function, and their physiologic regulation.

2. Neuropeptide multiplexing via isobaric tagging

The expression of neuropeptides is highly regulated by different synthesis and degradation pathways. These pathways can be perturbed by a multitude of different diseases (Hulme et al., 2020a, 2020b; Li et al., 2019), stressors (Buchberger et al., 2020; Liu et al., 2019; Sauer & Li, 2021), or even normal biological functions (DeLaney et al., 2021). Quantifying the expression changes of neuropeptides over time, after a biological event of interest, in different tissue regions, etc. can provide many insights into the regulation, function, and interplay of neuropeptides (DeLaney, Buchberger, & Li, 2018). Many methods of quantifying neuropeptides by MS exist, with the simplest being label-free quantification (LFQ). These workflows process each sample separately and compare signal intensity, area under the curve, etc. to quantify neuropeptide expression changes across different samples. A more detailed summary of LFQ methods is discussed in a recent review (Sauer et al., 2021), however, it is worth noting that the major limitations of these methods are low throughput and the requirement for large amounts of samples. Alternatively, using chemical tagging, neuropeptides can be differentially labeled and analyzed in a single MS experiment. These label-based approaches decrease run-to-run variability, the total amount of MS instrument time required, and the number of data files to be processed (Sauer & Li, 2021).

2.1 Isobaric tagging

With most chemical tagging strategies, stable isotopes (^{13}C, ^{2}H, ^{18}O, ^{15}N) are incorporated into different channels of the tag, thus incorporating different isotopes in the labeled product depending on the channel used. In an isotopic workflow, there exists light and heavy channels that are separated by a few Da due to the heavy channel containing more/larger isotopes than the light channel. Labeled products (i.e., neuropeptides) are differentiated at the precursor mass level (MS^1) and quantitative information is gained from either area under the curve or peak intensity. Common examples include dimethyl labeling (Buchberger et al., 2020; Tashima & Fricker, 2017) and TMAB tags (Boonen et al., 2018) and are summarized in a recent review (Sauer et al., 2021). As each channel increases the number of peaks in the spectra multiplicatively, spectra quickly become complicated, limiting the capacity for multiplexing.

Isobaric tagging alleviates this by incorporating the same nominal mass, regardless of channel, for a specific tag. Isotopes are strategically placed along the tag such that reporter ions of different masses form upon fragmentation and MS/MS analysis (Frost, Greer, & Li, 2015). See Fig. 1. At the precursor mass level, there is no additional spectral complexity, as all analytes of interest have the same mass addition, and there is only a modest increase in spectral complexity at the MS^2 level. Furthermore, the added peaks from the reporter ions are often in the low mass region (<200 Da) so they do not interfere with the peptide backbone ions (*b*/*y* or *c*/*z*). These advantages enable multiplexing capabilities of up to 21-plex using *N*,*N*-dimethyl

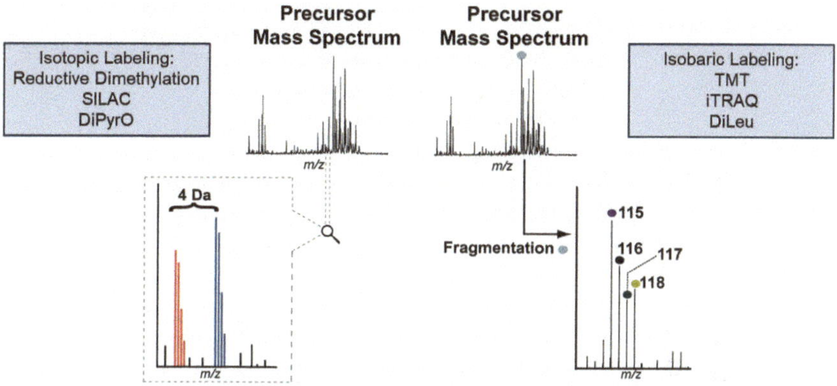

Fig. 1 Isotopic vs isobaric labeling. In an isotopic workflow, stable isotopes are incorporated to create a mass shift between differentially labeled samples (left). Isobaric tagging results in no mass shift between samples, but instead yields unique reporter ions at the MS^2 level that correlate to relative abundances from each sample (right).

leucine (DiLeu) tags developed in the Li Lab (Frost, Feng, & Li, 2020). Additionally, there are commercially available options, such as tandem mass tags (TMT) (Li et al., 2020) and isobaric tags for relative and absolute quantitation (iTRAQ) (Wiese, Reidegeld, Meyer, & Warscheid, 2007). These tags, however, are more expensive and lower throughput than DiLeu, thus we will be focusing on the use of DiLeu in this chapter.

The structure of the DiLeu tags is given in Fig. 2. Like many isobaric tags, DiLeu consists of a reporter region that, upon fragmentation, yields unique reporter ions for quantitation, a balance group to ensure the same nominal mass is incorporated at the precursor mass level, and a functional group that enables the tag to be covalently linked to the peptide. In the case of DiLeu, the functional group is an amine-reactive triazine ester that causes DiLeu to label primary amines, such as N-termini and lysine residues (Frost et al., 2015). The tags are synthesized readily by the reductive dimethylation of isotopic leucines, and the active triazine ester form is achieved by addition of N-methylmorpholine (NMM) and 4-(4,6-dimethoxy-1,3,5-triazin-2-yl)-4-methylmorpholinium tetrafluoroborate (DMTMM BF_4). Originally, four tags were conceived with reporter ion masses at *m/z* 115, 116, 117, and 118 (Xiang, Ye, Chen, Fu, & Li, 2010). With higher mass resolution, however, mDa mass differences arising from different mass additions from stable isotopes can be resolved. This has led to the expansion to a 12-plex labeling strategy (masses shown in Fig. 2) (Frost et al., 2015). Very recently, the synthesis of the tags has been altered to allow for more isotopologues and

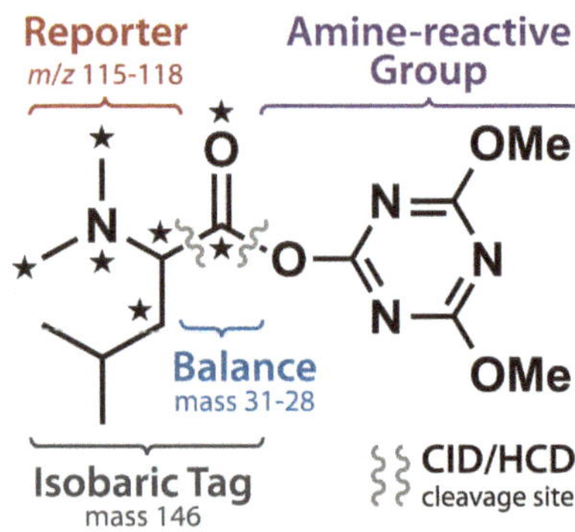

DiLeu Label	Stable Isotopes		Reporter Mass
	Reporter	Balance	
115a	^{15}N	^{13}C ^{18}O	115.12476
115b	^{13}C	^{13}C ^{18}O	115.13108
116a	^{13}C ^{15}N	^{18}O	116.12812
116b	^{13}C ^{13}C	^{18}O	116.13444
116c	^{2}H ^{2}H	^{18}O	116.14028
117a	^{13}C ^{13}C ^{15}N	^{13}C	117.13147
117b	^{2}H ^{2}H ^{15}N	^{13}C	117.13731
117c	^{13}C ^{2}H ^{2}H	^{13}C	117.14363
118a	^{13}C ^{13}C ^{13}C ^{15}N		118.13483
118b	^{13}C ^{2}H ^{2}H ^{15}N		118.14067
118c	^{13}C ^{13}C ^{2}H ^{2}H		118.14699
118d	^{2}H ^{2}H ^{2}H ^{2}H		118.15283

● ^{13}C ● ^{2}H ● ^{15}N ● ^{18}O

Fig. 2 DiLeu structure and masses. The chemical structure of the activated DiLeu tag is shown with stars denoting locations for stable isotope incorporation. The table beneath the structure shows the reporter ion masses for each of the 12-plex channels, along with the isotope placement for each tag. *Reprinted with permission from Frost, D. C., Greer, T., & Li, L. (2015). High-resolution enabled 12-plex DiLeu isobaric tags for quantitative proteomics.* Analytical Chemistry, 87*(3), 1646–1654. https://doi.org/10.1021/ac503276z.*

up to 21-plex analyses (Frost et al., 2020). This latest generation of DiLeu, however, has not yet been applied to neuropeptidomics, so we will be focusing our discussion on the 4- and 12-plex DiLeu tags.

2.2 Limitations for neuropeptidomics

DiLeu has been shown to be effective in analyzing protein digests, resulting in high proteome coverage and quantitative accuracy (Hao et al., 2019; Liu et al., 2020; Yu et al., 2020). The analysis of neuropeptides, however, presents some challenges when using isobaric tags, hence the limited publications featuring isobaric tagging of endogenous peptides. Although isobaric tagging offers numerous advantages, it does require that the analyte be selected for fragmentation and subsequent MS^2 analysis to form and measure the reporter ions. With proteins, not every constituent peptide needs to be identified to confidently quantify the starting protein from the sample, so long as there is sufficient sequence coverage. With endogenous peptides, however, the identification can come from only the single intact peptide.

Exacerbating this problem is the simple fact that neuropeptides are expressed with low *in vivo* abundance, and isobaric labeling workflows invariably use data-dependent acquisition (DDA) to collect mass spectra. In DDA analyses, a precursor scan is performed and the top ions (highest peak intensity) from that spectrum are selected for subsequent fragmentation and tandem MS analysis. After a preset number of MS^2 scans, another precursor scan is performed and the process repeats. DDA methods are widely used but are less useful for low level analytes. This is because it is often unlikely that conventional DDA methods will select low abundance analytes, like neuropeptides, for fragmentation and MS/MS analysis. As a result, the peptide backbone ions and reporter ions do not form, prohibiting accurate identification and quantitation of the analyte. Data-independent acquisition (DIA) methods (in which all ions are fragmented together) are gaining traction in the field of peptidomics (Delaney & Li, 2019; Zhong et al., 2020) but are incompatible with isobaric tagging as the reporter ion intensities cannot be deconvoluted from the source peptides. Inclusion lists and other targeted methods like parallel reaction monitoring (PRM) (Saidi, Kamali, & Beaudry, 2019) can be useful too, but they lack application for discovery-based experiments or when PTMs are unknown. Comparatively, optimization of DDA parameters can improve the number of neuropeptides identified in an isobaric workflow by increasing the number of unique MS^2 scan events and reducing redundant analyses of higher abundance analytes (Sauer & Li, 2021). In this chapter, we describe a protocol for the extraction, labeling, and LC-MS analysis of neuropeptides with discussion and details on how to optimize the analysis to greatly enhance neuropeptidomic coverage.

3. Protocol

An overview of the protocol is provided in Fig. 3.

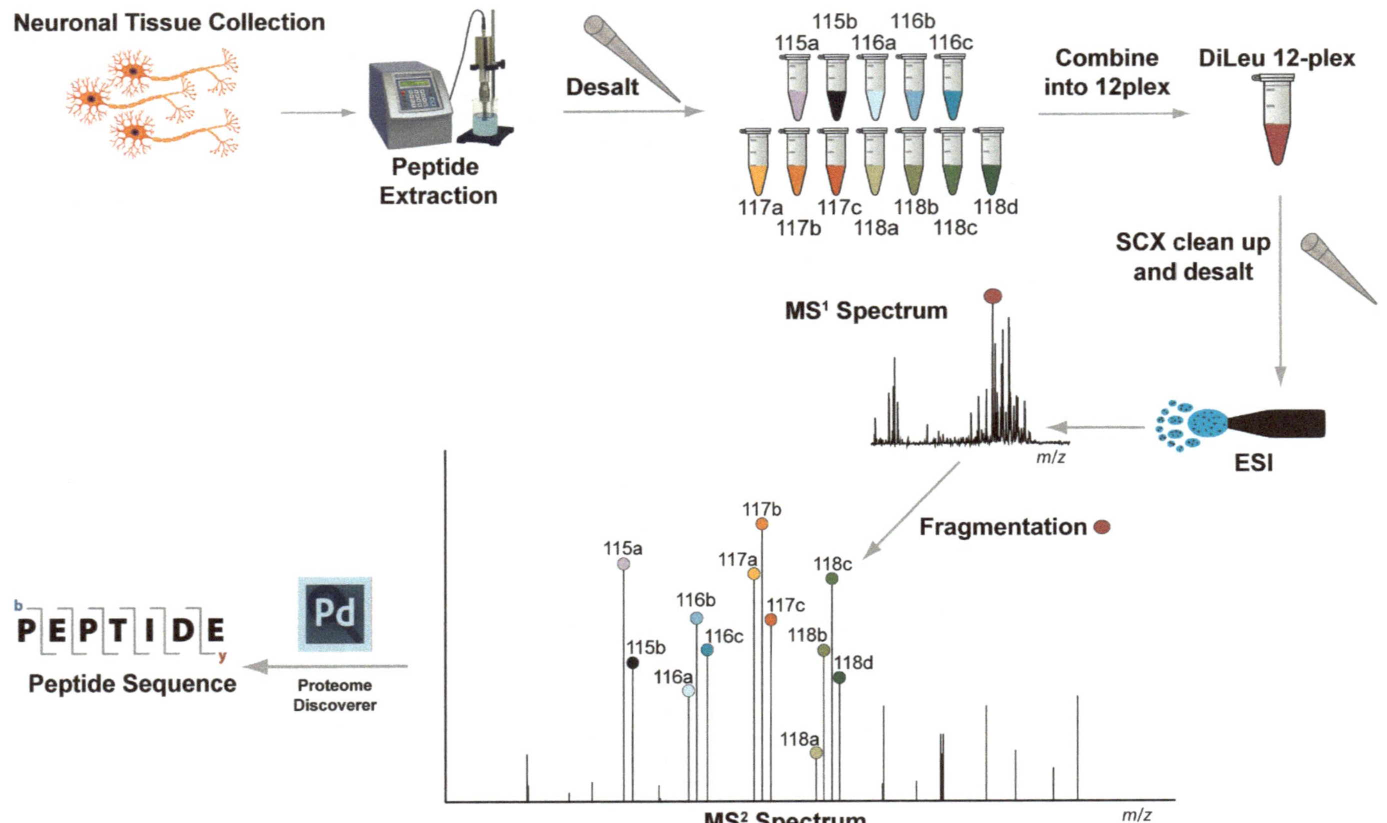

Fig. 3 Overall neuropeptidomic workflow. In brief, neuropeptides are extracted from tissues, cells, or fluids via ultrasonication. These neuropeptides are then desalted and purified so they can be labeled with DiLeu tags. The differentially labeled samples are then pooled and cleaned up further using SCX and C_{18} desalting before being analyzed by LC-MS. Data is then analyzed by Proteome Discoverer 2.1 to identify and quantify neuropeptides in the sample.

3.1 Extraction and purification of neuropeptides from tissue

3.1.1 Materials and equipment

1. Acidified methanol (90/9/1 methanol/water/acetic acid)
2. Resuspension solution: 0.1% formic acid (FA) in water
3. 10% FA in water
4. Elution solution 1: 0.1% FA in 25% acetonitrile (ACN) and 75% water
5. Elution solution 2: 0.1% FA in 50% acetonitrile (ACN) and 50% water
6. Elution solution 3: 0.1% FA in 75% acetonitrile (ACN) and 25% water
7. C_{18} resin-packed pipette tips (e.g., ZipTips or Omix Tips)
8. Ultrasonication probe dismembrator
9. Temperature controlled centrifuge (>20,000 x *g*)
10. Vacuum centrifuge drying system

3.1.2 Procedure

1. To the tissue sample, add enough acidified methanol to sufficiently cover the tissue. For crustacean brains, this is approximately 50–75 μL.
2. In a cold room (4 °C), use the ultrasonication probe dismembrator to homogenize the tissue and lyse cells. As the probe can generate heat, it is recommended to use a pulse setting with <10 s pulse and >15 s rest between pulses.
3. Centrifuge the homogenized samples at 20,000 x *g*, 4 °C for 20 min.
4. Transfer the supernatant to a clean tube and discard the pellet.
5. Dry down the samples in the vacuum centrifuge.
6. Resuspend samples in 15 μL resuspension buffer and ensure pH <4. Use the 10% FA solution to adjust the pH as needed. Sonicating in a water bath can help resolubilize sample if needed.
7. Purify samples using ZipTips per manufacturer's protocol. In brief, rinse the tip with 50/50 ACN/water and then equilibrate with the resuspension solution. Bind the sample and then wash with more resuspension solution to remove salts and other contaminants. Elute sequentially with the elution solutions listed above.
8. Pool the elution solutions and dry down under vacuum.

Note: After extracting the neuropeptides, it is important to purify them prior to labeling to ensure minimal interference from other contaminants. C_{18} desalting is an effective means of purification, and we recommend using resin-packed pipettes, such as Millipore ZipTips or Agilent Omix Tips.

3.2 Labeling and clean-up steps

3.2.1 Materials and equipment

1. Labeling Buffer (50/50 ACN/0.5 M triethylammonium bicarbonate, TEAB)
2. Anhydrous *N,N*-dimethylformamide (DMF)
3. *N*-methylmorpholine (NMM)
4. 4-(4,6-Dimethoxy-1,3,5-triazin-2-yl)-4-methylmorpholinium tetrafluoroborate (DMTMM)
5. 5% Hydroxylamine in water
6. DiLeu tags. The synthesis of these tags has been described in detail by Frost et al. (2015) and will not be described here
7. Pierce Peptide Assay Kit (or another means of quantifying peptide content)
8. Strong cation exchange (SCX) resin-packed tips (e.g., Millipore ZipTips or polyLC TopTips)
9. SCX loading buffer: 0.1% FA, 20% ACN
10. SCX elution buffer: 0.4 M ammonium formate, 20% ACN
11. C_{18} desalting pipette tips and buffers described in Section 3.1.1
12. Vacuum centrifuge

3.2.2 Procedure

1. Quantify peptide content of purified sample. We recommend using a simple and reliable method, such as a BCA assay or Pierce Peptide Assay kit and following the manufacturer's protocol.
2. Using information from the BCA or peptide assay, resuspend samples in a suitable solution dependent on your sample (e.g., water) and split into 25 μg aliquots. Dry down the aliquots via vacuum centrifugation and store at -80 °C until ready to proceed with labeling.
3. Create fresh activation solution with 495 μL anhydrous DMF, 15.5 mg DMTMM, and 5.18 μL NMM.
4. Add 50 μL activation solution to 1 mg of dry DiLeu tag and vortex at room temperature for 30 min. This yields 700 μg of active DiLeu tag capable of labeling primary amines.
5. While the tag is activating, resuspend the dry 25 μg aliquots of purified neuropeptides from Step 2 in 53.57 μL labeling buffer.
6. Add 35.71 μL (500 μg) of the activated label to the resuspended neuropeptides and vortex at room temperature for 1 h.

Note: The protocol described here is to label a set amount of neuropeptides at a 20:1 label:peptide ratio, but there is room for flexibility. The starting amount of neuropeptides is sample dependent and the rest can be scaled. Additionally, the label:peptide ratio and labeling time can be adjusted by the experimenter. It is important to keep the total organic content (coming from the ACN in the resuspension buffer and DMF in the activation solution) during the labeling reaction at approximately 70% to prevent hydrolysis of the tag and incomplete labeling.

7. Quench the reaction with 4.7 μL 5% hydroxylamine (0.25% final concentration).
8. Pool differentially labeled samples together at this time and dry down via vacuum centrifugation.

Note: After labeling, excess reagents need to be removed. SCX has proven useful for this and we recommend using pipette tips (e.g., Millipore SCX ZipTips), spin tips (e.g., polyLC SCX TopTips), or offline fractionation with a SCX column of choice. The protocol described here is for spin tips.

9. Remove excess labeling reagents using SCX spin tips per manufacturer's protocol. Briefly, resuspend the multiplexed samples in 100 μL SCX loading buffer. Equilibrate the resin with 100 μL loading buffer three times and apply the sample. It is helpful to reapply the sample after it has flown through the tip, up to three times, to ensure binding of sample. Wash the sample with 100 μL loading buffer five times. Elute with 100 μL SCX elution buffer three times. Pool the eluents together and dry down via vacuum centrifugation.
10. To remove the salt (ammonium formate) incorporated during SCX, desalting needs to be performed once again prior to LC-MS analysis. This is achieved using the same protocol as Step 7 of 3.1.2, but due to the increased sample size (from pooling multiple together) it is recommended to use a larger tip with greater binding capacity. Volumes should then be scaled accordingly. Additionally, it is important to use Optima-grade solvents and FA as this is the final purification step prior to LC-MS analysis.
11. Dry down the samples and store at -80 °C until ready for LC-MS analysis.

3.3 LC-MS analysis

3.3.1 Materials and equipment

1. C_{18}-packed nanoLC column, 75 μm internal diameter, 15 cm length, 300 Å pore size, either commercially available or self-packed.

2. Thermo Q Exactive HF with Dionex UltiMate3000 nanoLC system.
3. Mobile phase A: Optima-grade 0.1% FA in water.
4. Mobile phase B: Optima-grade 0.1% FA in ACN.

3.3.2 Procedure

1. Resuspend the samples in Mobile phase A so the concentration of neuropeptides is approximately 0.5–1 μg/μL.
2. Equilibrate the nanoLC column and set up the gradient for each sample injection: 0–120 min 3–35% B; 120–135 min 95% B; 135–150 min 3% B, all at a flow rate of 0.3 μL/min.
3. Set up a DDA-MS acquisition method with the following parameters: 30 k resolving power for both MS^1 and MS^2; 2 kV spray voltage; 200–2000 m/z precursor scan range; 100 m/z fixed first mass for MS^2 scans; normalized collision energy of 27; AGC target of 5e5 and 1e5 for MS^1 and MS^2 respectively; maximum injection time of 150 ms and 250 ms for MS^1 and MS^2 respectively; isolation window width of 2.0 m/z; 39 MS^2 scan events; 35 s dynamic exclusion window.
4. Inject 1–2 μL of sample with up to three technical replicates and a blank run between samples.

Note: The values listed above were those found to be optimal or sufficient for 4-plex labeled neuropeptides (Sauer & Li, 2021). The true optimal values are dependent on the sample, instrument, etc. Optimization techniques are described in Section 4.

3.4 Data analysis

There are different software packages that can be used to analyze multiplexed DiLeu data. There are commercially available options, such as Proteome Discoverer and PEAKS, that allow users to add custom isobaric tags to the search parameters. Open-source programs like MaxQuant can also be viable alternatives. In general, an analyst needs software that is compatible with the file type they collect (not all instruments produce the same data type) and that are reliable for quantitation in isobaric workflows. We chose to use Proteome Discoverer and detail a procedure below for the analysis of neuropeptide data.

3.4.1 Procedure

1. Import the data and database into Proteome Discoverer (we used version 2.1).

2. Create a data processing method for searching the data: Precursor min/max masses of 350/5000 Da; 25 ppm precursor mass tolerance; 0.2 Da fragment mass tolerance; 1% target false discovery rate (FDR); co-isolation threshold of 50%; static DiLeu modification of N-termini and Lys; C-terminal amidation for peptides known to be amidated endogenously; dynamic deamidation of Asn and Gln; dynamic dehydration of Ala, Asp, Glu, Ser, Thr, and Tyr; dynamic methylation of Asp, Glu, His, Arg, Ser, Thr; and three max modifications per peptide. A detailed overview of the Proteome Discoverer workflow used here can be found in the supplemental information.
3. Process the data to benchmark method performance (comparing LC-MS methods during optimization), or to quantify changes in a system (comparing experimental conditions to a control).
4. If performing statistical analyses, exporting the data to the proper software (e.g., Excel) is necessary. There the necessary statistical tests can be performed, such as t-tests, ANOVA, Dunnett's test, and other means of determining statistically significant changes between samples.

4. Optimization of MS data acquisition

Several instrument parameters influence the analytes selected by DDA analyses, and these can vary greatly depending on instrument. For our analyses, we utilize Orbitrap mass spectrometers, and our discussion will be focused on them. It is worth noting that the overall methods and workflows presented here are transferrable to other instrument types. In a DDA workflow, the user defines the number of tandem MS scans acquired per precursor scan (MS^2 scan events), the minimum amount of time required to pass before acquiring an MS^2 scan of an m/z previously acquired (dynamic exclusion window), the isolation window, the targeted amount of ions to be injected (automatic gain control target, or AGC target), and the amount of time allowed to accumulate ions prior to injection (maximum injection time), among others. Additionally, other parameters, such as the resolving power, do not directly contribute to DDA, but influence the duty cycle and total number of spectra that can be acquired.

The goal for optimizing these various parameters is to increase the number of acquired MS^2 scans for low abundance neuropeptides, decrease oversampling of high abundance analytes, and minimize the number of low-quality spectra. Optimizing the parameters listed above can be beneficial but is not so straightforward. For example, increasing the maximum injection time to allow more time for ions to accumulate from low abundance analytes could allow more neuropeptides to be identified from the acquired spectra, but it would also increase the cycle time and decrease the total number of acquired spectra from which to identify neuropeptides. Similarly, reducing the AGC target may allow more low abundance analytes to be sampled, but could also result in decreased signal-to-noise ratio (S/N) and therefore decreased confidence in the results.

Further complicating optimization process, there is interplay between parameters that is difficult to predict but needs to be considered. While some interplay can be more easily predicted, such as needing to increase the maximum injection time in order to reach a higher AGC target, others are less clear. Increasing the number of MS^2 scan events, for example, could result in sampling some neuropeptides too early (i.e., not at the highest peak intensity), and if the dynamic exclusion window is set too high, that neuropeptide may not be sampled again and not be identified/quantified due to poor signal. Due to the complex influence and co-dependence of various DDA parameters, thorough, simultaneous optimization of parameters is required to facilitate deeper neuropeptidomic coverage when using isobaric tags. In this section, we describe two different methods of optimization for DDA parameters.

4.1 Orthogonal array optimization

Optimization of parameters can be performed using complex modeling and statistical analysis, but developing these models is complex, requires many tests, and does not guarantee efficacy for experiments. Design of experiments (DoE) can be useful for efficiently evaluating different sets of conditions (Hecht, McCord, & Muddiman, 2015; Randall, Cardasis, & Muddiman, 2013). The simplest of these is the factorial DoE in which different levels (x) of each parameter (y) are tested in a combinatorial fashion, requiring y^x tests (Andrews, Dean, Hawkridge, & Muddiman, 2011). This is feasible for small numbers, but quickly becomes time and resource intensive to evaluate many conditions. Alternatively, using an orthogonal array optimization (OAO) or Taguchi array, greatly improves efficiency. Using this method, an array of values is generated that determine the level of each factor tested for a given experiment (Rao, Kumar, Prakasham, & Hobbs, 2008). The orthogonality of these arrays results in a table in which combinations of values from each column appear the same number of times across the array (Bolboacă & Jäntschi, 2007). This facilitates estimation of the same sample space as the more exhaustive factorial DoE but with far fewer total experiments. Different orthogonal arrays depending on the number of factors and levels to be explored have already been created and published, so the OAO method is readily accessible for new users. These arrays have notation of $L_n (x^y)$, where L is for Latin squares, n is the number of rows, y is the number of factors, and x is the number of levels of each factor (Bolboacă & Jäntschi, 2007).

In the context of optimizing DDA parameters, the OAO method is useful because it allows many parameters to be analyzed simultaneously while keeping the total number of experiments low. As an example, if seven parameters (number of MS^2 scans, dynamic exclusion window, isolation window, AGC target, maximum injection time, MS^1 resolution, and MS^2 resolution) are to be evaluated at two levels (low/high), a full factorial DoE would require 7^2 or 49 total experiments. Alternatively, by using an $L_8 (2^7)$ orthogonal array (Table 1), only 8 total experiments need to be performed. If performing technical replicates, the benefits of the OAO method are even greater. Reducing the number of experiments not only reduces the time required, but also reduces the total amount of sample needed and the run-to-run variability due to instrumental drift.

When evaluating the different experiments, the performance metrics are determined by the outcome the experimenter is trying to achieve. In an isobarically labeled neuropeptide workflow, this could be an increase in

Table 1 Orthogonal array optimization.

Experiment	MS^2 scan events	Dynamics exclusion window	Isolation window	AGC target	Maximum injection time	MS^1 resolution	MS^2 resolution
1	H	H	H	H	H	H	H
2	H	H	H	L	L	L	L
3	H	L	L	H	H	L	L
4	H	L	L	L	L	H	H
5	L	H	L	H	L	H	L
6	L	H	L	L	H	L	H
7	L	L	H	H	L	L	H
8	L	L	H	L	H	H	L

In this $L_8 (2^7)$ array, 7 parameters are evaluated at two levels each, high (H) and low (L).

the number of identifiable neuropeptides, the quantitative accuracy (determined from the observed reporter ion ratios compared to the known ratios based upon how samples were pooled together), etc. When optimizing for a specific outcome, the experimenter can use the information from the OAO to better understand the interplay between factors and which factors have a greater effect on the outcome of the experiments. These factors can then be the subject of more extensive optimization (demonstrated in Section 4.2).

4.2 Systematic simplex optimization

Design of experiments, like OAO, can provide experimenters with sufficient optimization to analyze neuropeptides, but more extensive optimization can offer further improvements. Using the knowledge gained from a DoE, or sometimes literature searches, the most impactful parameters can be determined. Systematically optimizing these parameters can then yield greater improvement in neuropeptidomic workflows as demonstrated in a recent publication (Sauer & Li, 2021). Similar to DoE, there are different options for systematically optimizing select parameters, but we will focus on the simplex optimization strategy in this section as it is simple and capable of simultaneous optimization.

In a simplex optimization strategy, an initial set of conditions (simplex) is assessed, and the optimization is visualized graphically. The conditions are then ranked by performance (determined by the desired outcome of optimization) and the next set of conditions is generated in the geometric opposition of the condition of worst performance (Bezerra et al., 2016). As more simplexes are evaluated, the conditions move toward optimization, and optimization is achieved experimentally when additional simplexes provide no increase in performance. An overview is shown in Fig. 4A. Reaching the optimum values for each parameter can be time consuming, so the simplex method can be modified to expedite the process. In a modified simplex algorithm, the geometric reflection (R) is evaluated along with the addition of an elongated reflection (E), contracted reflection (C_R), and a contraction toward the point of worst performance (C_W) (Bezerra et al., 2016) (Fig. 4B). By testing additional points in each simplex, the total number of simplexes that need to be tested can be reduced.

For the application of the simplex method, the experimenter must first generate an initial simplex of three vertices to be tested. These points should be based upon prior knowledge or published research for reasonable values

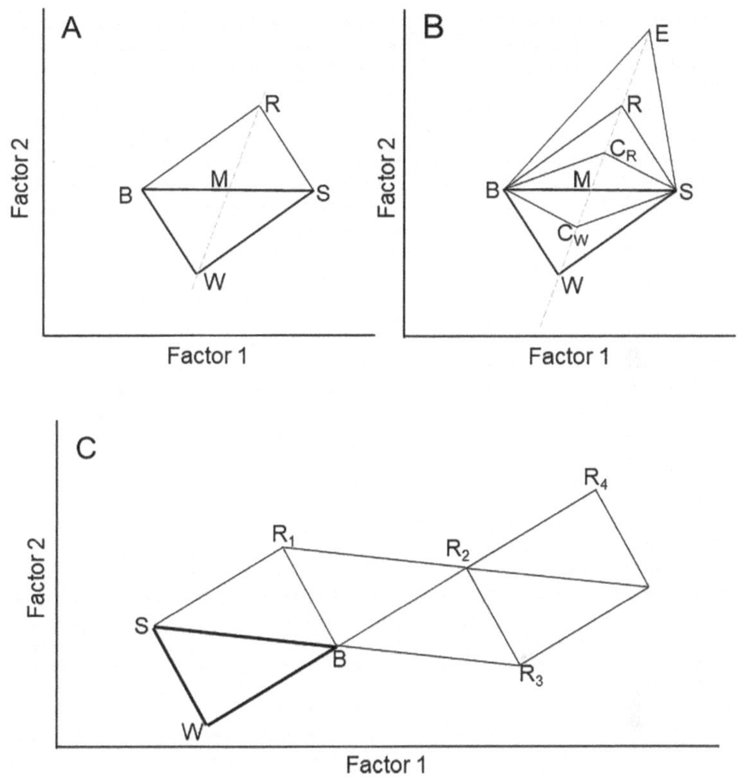

Fig. 4 Simplex optimization overview. The basic simplex is given in A and shows the reflection (R) across the midpoint (M) of the best (B) and second-best (S) vertices away from the worst vertex (W). Modifications to the workflow are given in B, where the next simplex also includes an elongated reflection (E), a contracted reflection (C_R), and a contracted reflection towards W (C_W). The process of reflecting away from the worst performing points is repeated until an optimum is reached. This is shown in C where the reflected vertices are sequentially labeled R_1, R_2, etc.

of the selected parameters. After evaluating the initial simplex, the vertices are ranked by performance as best (B), second-best (S), and worst (W). The next simplex is generated by reflecting W across the midpoint of BS. The equations for these reflections are summarized in Table 2. After evaluating the new points, the top performing condition of R, E, C_R, and C_W is ranked along with the best and second-best conditions from the prior simplex to generate a new BSW simplex. Again, W is reflected across the midpoint of BS as dictated by the equations in Table 2, the new points are evaluated, and the process repeats itself until an optimum is reached. A recent

Table 2 Simplex modification equations.

Vertex	Geometric translation	Equation
R	Reflection	2A–W
E	Elongated reflection	3A–2W
C_R	Contracted reflection	1.5A–0.5W
C_W	Contracted reflection towards W	0.5A + 0.5W

These equations are used to determine the new vertices to test. A is the average of points B and S (best and second-best), and W is the worst performing point. These equations are to be used for each factor being tested–if two factors are being tested, there will be two values making each coordinate point, and the equations need to be used twice to generate a new coordinate in the x,y grid.

application of this method can be found in a publication by Sauer and Li. Briefly, two parameters were selected for simultaneous optimization–the dynamic exclusion window and the number of MS^2 scan events. Keeping other factors constant, 3 paired sets of conditions were evaluated based upon the number of neuropeptides that could be identified from a constant sample. After the first three conditions were evaluated, a second set of conditions was selected in the geometrically opposed region of the worst performing point. These second conditions were evaluated, and a third set of conditions was selected, again moving away from low performing points. In the initial points, only 25–30 neuropeptides were able to be identified. By the end of the optimization (reached when additional points provide no further improvement), 88 neuropeptides were able to be identified. This approximate 3-fold increase in identifications allowed the method to be used for the study of neuropeptide expression changes resulting from copper toxicity.

Although effective, the simplex optimization does have limitations. Compared to the OAO, one of the major disadvantages to the simplex method is the time required for optimization. Data must be collected and analyzed between simplexes so the next simplex can be generated, instead of comparing all conditions at once. Additionally, although the method is amenable to simultaneous optimization of several factors, it becomes impossible to visualize graphically with more than three factors. Equations for determining the geometric reflections in extended dimensions exist, but these do not mitigate the number of experiments that need to be performed. As each additional factor for optimization increases the number of vertices in each simplex, it is less feasible to use the simplex method to optimize more than 2–3 parameters.

5. Applications of isobaric tagging in neuropeptidomics

The methods described in this chapter can help facilitate the application of isobaric labeling to neuropeptidomic workflows. One of the first published works in this area compared neuropeptides in lobster brains at different growth stages using 4-plex DiLeu tags (Jiang, Xiang, Jia, Buchberger, & Li, 2018). This work by Jiang et al. did not employ extensive optimization, but nonetheless showed the feasibility in using isobaric tags to study neuropeptide expression changes. Building on this work, Sauer and Li reported the benefits of the modified simplex optimization described in Section 4.2 and increased the neuropeptidomic coverage by approximately three-fold in a single LC-MS run using 4-plex DiLeu tags (Sauer & Li, 2021). Using these methods, neuropeptide expression changes were able to be quantified over time during acute copper exposure. Furthermore, several statistically significant changes were observed, demonstrating the potential role of allatostatin neuropeptides and others in the stress response to heavy metals like copper (Sauer & Li, 2021). In addition to using DiLeu to measure neuropeptides, isobaric tagging can influence neuropeptidomic studies by studying the associated neuroproteome. Liu et al. demonstrated the benefits of extracting both proteins and neuropeptides from the mouse brain and quantifying them using DiLeu and LFQ respectively (Liu et al., 2020). Although the isobaric tagging approach is not used for neuropeptides, this work highlights the use of using DiLeu in a multi-omic workflow to better understand the neuroendocrine system. Isobaric tagging of neuropeptides is still relatively new, but we expect increased application in the future as the need for high-throughput, quantitative neuropeptidomics continues to increase.

6. Summary

The investigation of neuropeptides and their roles in signaling pathways has great potential to improve understanding of many biological processes, disease, etc., but requires sensitive, high-throughput analyses. LC-MS analysis of isobarically tagged neuropeptides enables robust quantitation with higher orders of multiplexing than isotopic labeling and LFQ methods. As neuropeptides are of low abundance *in vivo*, they are often omitted from tandem MS analyses and go without identification or quantitation. Design of experiments and simplex optimization require initial time investment but can greatly improve the depth of coverage of the neuropeptidome, enabling future studies of the neuroendocrine system.

Acknowledgments

This work was supported by National Science Foundation (CHE-1710140 and CHE-2108223) and National Institutes of Health through grants (R01DK071801, RF1 AG052324, and P41GM108538). C.S.S. acknowledges a National Institute of Environmental Health Sciences fellowship as part of the National Ruth L. Kirschstein Research Service Award fellowship program (F31ES031859). L.L. acknowledges a Vilas Distinguished Achievement Professorship and Charles Melbourne Johnson Professorship with funding provided by the Wisconsin Alumni Research Foundation and University of Wisconsin-Madison School of Pharmacy.

References

Andrews, G. L., Dean, R. A., Hawkridge, A. M., & Muddiman, D. C. (2011). Improving proteome coverage on a LTQ-orbitrap using design of experiments. *Journal of the American Society for Mass Spectrometry*, *22*(4), 773–783. https://doi.org/10.1007/s13361-011-0075-2.

Bezerra, M. A., dos Santos, Q. O., Santos, A. G., Novaes, C. G., Ferreira, S. L. C., & de Souza, V. S. (2016). Simplex optimization: A tutorial approach and recent applications in analytical chemistry. *Microchemical Journal*, *124*, 45–54. https://doi.org/10.1016/j.microc.2015.07.023.

Bolboacă, S. D., & Jäntschi, L. (2007). Design of experiments: Useful orthogonal arrays for number of experiments from 4 to 16. *Entropy*, *9*(4), 198–232. https://doi.org/10.3390/e9040198.

Boonen, K., Haes, W. D., Van Houtven, J., Verdonck, R., Baggerman, G., Valkenborg, D., et al. (2018). Chapter 9 Quantitative peptidomics with isotopic and isobaric tags. In *1719. Peptidomics: methods and strategies, methods in molecular biology* (pp. 141–159).

Buchberger, A. R., Sauer, C. S., Vu, N. Q., DeLaney, K., & Li, L. (2020). A temporal study of the perturbation of crustacean neuropeptides due to severe hypoxia using 4-plex reductive dimethylation. *Journal of Proteome Research*, *19*, 1548–1555. https://doi.org/10.1021/acs.jproteome.9b00787.

Buchberger, A., Yu, Q., & Li, L. (2015). Advances in mass spectrometric tools for probing neuropeptides. *Annual Review of Analytical Chemistry*, *8*(1), 485–509. https://doi.org/10.1146/annurev-anchem-071114-040210.

Cao, L., Qu, Y., Zhang, Z., Wang, Z., Prykova, I., & Wu, S. (2017). Intact glycopeptide characterizatoin using mass spectrometry. *Physiology & Behavior*, *176*(3), 139–148. https://doi.org/10.1586/14789450.2016.1172965.Intact.

Christie, A. E., Stemmler, E. A., & Dickinson, P. S. (2010). Crustacean neuropeptides. *Cellular and Molecular Life Sciences*, *67*(24), 4135–4169. https://doi.org/10.1007/s00018-010-0482-8.

DeLaney, K., Buchberger, A. R., & Li, L. (2018). Chapter 17: Identification, quantitation, and imaging of the crustacean peptidome. In *1719. Methods in molecular Biology* (pp. 247–269). https://doi.org/10.1007/978-1-4939-7537-2_18.

DeLaney, K., Hu, M., Hellenbrand, T., Dickinson, P. S., Nusbaum, M. P., & Li, L. (2021). Mass spectrometry quantification, localization, and discovery of feeding-related neuropeptides in cancer borealis. *ACS Chemical Neuroscience*. https://doi.org/10.1021/acschemneuro.1c00007.

Delaney, K., & Li, L. (2019). Data independent acquisition mass spectrometry method for improved neuropeptidomic coverage in crustacean neural tissue extracts [Research-article]. *Analytical Chemistry, 91*(8), 5150–5158. https://doi.org/10.1021/acs.analchem.8b05734.

Frost, D. C., Feng, Y., & Li, L. (2020). 21-plex dileu isobaric tags for high-throughput quantitative proteomics. *Analytical Chemistry, 92*(12), 8228–8234. https://doi.org/10.1021/acs.analchem.0c00473.

Frost, D. C., Greer, T., & Li, L. (2015). High-resolution enabled 12-plex DiLeu isobaric tags for quantitative proteomics. *Analytical Chemistry, 87*(3), 1646–1654. https://doi.org/10.1021/ac503276z.

Han, B., Fang, Y., Feng, M., Hu, H., Qi, Y., Huo, X., et al. (2015). Quantitative neuropeptidome analysis reveals neuropeptides are correlated with social behavior regulation of the honeybee workers. *Journal of Proteome Research, 14*(10), 4382–4393. https://doi.org/10.1021/acs.jproteome.5b00632.

Hao, L., Thomas, S., Greer, T., Vezina, C. M., Bajpai, S., Ashok, A., et al. (2019). Quantitative proteomic analysis of a genetically induced prostate inflammation mouse model via custom 4-plex DiLeu isobaric labeling. *American Journal of Physiology - Renal Physiology, 316*(6), F1236–F1243. https://doi.org/10.1152/ajprenal.00387.2018.

Hecht, E. S., McCord, J. P., & Muddiman, D. C. (2015). Definitive screening design optimization of mass spectrometry parameters for sensitive comparison of filter and solid phase extraction purified, INLIGHT plasma N-glycans. *Analytical Chemistry, 87*(14), 7305–7312. https://doi.org/10.1021/acs.analchem.5b01609.

Hook, V., Lietz, C. B., Podvin, S., Cajka, T., & Fiehn, O. (2018). Diversity of neuropeptide cell-cell signaling molecules generated by proteolytic processing revealed by neuropeptidomics mass spectrometry. *Journal of the American Society for Mass Spectrometry, 29*(5), 807–816. https://doi.org/10.1007/s13361-018-1914-1.

Hu, M., Helfenbein, K., Buchberger, A. R., Delaney, K., Liu, Y., & Li, L. (2021). Exploring the sexual dimorphism of crustacean neuropeptide expression using callinectes sapidus as a model organism. *Journal of Proteome Research, 20*, 2739–2750. https://doi.org/10.1021/acs.jproteome.1c00023.

Hulme, H., Fridjonsdottir, E., Gunnarsdottir, H., Vallianatou, T., Zhang, X., Wadensten, H., et al. (2020a). Simultaneous mass spectrometry imaging of multiple neuropeptides in the brain and alterations induced by experimental parkinsonism and L-DOPA therapy. *Neurobiology of Disease, 137*(December 2019), 104738. https://doi.org/10.1016/j.nbd.2020.104738.

Hulme, H., Fridjonsdottir, E., Gunnarsdottir, H., Vallianatou, T., Zhang, X., Wadensten, H., et al. (2020b). Neurobiology of disease simultaneous mass spectrometry imaging of multiple neuropeptides in the brain and alterations induced by experimental parkinsonism and L-DOPA therapy. *Neurobiology of Disease, 137*(December 2019), 104738. https://doi.org/10.1016/j.nbd.2020.104738.

Jiang, X., Xiang, F., Jia, C., Buchberger, A. R., & Li, L. (2018). Relative quantitation of neuropeptides at multiple developmental stages of the American Lobster using *N,N*-dimethyl leucine isobaric tandem mass tags. *ACS Chemical Neuroscience*, acschemneuro.7b00521. https://doi.org/10.1021/acschemneuro.7b00521.

Li, L., & Sweedler, J. V. (2008). Peptides in the brain: Mass spectrometry–based measurement approaches and challenges. *Annual Review of Analytical Chemistry, 1*(1), 451–483. https://doi.org/10.1146/annurev.anchem.1.031207.113053.

Li, J., Van Vranken, J. G., Vaites, L. P., Schweppe, D. K., Huttlin, E. L., Etienne, C., et al. (2020). TMTpro reagents: A set of isobaric labeling mass tags enables simultaneous proteome-wide measurements across 16 samples. *Nature Methods, 17*(April). https://doi.org/10.1038/s41592-020-0781-4.

Li, C., Wu, X., Liu, S., Zhao, Y., Zhu, J., Liu, K., et al. (2019). Roles of neuropeptide Y in neurodegenerative and neuroimmune diseases. *Frontiers in Neuroscience*, *13*, 1–11. https://doi.org/10.3389/fnins.2019.00869.

Liu, Y., Buchberger, A. R., Delaney, K., Li, Z., & Li, L. (2019). Multifaceted mass spectrometric investigation of neuropeptide changes in atlantic blue crab, callinectes sapidus, in response to low pH stress. *Journal of Proteome Research*, *18*, 2759–2770. https://doi.org/10.1021/acs.jproteome.9b00026.

Liu, R., Wei, P., Keller, C., Salvatore, N., Shi, Y., Li, Z., et al. (2020). Integrated label-free and 10-plex dileu isobaric tag quantitative methods for pro fi ling changes in the mouse hypothalamic neuropeptidome and proteome: Assessment of the impact of the gut microbiome. *Analytical Chemistry*, *92*, 14021–14030. https://doi.org/10.1021/acs.analchem.0c02939.

Maes, K., Van Liefferinge, J., Viaene, J., Van Schoors, J., Van Wanseele, Y., Béchade, G., et al. (2014). Improved sensitivity of the nano ultra-high performance liquid chromatography-tandem mass spectrometric analysis of low-concentrated neuropeptides by reducing aspecific adsorption and optimizing the injection solvent. *Journal of Chromatography A*, *1360*, 217–228. https://doi.org/10.1016/j.chroma.2014.07.086.

Mast, D. H., Checco, J. W., & Sweedler, J. V. (2020). Differential post-translational amino acid isomerization found among neuropeptides in Aplysia californica. *ACS Chemical Biology*, *15*(1), 272–281. https://doi.org/10.1021/acschembio.9b00910.

Randall, S. M., Cardasis, H. L., & Muddiman, D. C. (2013). Factorial experimental designs elucidate significant variables affecting data acquisition on a quadrupole orbitrap mass spectrometer. *Journal of the American Society for Mass Spectrometry*, *24*(10), 1501–1512. https://doi.org/10.1007/s13361-013-0693-y.

Rao, R. S., Kumar, C. G., Prakasham, R. S., & Hobbs, P. J. (2008). The Taguchi methodology as a statistical tool for biotechnological applications: A critical appraisal. *Biotechnology Journal*, *3*(4), 510–523. https://doi.org/10.1002/biot.200700201.

Romanova, E. V., & Sweedler, J. V. (2015). Peptidomics for the discovery and characterization of neuropeptides and hormones. *Trends in Pharmacological Sciences*, *36*(9), 579–586. https://doi.org/10.1016/j.tips.2015.05.009.

Saidi, M., Kamali, S., & Beaudry, F. (2019). Neuropeptidomics: Comparison of parallel reaction monitoring and data-independent acquisition for the analysis of neuropeptides using high-resolution mass spectrometry. *Biomedical Chromatography*, *33*(7), 1–11. https://doi.org/10.1002/bmc.4523.

Sauer, C. S., & Li, L. (2021). Mass spectrometric profiling of neuropeptides in response to copper toxicity via isobaric tagging. *Chemical Research in Toxicology*. https://doi.org/10.1021/acs.chemrestox.0c00521.

Sauer, C. S., Phetsanthad, A., Riusech, O. L., & Li, L. (2021). Developing mass spectrometry for the quantitative analysis of neuropeptides. *Expert Review of Proteomics*. https://doi.org/10.1080/14789450.2021.1967146.

Tashima, A. K., & Fricker, L. D. (2017). Quantitative peptidomics with five-plex reductive methylation labels. *Journal of the American Society for Mass Spectrometry*, *29*, 866–878. https://doi.org/10.1007/s13361-017-1852-3.

Wiese, S., Reidegeld, K. A., Meyer, H. E., & Warscheid, B. (2007). Protein labeling by iTRAQ: A new tool for quantitative mass spectrometry in proteome research. *Proteomics*, 7(3), 340–350. https://doi.org/10.1002/pmic.200600422.

Xiang, F., Ye, H., Chen, R., Fu, Q., & Li, N. (2010). N,N-dimethyl leucines as novel Isobaric tandem mass tags for quantitative proteomics and peptidomics. *Analytical Chemistry*, *82*(7), 2817–2825. https://doi.org/10.1021/ac902778d.

Yu, Q., Zhong, X., Chen, B., Feng, Y., Ma, M., Diamond, C. A., et al. (2020). Isobaric labeling strategy utilizing 4-Plex N, N-dimethyl leucine (DiLeu) tags reveals proteomic changes induced by chemotherapy in cerebrospinal fluid of children with B-cell acute lymphoblastic leukemia. *Journal of Proteome Research*, *19*(7), 2606–2616. https://doi.org/10.1021/acs.jproteome.0c00291.

Zhong, X., Frost, D. C., Yu, Q., Li, M., Gu, T. J., & Li, L. (2020). Mass defect-based dileu tagging for multiplexed data-independent acquisition. *Analytical Chemistry*, *92*(16), 11119–11126. https://doi.org/10.1021/acs.analchem.0c01136.

CHAPTER ELEVEN

Revealing AMP mechanisms of action through resistance evolution and quantitative proteomics

Samantha J. Balboa and Leslie M. Hicks*

Department of Chemistry, University of North Carolina at Chapel Hill, Chapel Hill, NC, United States

*Corresponding author: e-mail address: lmhicks@unc.edu

Contents

Abstract

Antimicrobial resistance (AMR) is a significant public health issue that threatens our ability to treat common infections. AMR often emerges in bacteria through upregulation of proteins that allow a subpopulation of resistant bacteria to proliferate through natural selection. Identifying these proteins is crucial for understanding how AMR develops in bacteria and is essential in developing novel therapeutics to combat the threat of widespread AMR. Mass spectrometry-based proteomics is a powerful tool for understanding the biochemical pathways of biological systems, lending remarkable insight into AMR mechanisms in bacteria through measuring the changing protein abundances as a result of

Methods in Enzymology, Volume 663

ISSN 0076-6879

https://doi.org/10.1016/bs.mie.2021.09.008

antibiotic treatment. Here, we describe a serial passaging method for evolving resistance in bacteria that implements quantitative proteomics to reveal the differential proteomes of resistant bacteria. The focus herein is on antimicrobial peptides (AMPs), but the approach can be generalized for any antimicrobial compound. Comparative proteomics of sensitive vs. resistance strains in response to AMP treatment reveals mechanisms to survive the bioactive compound and points to the mechanism of action for novel AMPs.

1. Introduction

Antibiotic-resistant infections are a rising global health crisis with more than 2.8 million antibiotic resistant infections occurring each year in the United States, resulting in >35,000 deaths (Centers for Disease Control, 2019). The World Health Organization (WHO) has published a list of priority pathogens, dubbed the ESKAPE (*Enterococcus faecium, Staphylococcus aureus, Klebsiella pneumoniae, Acinetobacter baumannii, Pseudomonas aeruginosa,* and *Enterobacter* spp.) pathogens, for which new antimicrobial strategies are urgently needed (Tacconelli, Carrara, Savoldi, Kattula, & Burkert, 2017). While drugs of last resort are prescribed in severe cases of multidrug resistant (MDR) infections, the emergence of resistance to these last resort antibiotics is quickly depleting effective therapeutic strategies. Novel antibiotics are therefore essential to treat MDR infections.

Antimicrobial peptides (AMPs) are small peptides (<50 amino acids) that are expressed throughout all kingdoms of life and can inhibit the growth of parasites, fungi, viruses, and bacteria (Wang et al., 2019). AMPs are an attractive class of antimicrobials due to their ubiquitous nature, broad-spectrum activity, and slower emergence of resistance (Brogden, 2005; Magana et al., 2020). Despite these advantages, AMPs can still confer antibiotic resistance in bacteria, especially when applied in sublethal concentrations that allow resistant subpopulations of bacteria to rapidly develop. Well-characterized bacterial mechanisms to achieve resistance to AMPs include modulation of efflux pumps and uptake transporters to decrease intracellular AMP concentration, expression of peptidases to proteolytically degrade AMPs, and modification of the cellular membrane to reduce electrostatic interactions (Wang et al., 2019).

Antibiotic stress has enormous implications on the proteomic landscape and bacterial phenotype (Fernández-Reyes et al., 2009; Janssen et al., 2021; Nonejuie, Burkart, Pogliano, & Pogliano, 2013; Sun et al., 2020; Wong, Oliver, & Linington, 2012). As such, understanding the proteomic response of bacteria to AMP exposure is crucial to understanding mechanisms of action and potential routes for resistance evolution. This can be readily studied *in vitro* through serial passaging experiments in which bacteria cultures are initially exposed to sublethal concentrations of AMP and continually passaged into increasing AMP concentration to select for resistant bacteria (Sun et al., 2020). Mechanisms of action and resistance can then be informed through comparative quantitative proteomics of these resistant populations to those that are susceptible.

Herein, a method to induce AMR in bacteria and readout molecular phenotypic changes via quantitative proteomics is described. This includes a serial passaging method to induce AMP resistance in an example bacteria (*Escherichia coli*) followed by a detailed sample preparation and analysis protocol for label-free quantitative proteomics (Fig. 1A). Key considerations for

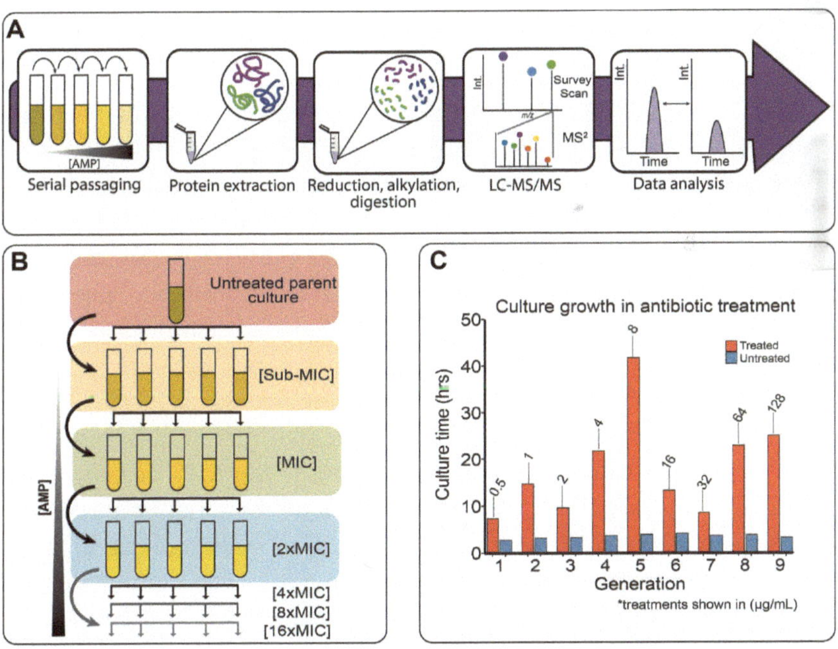

Fig. 1 (A) Overall workflow for *in vitro* evaluation of evolutionary antibiotic resistance via quantitative proteomics. Briefly, resistant cultures are generated through serial passaging in increasing concentrations of AMP. Then, the proteomes from the resistant cultures are harvested, extracted, reduced, alkylated, and digested with trypsin. The resulting peptides are then analyzed via LC-MS/MS. Data are then searched against a database to identify peptides. Peptides are assigned to proteins and quantified to determine changing protein abundances between conditions. (B) General overview of a serial passaging experiment in which a parent culture is used to generate 3–5 replicates. Cultures are then passaged into increasing concentrations of AMP until desired resistance is observed. (C) As the concentration of AMP increases in the bacterial culture, so do the incubation times to reach log-phase at $OD_{600} = 0.5–0.7$.

implementation of this method include: (1) requirement for sufficient antibiotic/AMP to induce resistance in multiple replicates across multiple generations with increasing concentrations of antimicrobial compound and (2) proteomics analysis will be most powerful with sequenced, well-annotated target bacterial strains.

2. Materials

2.1 Serial passaging

1. Bacterial strain: herein, *E. coli* ATCC® 25,922 will be used as an example
2. Mueller-Hinton agar/broth: Mueller-Hinton agar and broth can be readily purchased. To prepare agar plates, suspend 38 g of dehydrated agar medium in 1 L of Milli-Q water and sterilize by autoclaving. Cool media and pour approximately 20 mL into 100 × 15 mm petri dishes using aseptic technique. Allow plates to solidify overnight and store at 4 °C

   To prepare liquid Mueller-Hinton broth, dissolve 21 g of dehydrated medium in 1 L of Milli-Q water and sterilize by autoclaving.
3. Antimicrobial peptide sufficient for generation of at least three resistant lineages

   Tip: From a single parent culture, three biological replicates can be produced. Once each biological replicate reaches log-phase, five technical replicates can be collected for robust label-free quantitation. In a single passaging experiment comparing the proteomic differences between wild-type and three replicates of AMP-resistant strains, 20 total samples will be generated. It is important to note that there must be sufficient AMP material to produce resistant strains. If the MIC for an AMP is 1 μg/mL, the starting AMP concentration in the serial passage experiment could be 0.5 μg/mL, which is 0.5xMIC. If performing three passages, the final concentration for generating resistant strains is 4 μg/mL. Thus, there must be enough material to passage in 0.5, 1, 2, and 4 μg/mL of AMP for three biological replicates.
4. Shaking incubator at 37 °C, 250 rpm
5. 96-well plate UV/Vis spectrophotometer

2.2 Protein extraction

1. Lysis buffer: 100 mM Tris-HCl, pH 8, 0.1% Triton-X 100
   a. Stock solutions can be made for easy preparation of lysis buffer. When preparing the lysis buffer, stir slowly when fully dissolving contents to minimize agitation and bubble formation from the Triton-X 100

2. Covaris 2 mL milliTUBEs and 24 Place milliTUBE rack
   a. As an alternative, samples can be shaken for 15 min at RT, sonicated, or lysed using a French pressure cell press
3. 100 mM ammonium acetate in 100% methanol (MeOH), chilled
4. 100 mM TRIS, pH 8.0. Using a 1 M TRIS stock is recommended for ease of buffer preparation
5. Resuspension buffer: 8 M urea, 100 mM TRIS, pH 8.0
6. CB-X Protein Assay Kit (G-Biosciences, St. Louis, MO, USA) or equivalent protein quantification assay

2.3 Reduction, alkylation, and digestion

1. Reduction buffer: 500 mM dithiothreitol (DTT) in 100 mM TRIS, pH 8.0. Make fresh for each experiment and cover tube with aluminum foil or keep buffer in the dark to prevent degradation of light-sensitive DTT solution
2. Alkylation buffer: 500 mM iodoacetamide (IAM) in 100 mM TRIS, pH 8.0. Make fresh for each experiment and cover tube with aluminum foil or keep buffer in the dark to prevent degradation of light-sensitive IAM solution
3. Trypsin Gold, Mass Spectrometry grade, Promega (Madison, WI, USA), 0.5 μg/mL in 50 mM acetic acid
4. 20% trifluoroacetic acid (TFA).
5. ThermoMixer, Eppendorf

2.4 Desalting

1. Sep-Pak C18 1 cc Vac Cartridge, 50 mg, 55–105 μm particle size (Waters, Milford, MA, USA).
2. 0.1% TFA (LC-MS grade).
3. 80% acetonitrile (ACN, LC-MS grade), 0.1% TFA (LC-MS grade).
4. Vacuum manifold with 24-port cover (Phenomenex, Torrance, CA, USA) or equivalent setup
5. Vacuum centrifuge

2.5 LC-MS/MS

1. 5% ACN (LC-MS grade), 0.1% TFA (LC-MS grade).
2. LC-MS Total Recovery Vials.
3. Symmetry C18 trap column (100 Å, 5 μm, 180 μm × 20 mm; Waters).
4. HSS T3 C18 column (100 Å, 1.8 μm, 75 μm × 250 mm; Waters).
5. Mobile Phase A: 0.1% formic acid. Add 1 mL of Optima LC-MS grade formic acid to 1 L of Optima LC-MS grade water.

6. Mobile Phase B: 0.1% FA in ACN (LC-MS grade).
7. Acquity UPLC M-Class system (Waters).
8. Q Exactive HF-X Hybrid Quadrupole Orbitrap mass spectrometer (ThermoFisher, Waltham, MA, USA).

2.6 Data analysis

1. Progenesis QI for Proteomics v2.0 (Nonlinear Dynamics, Durham, NC, USA).

   Tip: Freeware options include MaxQuant, msInspect, and Skyline (Bellew et al., 2006; Maclean, Tomazela, Shulman, Chambers, & Finney, 2010; Tyanova, Temu, & Cox, 2016).
2. Mascot Daemon v3.5.1 (Matrix Science, Boston, MA, USA).
3. R script for processing proteomic data. The code used for processing these data is available on GitHub (https://github.com/hickslab/ProgenesisLFQ).

3. Methods

3.1 *In vitro* evolution of antibiotic resistance via serial passaging

1. Streak *E. coli* onto Mueller-Hinton agar plates and incubate for 12–18 h at 37 °C
2. Inoculate 5 mL of Mueller-Hinton broth with a single bacterial colony swabbed from the agar plate and incubate for 12–18 h at 37 °C, 250 rpm
3. From a single parent culture, back-dilute to an $OD_{600}=0.01$ into three different culture tubes containing 7 mL of Muller-Hinton broth and 0.5 μg/mL of AMP, or a sub-MIC value, to create three lineages and a fourth tube for the untreated culture as a control
4. Grow cultures at 37 °C, 250 rpm to an OD_{600} of 0.5–0.7 (log-phase) as determined by growth curves
5. Once log-phase has been reached, aliquot five replicates of 1 mL of bacterial culture and centrifuge for 5 min at 13,400 rcf, flash freeze, and store at −80 °C for proteomics analysis. Then, inoculate ~100 μL of the remaining log-phase culture into a fresh 5 mL of Mueller-Hinton broth containing 100% increased concentration of antibiotic to a final $OD_{600}=0.01$ (Fig. 1B).
6. Repeat steps 4–5 iteratively to the desired increase in the MIC of AMP is observed, indicating significant resistance

   Tip: Typically, a minimum of three passages of increasing AMP concentration is needed to induce AMP resistance from 0.5xMIC to 4xMIC, but this timeline is AMP and pathogen dependent as shown in several studies (Blanco, Hjort, Martinez, & Andersson, 2020; Samuelsen et al., 2005; Tambe, Sampath, & Modak, 2001). Sub-MIC treatment can also be adapted for this method by treating cultures with sub-MIC of AMP and measuring the proteomic differences. Note that AMP treatment of these cultures may increase incubation times to reach the desired OD_{600}, thus cultures must be continuously monitored (Fig. 1C).

3.2 Protein extraction

1. Resuspend bacterial pellets in 1 mL lysis buffer and transfer to Covaris 2 mL tubes. Keep samples on ice during resuspension

   Tip: Alternative cell lysis options include probe sonication, French cell press, manual grinding with a mortar and pestle, and freeze/thaw cycles. Reagent-based methods have advantages such as quick, mild, efficient, and reproducible results at the cost of introducing high concentrations of salt and detergents to the sample which are not compatible with mass spectrometry and may need further processing strategies. Physical disruption methods have the advantage of lysing cells without the addition of salts and detergents, but with the loss of ease and reproducibility.
2. Sonicate bacterial pellets for 3 min in a 4 °C water bath at 200 cycles/burst, 100 W power, and 13% duty cycle using an E220 focused ultrasonicator (Covaris, Woburn, MA, USA).
3. Transfer samples from Covaris tubes to 2-mL centrifuge tubes, keeping the samples on ice
4. Centrifuge cell lysates at 16,000 rcf for 10 min at 4 °C and collect the supernatant into a 15-mL conical tube
5. Precipitate proteins by adding 5 × volume (about 30 mL) of cold 100 mM ammonium acetate in MeOH
6. Incubate samples for 30 min at −20 °C
7. Collect protein pellet by centrifuging for 10 min at 3200 rcf, 4 °C. Decant the supernatant without disturbing the pellet
8. Allow protein pellets to dry for 20 min in a fume hood at room temperature
9. Resuspend the pellets in 400 μL of 4 M urea, 100 mM Tris, pH 8.0
10. Use a 10 μL aliquot of each replicate to perform protein quantification using the CB-X Protein Assay using manufacturer's protocol
11. Normalize each replicate to ~50–100 μg of total protein in 200 μL

3.3 Reduction, alkylation, and digestion

1. Reduce samples using 10 mM DTT. Incubate for 30 min at room temperature while shaking at 850 rpm
2. Alkylate samples using 40 mM IAM. Incubate for 45 min in the dark by covering samples with aluminum foil at RT while shaking
3. Precipitate proteins with 10 × volume of chilled 100% acetone and incubate for 30 min at −20 °C. Then, centrifuge for 10 min at 4 °C, 20,000 rcf. Remove supernatant and dry under N_2 in a fume hood
4. Resuspend protein pellets in 100 μL 8 M urea, 100 mM Tris, pH 8.0 and dilute to 2 M urea using 300 μL of 100 mM Tris pH 8.0
5. Perform overnight (12–18 h) digestion using mass spectrometry-grade trypsin (Trypsin Gold from Promega is recommended) at a protease to protein ratio of 1:50 (*w*/w) at 25 °C. Gently invert or shake the samples using a Thermomixer at 850 rpm during digestion
6. Following digestion, quench the reaction by adding 10% TFA to the samples until pH is <3 when measured with a pH test strip. Typically, 10–20 μL of 10% TFA is sufficient for 400 μL samples
7. (Optional) Samples can be stored at −80 °C until the next processing step is performed

3.4 Desalting

1. Set up one Waters™ Sep-Pak C18 SPE cartridge for each sample on a vacuum manifold using test tubes to collect the flow through from the cartridges
2. Wet cartridges by adding 1 mL of 80% ACN, 0.1% TFA
3. Equilibrate cartridges using 2 mL of 0.1% TFA
4. Load peptide samples onto the cartridge and recover the flow-through no faster than 1 drop/s in a new test tube
5. Reapply this flow through to the cartridge to ensure all peptides in the sample bind to the resin in the cartridge
6. Once the flow-through passes through, switch to a new test tube and flow 2 mL of 0.1% TFA through the cartridges to remove salts
7. Slowly elute desalted peptides at less than 1 drop/s into new 2-mL tubes by adding 1 mL of 50% ACN, 0.1% TFA to the cartridge. Once the samples finish eluting, apply vacuum for about 5 s to collect the remaining solvent from the packed bed
8. Following peptide elution, freeze the samples and vacuum centrifuge to dryness

3.5 LC-MS/MS

1. Resuspend peptide samples to 1 µg/µL in LC-MS water and dilute in LC-MS water +0.1% TFA to a final concentration of 0.25 µg/mL in Total Recovery Vials (Waters).
2. For each sample, inject 5 µL and perform LC-MS/MS analysis. Five technical replicates are injected for each biological replicate
   2.1. Acquity UPLC M-Class system (Waters): Inject the peptide mixture to a Symmetry C18 trap column (100 Å, 5 µm, 180 µm × 20 mm; Waters) with a flow rate of 5 µL/min for 3 min using 99% A and 1% B, then separate on a HSS T3 C18 column (100 Å, 1.8 µm, 75 µm × 250 mm; Waters) using a gradient of increasing mobile phase B at a flow rate of 300 nL/min for 120 min total. Increase mobile phase B from 5% to 35% in 90 min, ramp to 85% in 5 min, hold for 5 min, return to 5% mobile phase B in 2 min, and re-equilibrate for 13 min. This eluate is coupled directly to the subsequent ESI-MS/MS system
   2.2. The ESI-MS/MS system consists of a Q Exactive HF-X Hybrid Quadrupole Orbitrap mass spectrometer (ThermoFisher) with a Nanospray Flex Ion Source. Use the following MS parameters: use a tune file set with positive polarity, 2.2 kV spray voltage, 325 °C capillary temperature, and 40 S-lens RF level. Select full MS/DD-MS^2 scan type and set method duration to 120 min and default charge state to 2. Perform MS survey scan in profile mode across 350–2000 *m/z* at 120,000 resolution until 50 ms maximum IT or 3×10^6 AGC target is reached. Select the top 20 features above 5000 counts excluding ions with unassigned, +1, or >+8 charge state. Collect MS^2 scans at 30,000 resolution with NCE at 28 until 100 ms maximum IT or 1×10^5 AGC target. Set the dynamic exclusion window for precursor *m/z* to 10 s and an isolation window of 0.7 *m/z*

3.6 Data analysis

1. Upload the acquired LC-MS/MS spectral files (*.raw) into Progenesis QI for Proteomics (Nonlinear Dynamics). While the files are importing, start automatic processing to allow Progenesis to select an alignment reference to align the total ion chromatograms to minimize LC-MS/MS between-run differences in retention time and normalize peak abundances. Review the alignment score to ensure that all "Score" values are >80%. Design the experiment so that replicates are grouped together as one subject and export the aligned runs as a combined peak list (*.mgf).

Tip: Low alignment scores are often a result of drifting LC retention times and/or fluctuating MS mass accuracy between-runs during LC-MS/MS acquisition. Should a run fail to align during the automatic processing procedure, repeating the process with this failed run will often produce successful alignment but with overall lower alignment scores across the runs.

2. Perform Mascot (Matrix Science) database searching using the .mgf peak list to determine peptide sequence and identify proteins. Use the following search parameters: Search against the database containing the proteome for the organism of interest, in this case the *E. coli* (strain 25,922) proteome along with the sequence for common laboratory contaminants (https://www.thegpm.org/cRAP; 116 entries). Use a target decoy MS/MS search with trypsin protease specificity with up to two missed cleavages, a peptide mass tolerance of 15 ppm, and a fragment mass tolerance of 0.1 Da. Set a fixed modification of carbamidomethylation at cysteine and include the following variable modifications: acetylation at the protein N-terminus. After the search is complete, adjust the false discovery rate of the significant peptide identifications to be less than 1% using the embedded Percolator algorithm. Export matches the matches into a result file (*.xml).
3. Import the Mascot search results into Progenesis to perform peptide feature mapping. Export the "Protein Measurements" as an Excel document from the "Review Proteins" tab
4. Parse the "Protein Measurements" data file using the *QuantifyR* package developed by our lab (https://github.com/hickslab/ProgenesisLFQ) or using a similar parsing technique. This script groups together features identified in Progenesis and matches identical sequence, modifications, and score with differing protein accessions. These features are then represented by the protein accession with the highest number of unique peptides and largest confidence score assigned by Progenesis. Features duplicated by multiple peptide identifications are reduced to a single peptide with the highest Mascot ion score. Identifiers are made by joining the protein accession of each feature with the single-letter amino acid code of the modified residue and location of the modification. The data are then reduced to unique identifiers by summing the abundance of all contributing features (charge states, missed cleavages, etc.). Each identifier group is represented in the final dataset by the peptide with the highest Mascot score

5. To determine whether there is a significant difference in resistant and wild-type strains, the "Analyze" module of the *QuantifyR* package is used to perform two-sided, equal-variance *t*-tests and one-way ANOVA. The false discovery rate (FDR) is controlled in both analyses by using the Benjamini-Hochberg FDR correction and is used to determine proteins that were significantly changing between wild-type and resistant conditions (q-value <0.05). Using this workflow, a volcano plot displaying significantly changing proteins can be generated (Fig. 2) from the "Plot" function of the *QuantifyR* package. These changing protein abundances can then be used to inform further experiments to study the altered processes in the bacteria such as phenotypic assays, motility studies, and cell membrane permeability assays

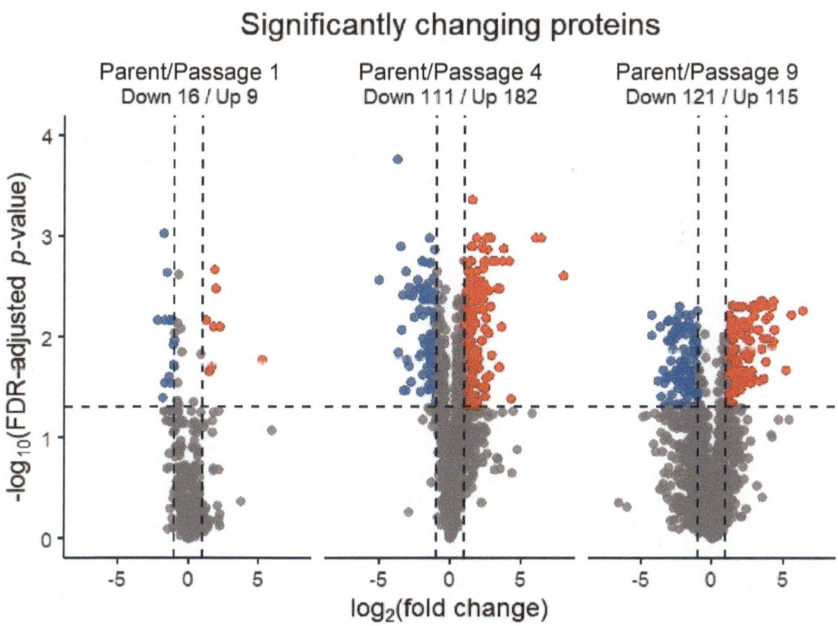

Fig. 2 Volcano plot showing changing protein abundances between parent cultures, serial passage 1, serial passage 4, and serial passage 9. The x-axis represents $\log_2$-transformed fold changes, with dashed lines indicating expected fold changes in *E. coli* background proteins. The y-axis represents the $-\log_{10}$-transformed FDR-adjusted *P*-values (q-values), with the dashed lines indicating the untransformed significance level, 0.05. Proteins in increasing abundance are shown in red while proteins decreasing in abundance are shown in blue.

4. Summary

Antimicrobial resistance is a challenging global health issue with enormous implications in human health. Recent studies highlight the importance of proteomic responses to AMP stress (Janssen et al., 2021; Sadecki et al., 2021; Sun et al., 2020). Interestingly, Sadecki et al. (2021) showed that polymyxin B, an AMP antibiotic, enhances the production of a genotoxic metabolite called colibactin in *E. coli* NC101, a commensal gut microbe, through the increase of proteins involved in the biosynthesis of colibactin. This study highlights the importance of understanding how AMP resistance can have immense implications to their target organism. This chapter outlines a method to induce AMR in bacteria through serial passaging and measuring the resulting proteomic changes via quantitative proteomics to inform further studies.

Acknowledgments

This research was supported by the NIH-NIGMS under award R01 GM125814 to L.M.H., and S.J.B. acknowledges NIH Biophysics training grant support (T32 GM008570) and P.W.S. for their development of this method.

References

Bellew, M., Coram, M., Fitzgibbon, M., Igra, M., Randolph, T., et al. (2006). A suite of algorithms for the comprehensive analysis of complex protein mixtures using high-resolution LC-MS. *Bioinformatics*, *22*, 1902–1909.

Blanco, P., Hjort, K., Martinez, J. L., & Andersson, D. I. (2020). Antimicrobial peptide exposure selects for resistant and fit Stenotrophomonas maltophilia mutants that show cross-resistance to antibiotics. *mSphere*, *5*.

Brogden, K. A. (2005). Antimicrobial peptides: Pore formers or metabolic inhibitors in bacteria? *Nature Reviews Microbiology.*, *3*, 238–250.

Centers for Disease Control. (2019). *Antibiotic resistance threats in the United States.*

Fernández-Reyes, M., Rodríguez-Falcón, M., Chiva, C., Pachón, J., Andreu, D., & Rivas, J. (2009). The cost of resistance to Colistin in *Acinetobacter baumannii*: A proteomic perspective. *Proteomics*, *9*(6), 1632–1645. https://doi.org/10.1002/pmic.200800434.

Janssen, A. B., Doorduijn, D. J., Mills, G., Rogers, M. R. C., Bonten, M. J. M., Rooijakkers, S. H. M., et al. (2021). Evolution of Colistin resistance in the *Klebsiella pneumoniae* Complex follows multiple evolutionary trajectories with variable effects on fitness and virulence characteristics. *Antimicrobial Agents and Chemotherapy.*, *65*.

Maclean, B., Tomazela, D. M., Shulman, N., Chambers, M., & Finney, G. (2010). Skyline: An open source document editor for creating and analyzing targeted proteomics experiments. *Bioinformatics*, *26*, 966–968.

Magana, M., Pushpanathan, M., Santos, A. L., Leanse, L., Fernandez, M., Ioannidis, A., et al. (2020). The value of antimicrobial peptides in the age of resistance. *The Lancet Infectious Diseases*, *20*, e216–e230.

Nonejuie, P., Burkart, M., Pogliano, K., & Pogliano, J. (2013). Bacterial cytological profiling rapidly identifies the cellular pathways targeted by antibacterial molecules. *PNAS*, *110*, 16169–16174.

Sadecki, P. W., Balboa, S. J., Lopez, L. R., Kedziora, K. M., Arthur, J. C., & Hicks, L. M. (2021). Evolution of Polymyxin resistance regulates Colibactin production in *Escherichia coli*. *ACS Chemical Biology*, *16*, 1243–1254.

Samuelsen, Ø., Haukland, H. H., Jenssen, H., Krämer, M., Sandvik, K., Ulvatne, H., et al. (2005). Induced resistance to the antimicrobial peptide Lactoferricin B in *Staphylococcus aureus*. *FEBS Letters.*, *579*, 3421–3426.

Sun, L., Rasmussen, P. K., Bai, Y., Chen, X., Cai, T., Wang, J., et al. (2020). Proteomic changes of Klebsiella pneumoniae in response to Colistin treatment and CrrB mutation-mediated Colistin resistance. *Antimicrobial Agents and Chemotherapy*, *6*.

Tacconelli, E., Carrara, E., Savoldi, A., Kattula, D., & Burkert, F. (2017). *Global priority list of Antibiotic-resistant bacteria to guide research, discovery, and development of new antibiotics*. World Health Organization.

Tambe, S. M., Sampath, L., & Modak, S. M. (2001). In vitro evaluation of the risk of developing bacterial resistance to antiseptics and antibiotics used in medical devices. *The Journal of Antimicrobial Chemotherapy*, *47*, 589–598.

Tyanova, S., Temu, T., & Cox, J. (2016). The MaxQuant computational platform for mass spectrometry-based shotgun proteomics. *Nature Protocols.*, *11*, 2301–2319.

Wang, J., Dou, X., Song, J., Lyu, Y., Zhu, X., Xu, L., et al. (2019). Antimicrobial peptides: Promising alternatives in the post feeding antibiotic era. *Medicinal Research Reviews*, *3*, 831–859.

Wong, W. R., Oliver, A. G., & Linington, R. G. (2012). Development of antibiotic activity profile screening for the classification and discovery of natural product antibiotics. *Chemistry & Biology*, *11*, 1483–1495.

CHAPTER TWELVE

Molecular networking-based strategies in mass spectrometry coupled with *in silico* dereplication of peptidic natural products and gene cluster analysis

Andrew C. McAvoy[a] and Neha Garg[a,b,*]

[a]School of Chemistry and Biochemistry, Georgia Institute of Technology, Atlanta, GA, United States
[b]Center for Microbial Dynamics and Infection, Georgia Institute of Technology, Atlanta, GA, United States
*Corresponding author: e-mail address: neha.garg@chemistry.gatech.edu

Contents

Abstract

Peptidic natural products (PNPs) are valuable sources for drug development as they are often associated with desirable bioactivities. Chemoenzymatic approaches coupled with description of structure-activity relationships have enabled the production of analogs to PNPs such as daptomycin and vancomycin which exhibit optimized bioactivities. The natural promiscuity of biosynthetic enzymes allows the producer organism to generate various analogs potentially serving as multiple hits for one target or to

Methods in Enzymology, Volume 663
ISSN 0076-6879
https://doi.org/10.1016/bs.mie.2021.09.012

diversify the bioactivity of a given natural product. Thus, approaches that allow rapid discovery of novel peptide natural products by mining their presence in crude extracts while providing insights into promiscuity of biosynthetic enzymes are sought out by natural product biochemists and chemists. In this chapter, we describe how two different molecular networking-based approaches combined with *in silico* tools can be used to dereplicate PNPs from untargeted mass spectrometry data acquired on crude extracts and how to propagate annotations to structurally related molecules. We further describe steps to gain knowledge of enzyme promiscuity by combining molecular networking analysis with gene cluster analysis.

1. Introduction

Microbially derived natural products are attractive targets for discovery of compounds with potential therapeutic value (Atanasov et al., 2021). Peptidic natural products (PNPs) are an important class of natural products with diverse biological functions, including antibacterial, antifungal, antiviral, immunosuppressant, and anticancer agents (Dang & Sussmuth, 2017). PNPs are either biosynthesized by non-ribosomal peptide synthetases (NRPSs) or are ribosomally synthesized and post-translationally modified (Liu, Rubin, Dhakal, Chen, & Ding, 2021). PNPs often exhibit additional structural modifications as well as incorporation of non-canonical amino acids to achieve chemical diversity (Baumann et al., 2017; Liu et al., 2021).

Ultra-high-performance liquid chromatography coupled to tandem mass spectrometry (UHPLC-MS/MS) is the analytical method of choice for high-throughput metabolomic screening of complex biological samples due to high sensitivity and selectivity. While this technology enables rapid acquisition of large amounts of data, interpretation of this data remains a major bottleneck for natural products researchers. Dereplication is a key step in natural product discovery pipelines, where previously reported natural products are rapidly identified. Additionally, samples often contain structural variants of known compounds that may initially appear as novel unknown compounds, requiring sophisticated algorithms to parse out unknown PNPs from their known variants (Gurevich et al., 2018). Many such algorithms and advanced dereplication methods have been developed in the last decade to decode the dark metabolome acquired on crude extracts with desired bioactive properties. The molecular networking approach detailed below is one such approach available through the Global Natural Product Social molecular networking (GNPS) platform, which has served as a resource for the bioinformatics community in development of advanced *in silico* dereplication strategies (Wang et al., 2016).

A molecular networking approach allows visualization of chemical structural similarity between detected metabolites based solely on their MS^2 spectral similarity, while searching for matching spectra in publicly available MS^2 spectral libraries (Wang et al., 2016; Yang et al., 2013). In a molecular network, identical spectra representing a unique metabolite is shown as a single node, and edges connecting nodes indicate spectral similarity between the two metabolites. In the first iteration of molecular networking (known as classical networking), clustering is based entirely on MS^2 similarity, and therefore two structural isomers featuring identical precursor masses and MS^2 spectra may incorrectly be combined into a single node despite displaying chromatographic separation. Feature-based molecular networking (FBMN) avoids this limitation by taking advantage of LC/MS data processing software such as MZmine 2 to incorporate MS^1 information (including isotope patterns, relative abundance and retention time) into networks, with the added benefit of allowing for relative quantification of metabolites (Nothias et al., 2020; Pluskal, Castillo, Villar-Briones, & Oresic, 2010). Molecular networking is especially effective when complemented with computational tools for PNP dereplication such as DEREPLICATOR (Mohimani et al., 2018), MS2LDA (van der Hooft, Wandy, Barrett, Burgess, & Rogers, 2016), VarQuest (Gurevich et al., 2018), or Sirius with CSI:FingerID (Dührkop et al., 2019; Dührkop, Shen, Meusel, Rousu, & Böcker, 2015). In this chapter, we describe use of DEREPLICATOR, which predicts MS^2 spectra of peptides within a chemical database and compares each experimental MS^2 spectrum against the predicted spectra of compounds within a specified mass range (Mohimani et al., 2017). These annotations can then be supported by linking PNPs to their biosynthetic gene clusters available in repositories such as MIBiG (Minimum Information about a Biosynthetic Gene cluster) (Kautsar et al., 2020). Gene cluster analysis leads to additional insights about the specific amino acids which are incorporated into NRPs and their potential modifications, guiding the use of structure-activity relationships to generate analogs with optimized bioactivity (Marahiel, Stachelhaus, & Mootz, 1997). In this chapter, we outline an analysis pipeline for PNP dereplication and variable dereplication using the siderophore ornibactin and related analogs detected in previously published data collected on culture extracts of various *Burkholderia* spp. (McAvoy et al., 2020).

2. Materials

2.1 Data

The converted data (centroided and lock-mass calibrated) in .mzXML file format used in the generation of molecular networks and analysis, a metadata table containing detailed sample information, a batch file with the settings employed for data pre-processing in MZmine 2, and the output files from MZmine 2 pre-processing can be downloaded at https://massive.ucsd.edu with the identifier MSV000087997 by clicking the "MassIVE Datasets" menu option on the top toolbar, or from https://gnps.ucsd.edu/ by clicking on MassIVE Datasets tab. The dataset in this chapter was acquired using an Agilent 1290 Infinity II UHPLC system (Agilent Technologies) coupled to an ImpactII ultra-high resolution Qq-TOF mass spectrometer (Bruker Daltonics, GmbH) equipped with ESI source. The classical molecular network analysis is available at: https://gnps.ucsd.edu/ProteoSAFe/status.jsp?task=304fb28f8ccc46b68645a07ae27790bb. The FBMN analysis is available at: https://gnps.ucsd.edu/ProteoSAFe/status.jsp?task=18084ed411c14c0d80a159496b43dc1b. The DEREPLICATOR analysis is available at https://gnps.ucsd.edu/ProteoSAFe/status.jsp?task=b9ac18b5db534196b0bad86f5727487c (for the classical network) and https://gnps.ucsd.edu/ProteoSAFe/status.jsp?task=b9ac18b5db534196b0bad86f5727487c (for the FBMN).

2.2 Computational resources

All software, data analysis resources, and online databases are publicly available at the following web pages.

1. Data conversion—Bruker data: https://gnps-quickstart.ucsd.edu/conversion, other vendor data: https://gnps-quickstart.ucsd.edu/
2. mzMine 2—http://mzmine.github.io/
3. ProteoSAFe—https://proteomics2.ucsd.edu/ProteoSAFe/index.jsp
4. GNPS—https://gnps.ucsd.edu/ProteoSAFe/static/gnps-splash.jsp
5. Cytoscape—https://cytoscape.org/
6. PubChem compound database—https://pubchem.ncbi.nlm.nih.gov/
7. Chemical editing software—ChemDraw: https://perkinelmerinformatics.com/products/research/chemdraw/, Marvin Suite: https://chemaxon.com/products/marvin

8. Exact *m*/*z* calculator—https://www.chemcalc.org/
9. MIBiG—https://mibig.secondarymetabolites.org/

3. Analysis pipeline

3.1 Running a classical molecular network

1. Navigate to the GNPS homepage, and log in using your GNPS username and password. If you do not have an account yet, create one using the "Register" link in the upper right corner of the page, below the login box.
2. Upload data to the FTP server at massive.ucsd.edu using your preferred FTP client program (such as WinSCP). Data must be centroided and in an open spectrum format (.mzXML, .mzML, and .mgf files are supported).
   - **(a)** Set up the FTP server connection using plain FTP file protocol with no encryption.
   - **(b)** Enter "massive.ucsd.edu" as the host/server name (port number: 21).
   - **(c)** Login to server using GNPS username and password.
   - **(d)** After establishing a connection, upload files from your local computer (left panel) to the server (right panel) by clicking and dragging. It is recommended to upload files into folders organized by experimental design to simplify data selection in subsequent steps.
3. Next, on the GNPS homepage, scroll down to the "Data Analysis" heading and click "Create Molecular Network" under "Molecular Networking." This will open the classical molecular network workflow input screen.
4. Under the "Basic Options" heading, select the files to be analyzed by clicking on the "Select Input Files" button next to "Spectrum Files (Required)." This will open the CCMS ProteoSAFe File/Resource Manager in a popup window. Select the files to analyze by clicking on the appropriate files in the left panel (under "Select Input Files"). An entire folder can also be selected to analyze all files within the folder. The files can be attributed to as many as six groups using this method. Group assignment can also be performed using a metadata file (https://ccms-ucsd.github.io/GNPSDocumentation/metadata/). In the current example, data files for solvent and media blanks were placed under G1, data files for *B. cenocepacia* culture extracts under G2, data files for *B. multivorans* culture extracts under G3, data files for *B. ambifaria*

culture extracts under G4, and data files for *B. dolosa* culture extracts under G5. The files and/or folders assigned to each group can be seen in the "Selected Files" panel. To remove a file from a group, select that file in the "Selected Files" panel and click the cross icon.

5. Once all data is organized into groups, scroll all the way down and choose "Finish Selection" to close the pop-up window. Set the networking workflow parameters, which are dependent on the instrument and the method used for data acquisition. Hovering on each parameter opens a window with detailed description of individual parameters. The settings used to analyze the sample data with classical molecular networking are summarized in Table 1. Although a comprehensive discussion on each of the workflow parameters available on GNPS is beyond the scope of this chapter, readers who are interested in further optimization are encouraged to refer to the Molecular Networking documentation (https://ccms-ucsd.github.io/GNPSDocumentation/networking/). After setting parameters as desired, ensure that the email address under "Workflow Submission" is correct before submitting the job *via* the "Submit" button in the bottom right corner.
6. Once networking job is complete, access it under the "Jobs" tab *via* the GNPS homepage where status will be listed as "Done." Click "Done" to access the results screen.

3.2 Generation of metabolite feature table

The first step for generation of feature-based molecular network is to extract a metabolite feature list which includes *m/z*, retention time, abundance *via* integration of area under the chromatographic peak from centroided data. In addition, each metabolite feature is linked to its representative MS/MS spectrum to generate spectral similarity-derived molecular network.

1. To generate a feature list with linked MS/MS spectra, download and open latest version of MZmine 2. The data processing in this protocol is performed using the latest MZmine 2 version available at the time of writing (v2.53). Note that other LC/MS data processing tools (including OpenMS, MS-DIAL, MetaboScape, XCMS, and Progenesis QI) are also supported for FBMN by GNPS (Rost et al., 2016; Smith, Want, O'Maille, Abagyan, & Siuzdak, 2006; Tsugawa et al., 2015). While this protocol will demonstrate the use of MZMine 2, additional information on MZmine 2 and other supported preprocessing software can be found in the FBMN documentation available on the GNPS website (https://ccms-ucsd.github.io/GNPSDocumentation/featurebasedmolecularnetworking/).

Table 1 Classical Molecular Networking workflow parameters used to analyze example dataset.

Parameter	Setting
Workflow selection	
Basic options	
Precursor ion mass tolerance	0.01 Da
Fragment ion mass tolerance	0.05 Da
Advanced network options	
Min pairs cos	0.7
Minimum matched fragment ions	4
Network TopK	10
Minimum cluster size	2
Run MSCluster	yes
Maximum connected component size	100
Advanced library search options	
Spectral library	Speclibs
Library search min matched peaks	4
Score threshold	0.7
Search analogs	Don't search
Maximum analog search mass difference	100
Advanced filtering options	
Filter below Std Dev	0
Minimum peak intensity	0
Filter precursor window	Don't filter
Filter library	Filter library
Filter peaks in 50 Da window	Don't filter
Filter Spectra from G6 as Blanks Before Networking	Don't filter
Advanced GNPS repository search options	
Find Related Datasets (Temporarily Deprecated)	Don't
Advanced output options	
Create cluster buckets and qiime2/Biom/PCoA plots output	Create
Create Ili mapping output	Don't create

2. Import raw data files. On the top toolbar, select "Raw data methods" → "Raw data import" and choose which data files which will be analyzed.
3. MZmine 2 subjects raw LC/MS data to a series of steps to perform feature detection, where "features" are defined as distinct LC/MS signals with unique *m/z* and retention times. These features are then aligned across all samples in the dataset of interest, and the abundances of features in each sample are calculated for downstream analysis. The final output is a matrix containing information about the *m/z* and retention time of each feature and the abundance of each feature in every data file. The MZmine 2 workflow used to preprocess current dataset is outlined in Fig. 1, and exact parameters are available in batch file included with the MassIVE dataset. It should be noted that these parameters should be optimized for each dataset based on the data acquisition parameters and instrument used. Upload the batch file by selecting "Project" → "Batch mode" on the toolbar and configure each step appropriately for the method and instrument used.
4. Alternatively, the LC_MZMINE2 workflow available on the UCSD CCMS ProteoSAFe web server (https://proteomics3.ucsd.edu/) can be used for rapid processing of data with MZMine2 using batch presets provided, which are optimized for several instrumental setups. After selecting the LC_MZMINE2 workflow, upload data files in mzXML format to the FTP server at massive.ucsd.edu (see steps 5–6 in Section 3.1) and process using a preset series of parameters appropriate for the dataset of interest. Advanced parameters including ppm error threshold, noise threshold, retention time range, and retention time tolerance can be adjusted as appropriate. The resulting output is analyzed *via* feature based molecular networking workflow by clicking on "Analyze with FBMN" under the "Follow up Analysis" header on the output screen. In the author's experience, this tool is often a useful starting point for evaluating which feature detection parameters are appropriate for a particular dataset, especially since an MZmine 2 batch file (named "params.xml") containing all parameters is accessible by clicking "Download Data" on the results screen, unzipping the resulting .zip file, loading into MZmine 2 and modifying as appropriate.

3.3 Creation of metadata table for feature-based molecular networking

The second step for generation of FBMNs involves creation of a metadata file that allows analysis of networks in relation to a physiological or

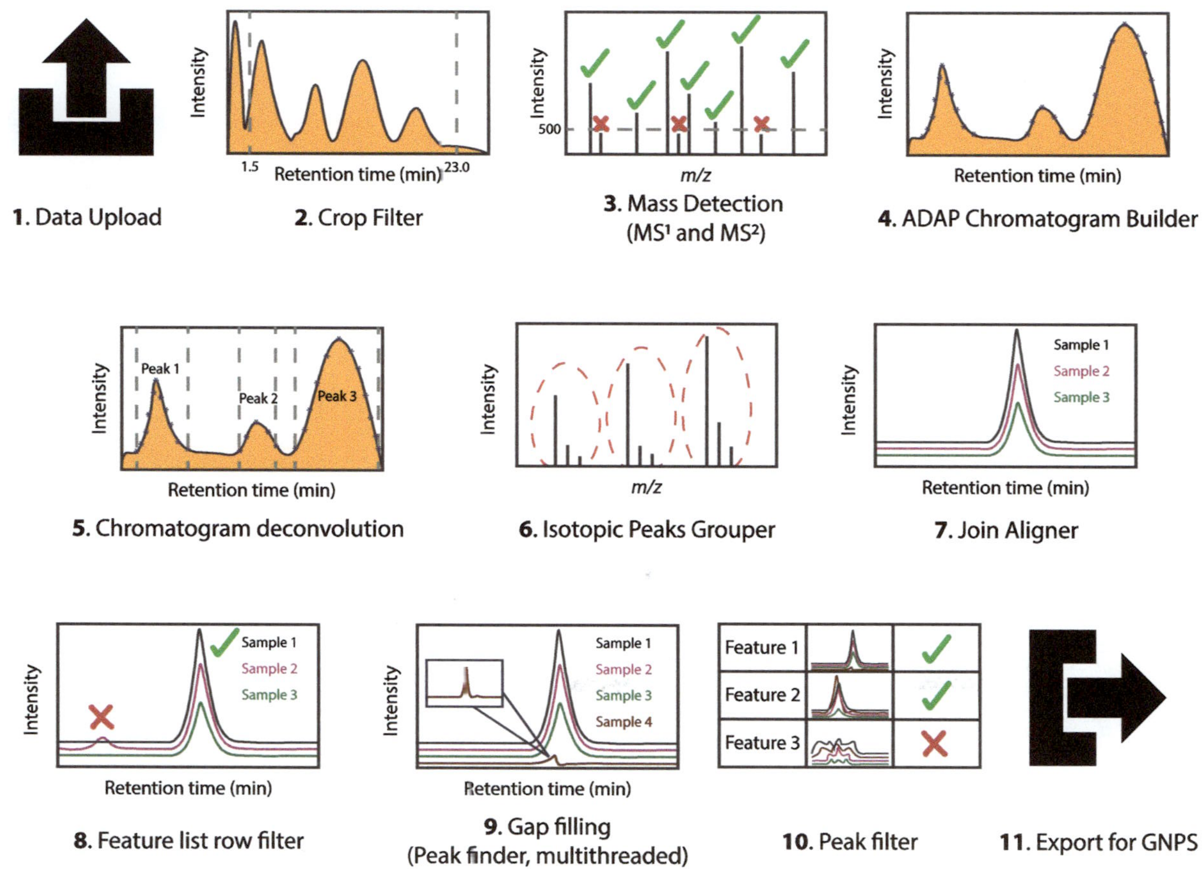

Fig. 1 Outline of MZmine 2 workflow used to preprocess mass spectrometry data for FBMN analysis.

experimental condition under which a given metabolite feature is observed. Thus, the metadata file facilitates downstream analysis of metabolomic trends in relation to experimental design. For FBMN, metadata files must be in a tab separated text file format (.txt), which can be generated using Excel or an equivalent text editing software. When creating the metadata file, it is important to ensure the file is correctly formatted. The title of the first column must be "filename" in all lowercase. Each cell in this column must contain the full filename for all files included in MZmine2 processing. Additional metadata columns containing experimental information can be incorporated as needed. The headers for these metadata columns must begin with the prefix "ATTRIBUTE_," followed by a short, descriptive name specific to each property. The contents of each metadata table should contain a short text descriptor, making sure to not leave any blank cells. In this example, the "ATTRIBUTE_Species" column contains six attributes which denote each of the four *Burkholderia* species being compared, the solvent blank and the growth media control. An example of the correct metadata format is shown in Fig. 2.

3.4 Generate feature-based molecular network

1. Logon to the GNPS website (https://gnps.ucsd.edu/ProteoSAFe/static/gnps-splash.jsp).
2. On the homepage, scroll down to the header called "Advanced Analysis Tools" and click on the "Analyze" button under "Feature Networking." As mentioned earlier, if the LC_MZMINE2 workflow on the UCSD CCMS ProteoSAFe web server was used for feature detection, the output will be directly imported for the FBMN analysis by selecting "Analyze with FBMN" on the results screen.
3. Provide a descriptive title in the "Title" box under "Workflow Selection."
4. Upload MS2 file (.mgf format if MZmine 2 was used), feature quantification table (.csv file), and sample metadata table (tab-separated .txt file)

	A	B	C	D	E	F
1	filename	ATTRIBUTE_Species	ATTRIBUTE_Strain	ATTRIBUTE_Trimethoprim	ATTRIBUTE_Extraction	ATTRIBUTE_Timepoint
2	BcC5424_LB_EA_24_1_P1-C-1_01_4077.mzXML	cenocepacia	C5424	No	EA	24
3	BcC5424_LB_EA_24_2_P1-C-1_01_4078.mzXML	cenocepacia	C5424	No	EA	24

Fig. 2 Template format for metadata table file. This file contains sample metadata which is used by FBMN to link metabolomic differences between samples to experimental conditions.

to GNPS by clicking "Select Input files," choosing the "Upload Files" tab, selecting the appropriate upload folder on the left panel, and dragging files to the "Drop Files Here To Upload" area. Once uploaded, navigate back to the "Select Input Files" tab and locate uploaded files. Select .mgf file and click "MS2 File MGF/MSP (Progenesis QI)/ mzXML(MzTAab-M)" button. Repeat for "Feature Quantification Table" with the .csv file and "Sample Metadata Table" with the .txt file. When finished, close window by selecting "Finish Selection."

5. Change FBMN workflow parameters as needed. The settings used to analyze the sample data are summarized in Table 2. Readers who are interested in optimizing these parameters further are encouraged to refer to the FBMN documentation (https://ccms-ucsd.github.io/GNPSDocumentation/featurebasedmolecularnetworking), and the original FBMN manuscript (Nothias et al., 2020). After setting parameters as desired, confirm your email address under "Workflow Submission" and click "Submit" to begin running job.
6. Access the output of FBMN workflow under the "Jobs" tab on GNPS webpage.

3.5 Visualization of molecular networks in cytoscape

1. Export molecular networks (both classical and feature-based) for visualization in Cytoscape *via* the GNPS Cytoscape Downloader. This tool is accessible on the results screen of a molecular networking job under the "Advanced Views—External Visualization" header by clicking on "Direct Cytoscape Preview/Download". Directly download the Cytoscape file (.cys) by clicking on "Download Cytoscape File." Prior to downloading, networks can also be rendered with alternative visualization styles.
2. Open the Cytoscape file in Cytoscape by navigating to "File" → "Open..." and then selecting the appropriate file.
3. In the "Styles" tab to the left, properties of nodes and edges (such as size, color, label, transparency *etc.*) can be mapped to various data attributes. Readers should refer to the following links to GNPS documentation to learn more about visualization of classical (https://ccms-ucsd.github.io/GNPSDocumentation/cytoscape/) and feature-based (https://ccms-ucsd.github.io/GNPSDocumentation/featurebasedmolecularnetworking-cytoscape/) molecular networks with Cytoscape.

Table 2 FBMN workflow parameters used to analyze example dataset.

Parameter	**Setting**
Workflow selection	
File selection	
MS2 MGF file	MZMine_output_FBMN.mgf
Feature quantification table	MZMine_output_FBMN_quant.csv
Sample metadata table	Metadata_Burk_example.txt
Basic options	
Quantification table source	Mzmine
Precursor ion mass tolerance	0.01 Da
Fragment ion mass tolerance	0.05 Da
Advanced network options	
Min pairs cos	0.7
Minimum matched fragment ions	4
Maximum shift between precursors	500 Da
Network TopK	10
Maximum connected component size (beta)	130
Advanced library search options	
Library search min matched peaks	4
Score threshold	0.7
Search analogs	Don't Search
Maximum analog search mass difference	100.0 Da
Top results to report per query	1
Advanced filtering options	
Maximum peak intensity	0.0
Filter precursor window	Don't filter
Filter library	Filter library
Filter peaks in 50 Da window	Don't filter

Table 2 FBMN workflow parameters used to analyze example dataset.—cont'd

Parameter	Setting
Advanced quantification options	
Normalization Per File	No norm
Aggregation Method For peak abundances per group	Mean
Advanced multivariate statistics options	
PCoA distance metric	Cosine
Advanced univariate statistics options	
Metadata column to compare	None
Run stats and plots	No
Metadata field to compare	None
Metadata field to compare	None
Metadata column to facet	None
Advanced external tools	
Run dereplicator	Run

3.6 DEREPLICATOR analysis

1. The DEREPLICATOR *in silico* tool can be run on both classical and feature-based molecular networks to facilitate identification of known PNPs in large-scale tandem mass spectrometry datasets. DEREPLICATOR first creates a database of decoy PNPs by randomly shuffling the amino acid sequence of each PNP present in the publicly available databases. Next, *in silico* MS^2 spectra are predicted for each target and decoy PNP. Experimental MS^2 spectra referred to *Spectrum* and *in silico* spectra referred as *Peptide* are scored using Mass Spectrometry Generating Function. A peptide-spectrum match (PSM) is further supported by the matching of experimental precursor mass and the exact mass of the candidate peptide up to a predefined tolerance threshold. The significance of high-scoring matches are evaluated using the MS-DPR algorithm to calculate a *p*-value, which is further adjusted by the false-discovery rate estimated by querying the experimental spectrum against target and decoy reference spectra. The mass spectral databases for PNPs are limited. Thus, use of tools such as DEREPLICATOR

allows the user to overcome this limitation by using compound databases lacking mass spectra. Furthermore, DEREPLICATOR is run periodically on datasets available publicly in GNPS and so the user can subscribe to new annotations as they become available. Lastly, the user can also search for the presence of PNPs across different datasets, gaining insights into their prevalence in biological systems. For the purpose of this analysis, DEREPLICATOR is run on both networks. For further guidance, readers are encouraged to reference the documentation available on GNPS (https://ccms-ucsd.github.io/GNPSDocumentation/dereplicator/), as well as the manuscript describing this tool (Mohimani et al., 2017).

(a) To run on a classical molecular network, navigate to the results screen of the appropriate GNPS job. Under the "Advanced Views—Experimental Views" heading, select "Annotate with DEREPLICATOR." This selection will bring the investigator to a page where the desired workflow parameters for DEREPLICATOR analysis are selected.

(b) To analyze FBMN with DEREPLICATOR, simply set the "Run Dereplicator" option under "Advanced External Tools" to "Run" when setting up the FBMN parameters. This analysis will automatically be run alongside FBMN job. Access the DEREPLICATOR results under the "Advanced Views—External Tools" heading on the FBMN results page by clicking "View Dereplicator Results." The individual workflow parameters can be adjusted as needed by cloning the DEREPLICATOR job. To do so, select the "Clone to Latest Version" option on the DEREPLICATOR results page. The DEREPLICATOR analysis of both classical and feature-based molecular networks was performed using the same parameters, as described in Table 3.

2. View the list of unique candidate annotations by clicking on "View Unique Peptides" under the "Significant Matches" heading. Select "View All PSM" for a detailed view of each spectrum match to a peptide.
3. Examine the resulting list of candidate annotations. Annotations can be sorted based on each column (typically either "Score" or "*P*-value"), or searched by compound name using the filter function. To use the filter function, type the desired search parameters in the box under the column of interest and hit "Enter" on the keyboard. It should be noted that the "Accurate *P*-values" setting must be checked prior to running DEREPLICATOR analysis for *p*-values to be meaningful.

Table 3 Parameters used for DEREPLICATOR analysis of example dataset.

Parameter	**Setting**
Workflow selection	
Basic options	
Search analogs (VarQuest)	Unchecked
Precursor ion mass tolerance	0.02 Da
Fragment ion mass tolerance	0.02 Da
Advanced options	
PNP database	PNPdatabase
Max charge	2
Accurate P-values	Checked (Must be checked for p-values to be meaningful)
Min number of AA	5
Advanced dereplicator options	
Max isotopic shift	0 (none)
Adducts	Na unchecked, K unchecked
Advanced VarQuest options	
Max allowed modification mass	150 Da
Min matched peaks with known compound	5

4. One predicted annotation appearing in the DEREPLICATOR output for both classical and feature-based molecular networks is "Ornibactin_C4_N5-Deacyl,_N5-(3?-hydroxyoctanoyl)." Although the compound names provided by DEREPLICATOR can be ambiguous, the SMILES identifier associated with this peptide can be found by clicking the "Show" button in the "SMILES" column. A PubChem search of this SMILES ID reveals that the predicted compound is ornibactin C8. Herein, we will use ornibactin as an illustrative example to show how to propagate annotations using a molecular network and link observed structural modifications to the gene cluster analysis.
5. To view the experimental MS^2 spectrum and the peaks explained by *in silico* fragmentation of the predicted molecule, choose the "Show Annotation" button on the left side of the table for the annotation of

interest. The blue peaks represent fragment ions which match with predicted *in silico* fragments for the peptide in question (Fig. 3). Click on each of these peaks to highlight the predicted corresponding fragment on the molecular structure to the right, along with the properties in the list of annotated peaks (such as experimental m/z, mass error, charge, and intensity). In this case, several peaks observed in the experimental MS^2 spectra are explained by the *in silico* spectra generated for ornibactin C8. While this analysis supports the proposed annotation, further verification steps are still needed to confirm the accuracy of this prediction.

6. To meet the standards outlined by the Chemical Working Group Metabolomics Standards Initiative, at least two orthogonal analytical techniques (such as liquid chromatography and mass spectrometry) should be used to compare compounds against an analytical standard for level 1 validation of a metabolite annotation (Sumner et al., 2007). However, putative annotations (level 2) can also be achieved by comparing experimental MS^2 spectra with published MS^2 spectra (Sumner et al., 2007). In this case, a published MS^2 spectra is available for ornibactin C8 in the GNPS public MS/MS libraries. The spectra for specific compounds are accessible in various databases such as NIST, MarinLit, Metlin, MoNA, and HMDB, as well as in published literature.
   - **(a)** To search GNPS public spectral libraries, navigate to the GNPS homepage, scroll down to the "GNPS Spectrum Library Knowledgebase" header, and click the "Browse Libraries" link under "MS/MS Libraries."
   - **(b)** To view spectra for all GNPS libraries, choose the "View" button next to "All GNPS Library Spectra."
   - **(c)** To search for a specific metabolite, use the filter function to search the "Name" column. Search for "ornibactin" to return three library spectra for ornibactin C4, C6, and C8.
   - **(d)** Open the library spectra for ornibactin C8 by clicking the Spectrum ID link in the "View" column (CCMSLIB00001059073). This returns the library spectra for the molecule along with chemical properties and experimental conditions used for data collection (such as ion source, ion mode, and instrument).
   - **(e)** To compare this published MS^2 spectra against the experimental MS^2 spectra, click on the "Metabolomics Spectrum Resolver" link. The GNPS Metabolomics Spectrum Identifier (MSI) Resolver will open and automatically load the library Universal Spectrum Identifier (USI) for visualization.

Fig. 3 Example of DEREPLICATOR output. Some columns are omitted from the view for clarity. The peaks in blue correspond to potential substructure annotations. Clicking these peaks labels the predicted substructure in blue on the interactive substructure.

(f) The USI associated with the experimental MS^2 spectra is found in the molecular network for a node of interest by navigating to the results screen for the appropriate GNPS job in a separate browser window. Select "View All Clusters With IDs" under the "Default Molecular Networking Results Views" heading. The number in the "Scan" column in the DEREPLICATOR output (5577 in the classical network, 1504 in the FBMN analysis) represents the "ClusterIdx" in the molecular network. Search this node number in the "ClusterIdx" column by using the filter function.

(g) Next, click on the spectra icon in the leftmost column, to the left of "View Analogs." Upon clicking, an interactive spectrum for the node of interest will open. The USI is listed above this spectrum. Copy this USI to the clipboard by selecting the "(Copy USI String)" option.

(h) Navigate back to the MSI Resolver, and paste the USI for the second experimental MS^2 spectra in the second "Spectrum USI" box under "USI Data Selection" to generate a mirror plot for comparisons of the two spectra, as exemplified in Fig. 4. It should be noted that in our case molecular networking analysis did not result in automated annotation of nodes corresponding to ornibactin, even though experimental spectra are present in the GNPS library. The spectra in the GNPS library was acquired using linear FT, whereas we used a qTOF to acquire the experimental data. Thus, the spectral similarity did not meet the cosine score threshold of 0.7 set for library search matches, further highlighting the utility of *in silico* tools such as DEREPLICATOR for dereplication of PNPs.

3.7 Propagation of annotations in molecular network

1. After confirming the annotation as ornibactin, use the molecular network to propagate annotations for connected nodes in the ornibactin cluster. To do so, navigate to the networking results page, "View All Clusters With IDs," and filter the ornibactin node by searching the network by "ClusterIdx," as described in step 6f in the previous subsection.
2. Under the "Spec Family" column, select "View Network" to view the cluster associated with the node of interest. This node will be highlighted at first. To deselect, click elsewhere in the white space in the network viewer. A screenshot of the ornibactin cluster in the classical molecular network is shown in Fig. 5.

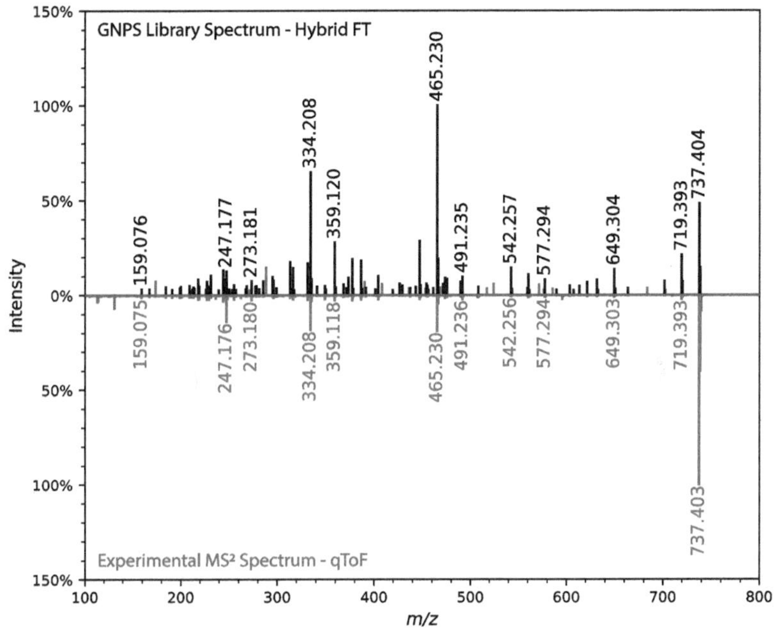

Fig. 4 Mirror plots between GNPS library (top) and experimental (bottom) MS^2 spectra of ornibactin C8, supporting the level 2 identification is shown. The Spectrum ID of this library spectrum is CCMSLIB00001059073.

3. By default, nodes will be labeled by cluster index. To label nodes by *m/z*, choose the "parent mass" option under "Node Labels." Nodes exhibiting database matches to the GNPS spectral libraries can also be labeled by "LibraryID."
4. The options under "Edge Labels" allow for labeling of the edges that connect nodes. Select "DeltaMZ" to label edges by the mass differences between the connected nodes. This feature allows users to relate common exact mass differences to the gain or loss of particular functional groups. Pick the "EdgeAnnotate" option to label edges with the most likely molecular formula corresponding to the observed mass differences. While helpful, these annotations are not always accurate and therefore the exact theoretical mass of the predicted formula should also be compared against the observed mass difference. A molecular formula calculator such as the one available on ChemCalc.org allows rapid annotations of DeltaMZ (Patiny & Borel, 2013).
5. The "Draw Pies" function under the "Node Coloring" panel allows users to visualize relative abundances of each metabolite feature across

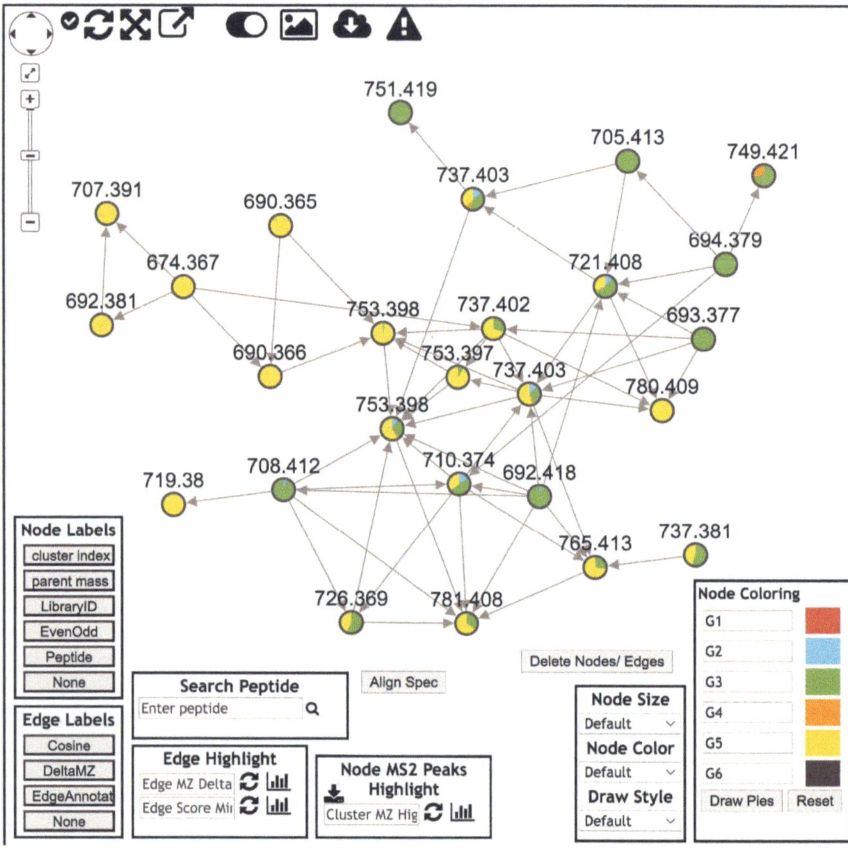

Fig. 5 Ornibactin cluster in classical molecular network visualized in the GNPS browser.

experimental groups (which are either defined during step 4 in Section 3.1, or indicated in the metadata file). For classical networks greater than six groups or for FBMNs, the following steps are required.

(a) To view pies for a FBMN in the GNPS browser, replace the group names in the "Node Coloring" panel with "GNPSGROUP:X," where X is the metadata descriptor to be plotted. For example, to plot the relative abundance of all *B. cenocepacia* samples in blue, clear "G2" out of the first textbox in the "Node Coloring" panel and replace with "GNPSGROUP:cenocepacia." Repeat for all groups with the desired metadata descriptors (for the purposes of this analysis, assign G1 to media blanks, G2 to *B. cenocepacia* culture extracts, G3 to *B. multivorans* culture extracts, G4 to *B. ambifaria* culture extracts, and G5 to *B. dolosa* culture extracts).

(b) Such visualization can also be carried out in Cytoscape. Select the "Style" tab in the toolbar to the left of the screen. Make sure to view node properties by clicking the "Node" tab at the bottom of the properties window. Next, under "Properties," click on the left most of the three boxes to the left of "Image/Chart 1". Doing so will open the "Graphics" window. Select the "Charts" tab, and then the pie chart icon. In the "Data" tab on the left side of the window, select the desired metadata columns under "Available Columns," choose the columns of interest, and click the ">" button to add to "Selected Columns." Open the "Options" tab on the left side of the "Graphics" window to view and change the colors representing each group.

6. To begin propagating annotations to surrounding nodes, start by locating a known node in the network (in this case, ornibactin C8). Click on this node to highlight it and all other nodes connected by an edge. Label edges by either "DeltaMZ" or "EdgeAnnotations."

7. Start by annotating nodes connected to the known node by a recognizable mass difference. In both the classical and feature-based molecular networks, ornibactin is connected to a node with *m/z* 753.398. Therefore, this molecule represents 737.404 (mass of ornibactin C8) + 15.995, indicating a gain of a hydroxy group.

(a) Note that the three nodes (two with *m/z* 753.397 and one with *m/z* 753.398) appear in the classical molecular network, while only one node with *m/z* 753.3987 is observed in FBMN. Further manual inspection of data reveals that the three nodes appearing in the classical network share similar retention times and identical MS^2 spectra, therefore representing duplicate nodes of the same feature. This example highlights one of the key advantages of FBMN over classical networks; that classical networking does not incorporate retention time which can lead to false duplicate nodes. The MS^2 spectra acquired at varied abundances of the same metabolite can be represented as multiple nodes in the classical network if their spectral similarity threshold is below the threshold used for generation of network. On the other hand, the structural isomers with different retention times are clumped into a single node in classical molecular network whereas they are observed as separate nodes in FBMN.

8. The next step in propagating annotations is to locate the position where gain of hydroxy group occurs in ornibactin C8. By clicking the edge linking the two nodes, an interactive MS^2 spectrum will appear for each compound. Hover the cursor over specific peaks to show the corresponding m/z and intensity. Using this method, the user can identify peaks that have shifted by the same m/z as the precursor molecule.

   (a) It should be noted that while analysis in the GNPS browser is certainly convenient, readers are urged to always inspect raw data, as the consensus spectra representing nodes may overlook low intensity fragment peaks which contain valuable structural information. Additionally, m/z values are usually more precise in the raw data.

9. As a PNP, ornibactin C8 exhibits fragmentation along peptide bonds, resulting in characteristic b and y ions (Hunt, Yates, Shabanowitz, Winston, & Hauer, 1986; Roepstorff & Fohlman, 1984). These fragmentation products are detected in the mass spectrometer as either protonated amine (y ions) or oxazolone (b ions) cations.

   (a) Draw the structure of ornibactin C8 and cleave each peptide bond to predict b and y fragment ions. The user can use various softwares for this purpose, such as ChemDraw or Marvin Suite, as well as with the online PubChem Sketcher (Ihlenfeldt, Bolton, & Bryant, 2009). The b and y ions can also be generated using vendor software or ProSight available at https://prosightptm.northwestern.edu/ionpredictor/. All structures must be drawn in their protonated form, and with exact monoisotopic masses used for annotation of fragment ions. The interactive ornibactin MS^2 spectrum in DEREPLICATOR is also helpful for this process. Theoretical masses of expected fragments for ornibactin C8 are provided in Table 4.

   (b) Locate these potential fragments in the experimental MS^2 spectrum of ornibactin C8 to determine whether are detected. In this dataset, all five y ions and all b ions except b_1 are detected. The b_1 ion is not detected at any appreciable intensity, as expected according to the currently accepted model of b-ion formation *via* an oxazolone intermediate (Schlosser & Lehmann, 2000).

   (c) Identify peaks in the MS^2 spectrum of m/z 753.399 displaying a mass shift from the annotated fragments in the ornibactin C8 spectrum as shown in Fig. 6. This procedure is facilitated using the "Align Spec" option located beneath the network in the GNPS browser. Peaks shared between spectra will be colored in maroon while those shifted by the same m/z difference separating the precursor masses will be colored in blue. As seen in Fig. 6, all y-ions

Table 4 Theoretical monoisotopic masses of predicted b and y ions formed during fragmentation of ornibactin C8.

	b Ions	y Ions
1	143.107	89.107
2	273.181	247.177
3	404.203	334.209
4	491.235	465.230
5	649.304	595.305

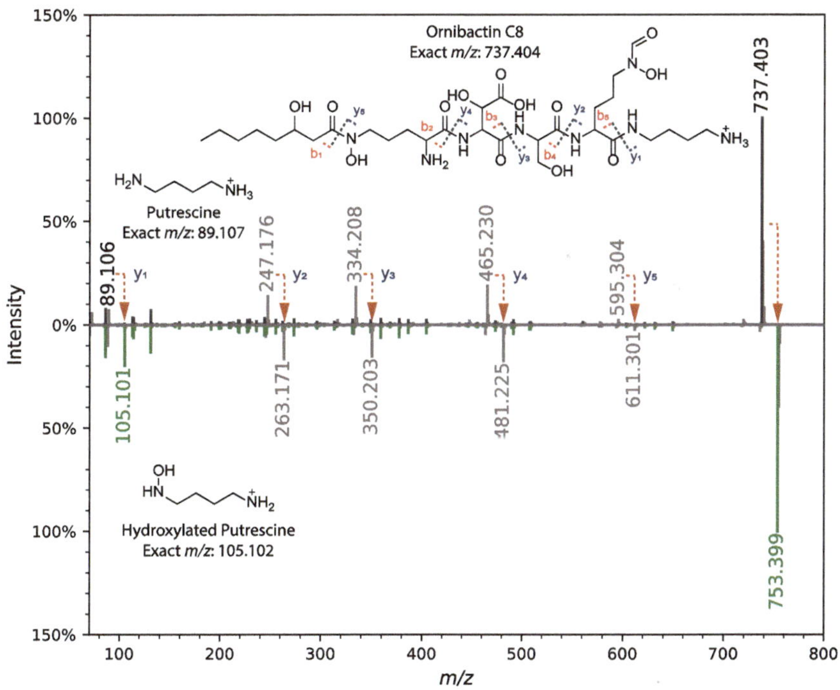

Fig. 6 Mirror spectra comparing MS^2 spectra of ornibactin C8 and *m/z* 753.399, highlighting mass shifts on y-ions.

detected in ornibactin C8 display a mass gain of 15.995. Therefore, we can conclude that the hydroxylation occurs on the y_1 fragment corresponding to the putrescine residue. Further experiments are required to determine where the putrescine residue is hydroxylated, although knowledge of the mechanism for biosynthesis of biological molecules can guide such annotations.

10. As another annotation example, a node with *m/z* 710.376 is connected to the node for ornibactin C8 in both the classical and FBMN. These nodes are separated by a mass difference of 27.029. Labeling edges with the "EdgeAnnotate" option predicts C_2H_3, which has an exact mass of 27.023. This mass difference is less than the observed mass difference of 27.029. Since this is not an immediately recognizable mass difference for commonly observed functional groups, further analysis is needed to annotate this molecule.
11. Using the approach outlined in steps 9–10, it is apparent that the structural differences from ornibactin C8 occur entirely on the N^5-formyl-N^5-hydroxyornithine residue (Fig. 7). The y_1 ion at *m/z* 89.107 is found in the spectra of ornibactin C8 and node with *m/z* 710.376, but the y_2 ion at *m/z* 247.176 has shifted to 220.148 (mass difference of 27.028). A similar mass shift is also observed for y_3–y_5 fragment ions. Furthermore, the b_4 ion at *m/z* 491.235 is detected in both spectra.

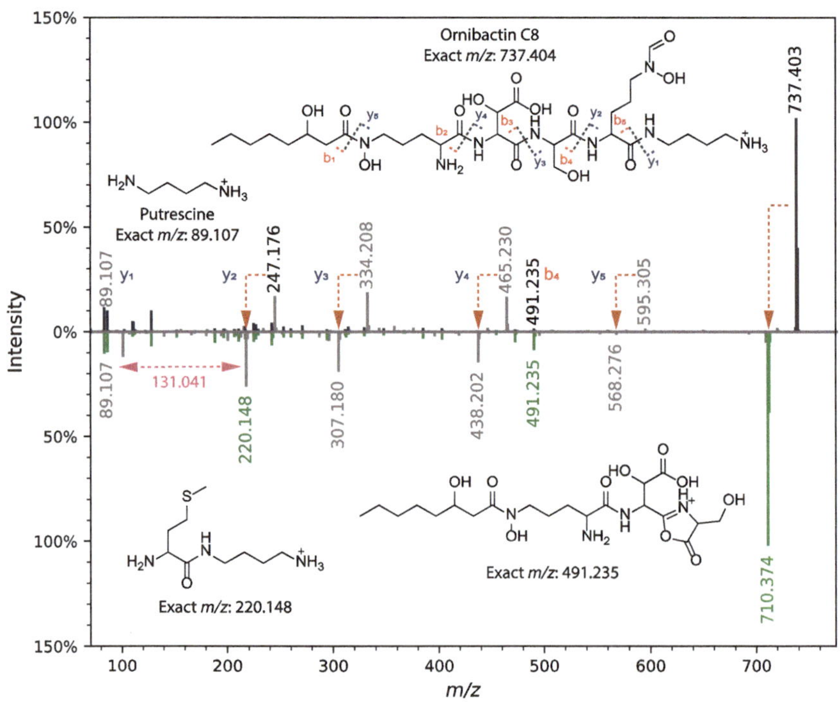

Fig. 7 Mirror spectra comparing MS^2 spectra of ornibactin C8 and *m/z* 710.374, highlighting mass shifts on y-ions.

12. While the observed mass difference between precursor molecules defies annotation, it is possible to determine the structural difference using another approach after identifying where the modification occurs. The mass difference between the y_2 and y_1 ion of *m/z* 710.374 is 131.041 (220.148–89.107), which corresponds to a methionine residue. A table containing the monoisotopic masses of canonical amino acid residues is available for reference at https://ionsource.com/Card/aatable/aatable.htm. Therefore, *m/z* 710.374 represents an analog of ornibactin C8 where the N^5-formyl-N^5-hydroxyornithine residue is replaced by methionine. Applying this same logic to the node with *m/z* 692.418 leads to identification of a similar substitution of leucine/isoleucine for the N^5-formyl-N^5-hydroxyornithine residue. Since both leucine and isoleucine have the same exact mass, it is not possible to distinguish these possibilities using the available data.

13. Extend this approach to annotate the remaining nodes within the ornibactin cluster. Annotated networks are shown in Fig. 8. Spectral evidence for these annotations has been published in McAvoy et al. (2020). Note that while the ornibactin cluster appears similar in the classical and feature-based molecular networks, there are a few subtle differences, which warrant use of both strategies for comprehensive analysis of potential ornibactin analogs.

   (a) As mentioned earlier, several duplicate nodes are found in the classical network with identical masses, MS^2 spectra, and retention times.

   (b) Several nodes in the ornibactin cluster are exclusively found in either the classical (such as *m/z* 780.409) or FBMN (such as *m/z* 723.3890).

   (c) Finally, the pie charts representing abundance appear differently for select compounds in classical and feature-based networks. While FBMN is accurate for comparing the relative abundance of a compound across samples, these networks are heavily dependent on the parameters used for preprocessing. This can be exemplified by the nodes with *m/z* 719.3819 and 765.4137, which are reported to be detected in media blanks in the FBMN but not the classical network. Further investigation reveals that detection in the blanks was artificially introduced during the gap-filling steps during MZmine 2 preprocessing. Readers are cautioned to consider this when selecting parameters, and to carefully examine data after each preprocessing step.

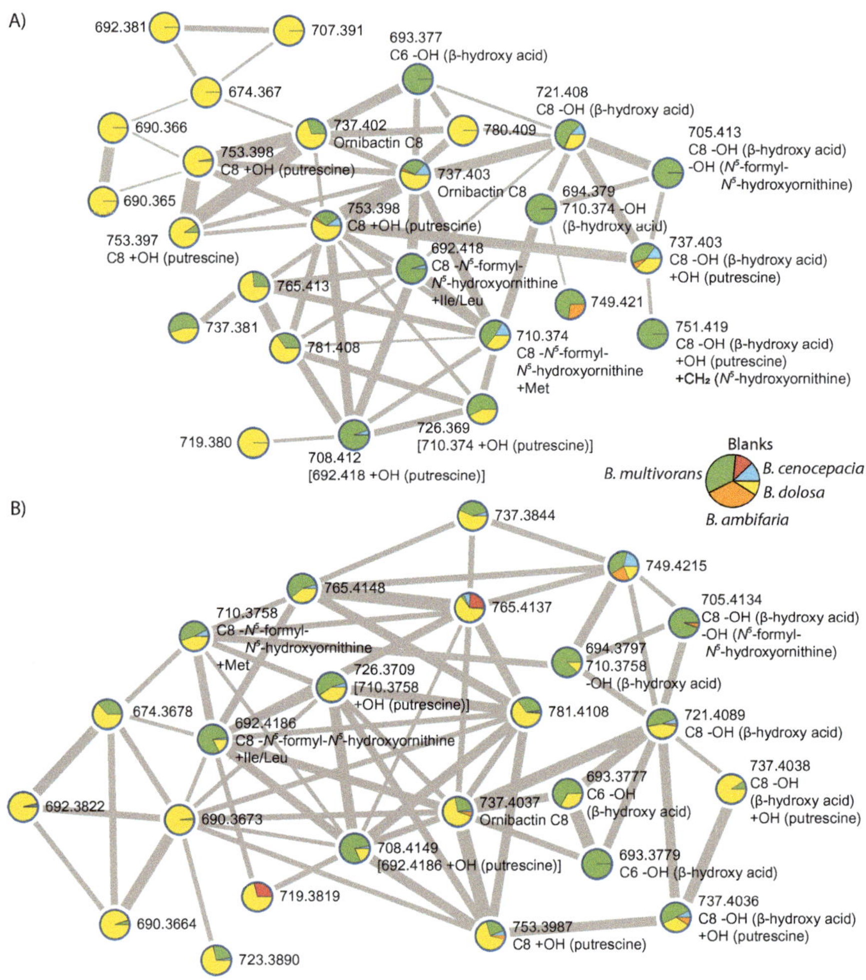

Fig. 8 Annotated (A) classical and (B) feature-based molecular networks.

3.8 Gene cluster analysis

Using the molecular networking-based strategies above, several structural analogs of ornibactin C8 are described in *Burkholderia* spp., with some of these analogs exhibiting species and strain specific production (McAvoy et al., 2020). The Minimum Information about a Biosynthetic Gene Cluster (MIBiG) Data Standard and Repository is a useful tool to further explore the biosynthetic basis of secondary metabolite production (Kautsar et al., 2020). To probe the promiscuity of specific biosynthetic enzymes, gene cluster analysis can be coupled to network analysis to identify candidates for pathway engineering or for use in semi-synthesis of candidate natural products.

1. Navigate to the MIBiG homepage (https://mibig.secondarymetabolites.org/), and click on "Search" on the right side of the top toolbar.
2. Enter "ornibactin" as the search term, and click "Search" to find biosynthetic gene clusters (BGCs) associated with ornibactin production.
3. The search results will appear, along with a basic description of each gene cluster hit. Click on the row for a particular cluster to open the BGC overview page. A screenshot of this overview page is pictured in Fig. 9. All genes within the cluster are displayed in the locus overview section on the top left.
4. Select an individual gene in the locus overview section to display the relevant gene information in the "Gene details" section on the top right. This includes the gene name, genomic location, functional annotation, as well as links to the amino acid and nucleotide sequences.
5. To access additional information about the biosynthetic cluster including literature references supporting genomic annotations, browse through the tabs in the lower section.
   (a) Click on the "NRPS/PKS domains" tab to see detailed domain annotations for the non-ribosomal peptide synthetases OrbI and OrbJ. Clicking on a particular domain displays the function, genomic location, and links to the corresponding amino acid and nucleotide sequences for that domain.

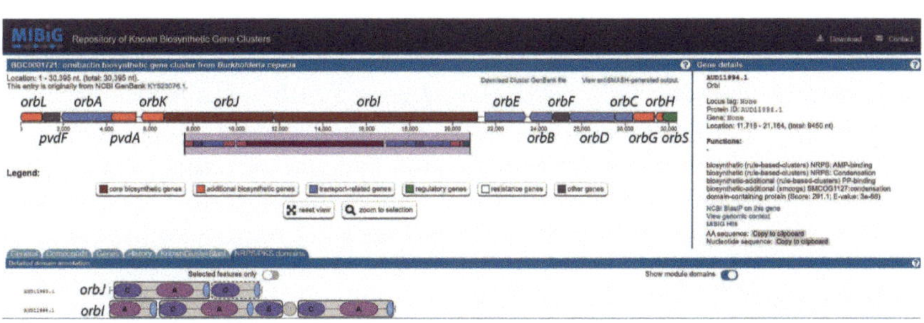

Fig. 9 MIBiG overview page for the ornibactin biosynthetic gene cluster. Gene names are manually labeled in screenshot. In the "NRPS/PKS domains" tab on the bottom section, "A" represents AMP-binding domains, "C" represents condensation domains, and unlabeled ovals represent Peptidyl-carrier protein (PCP) domains. The "Gene details" section is not shown for clarity.

(b) Select "KnownClusterBlast" to view other known biosynthetic gene clusters sharing similar genes. This knowledge provides clues about the potential functions of the biosynthetic products of a gene cluster, as well as insight into their probable biosynthetic mechanism.

6. Click on "View antiSMASH-generated output" to view the results of antiSMASH analysis of this cluster. antiSMASH enables more in-depth analysis of multimodular enzymes like NRPSs, including module detection to predict monomers which are integrated into the product compound (Blin et al., 2021). Unselect the "Show module domains" option to see predicted products of each module. Select "Show module domains" to see the protein domains detected in each module.

4. Summary

Discovery and annotation of PNPs from mass-spectrometry based metabolomics data presents several challenges owing to their structural complexity and diversity. In this chapter, we use the ornibactin molecular family as an example to illustrate a data analysis workflow that exploits molecular networking in combination with the *in silico* DEREPLICATOR tool to aid with the processes of dereplication and variable dereplication of PNPs in large datasets which would overwhelm manual curation. We also describe how gene cluster analysis can complement this metabolomics workflow, providing additional confidence in annotation while also providing biochemical insights into their biosynthesis. Furthermore, we provide discussion on the relative merits of classical and feature-based molecular networking, highlighting strengths and weakness offered by each method. Prudent implementation of these tools for analysis of high-throughput metabolomics data can help researchers bypass bottlenecks in data interpretation during the natural product discovery pipeline and effectively exploit structure-activity relationships to generate analogs with optimal pharmaceutical properties.

References

Atanasov, A. G., Zotchev, S. B., Dirsch, V. M., Orhan, I. E., Banach, M., Rollinger, J. M., et al. (2021). Natural products in drug discovery: Advances and opportunities. *Nature Reviews Drug Discovery*, *20*(3), 200–216. https://doi.org/10.1038/s41573-020-00114-z.

Baumann, T., Nickling, J. H., Bartholomae, M., Buivydas, A., Kuipers, O. P., & Budisa, N. (2017). Prospects of *in vivo* incorporation of non-canonical amino acids for the chemical diversification of antimicrobial peptides. *Frontiers in Microbiology*, *8*, 124. https://doi.org/10.3389/fmicb.2017.00124.

Blin, K., Shaw, S., Kloosterman, A. M., Charlop-Powers, Z., van Wezel, G. P., Medema, M. H., et al. (2021). antiSMASH 6.0: Improving cluster detection and comparison capabilities. *Nucleic Acids Research*, *49*(W1), W29–W35. https://doi.org/10.1093/nar/gkab335.

Dang, T., & Sussmuth, R. D. (2017). Bioactive peptide natural products as lead structures for medicinal use. *Accounts of Chemical Research*, *50*(7), 1566–1576. https://doi.org/10.1021/acs.accounts.7b00159.

Dührkop, K., Fleischauer, M., Ludwig, M., Aksenov, A. A., Melnik, A. V., Meusel, M., et al. (2019). SIRIUS 4: A rapid tool for turning tandem mass spectra into metabolite structure information. *Nature Methods*, *16*(4), 299–302. https://doi.org/10.1038/s41592-019-0344-8.

Dührkop, K., Shen, H., Meusel, M., Rousu, J., & Böcker, S. (2015). Searching molecular structure databases with tandem mass spectra using CSI:FingerID. *Proceedings of the National Academy of Sciences of the United States of America*, *112*(41), 12580–12585. https://doi.org/10.1073/pnas.1509788112.

Gurevich, A., Mikheenko, A., Shlemov, A., Korobeynikov, A., Mohimani, H., & Pevzner, P. A. (2018). Increased diversity of peptidic natural products revealed by modification-tolerant database search of mass spectra. *Nature Microbiology*, *3*(3), 319–327. https://doi.org/10.1038/s41564-017-0094-2.

Hunt, D. F., Yates, J. R., 3rd, Shabanowitz, J., Winston, S., & Hauer, C. R. (1986). Protein sequencing by tandem mass spectrometry. *Proceedings of the National Academy of Sciences of the United States of America*, *83*(17), 6233–6237. https://doi.org/10.1073/pnas.83.17.6233.

Ihlenfeldt, W. D., Bolton, E. E., & Bryant, S. H. (2009). The PubChem chemical structure sketcher. *Journal of Cheminformatics*, *1*(1), 20. https://doi.org/10.1186/1758-2946-1-20.

Kautsar, S. A., Blin, K., Shaw, S., Navarro-Munoz, J. C., Terlouw, B. R., van der Hooft, J. J. J., et al. (2020). MIBiG 2.0: A repository for biosynthetic gene clusters of known function. *Nucleic Acids Research*, *48*(D1), D454–D458. https://doi.org/10.1093/nar/gkz882.

Liu, D., Rubin, G. M., Dhakal, D., Chen, M., & Ding, Y. (2021). Biocatalytic synthesis of peptidic natural products and related analogues. *iScience*, *24*(5), 102512. https://doi.org/10.1016/j.isci.2021.102512.

Marahiel, M. A., Stachelhaus, T., & Mootz, H. D. (1997). Modular peptide synthetases involved in nonribosomal peptide synthesis. *Chemical Reviews*, *97*(7), 2651–2674. https://doi.org/10.1021/cr960029e.

McAvoy, A. C., Jaiyesimi, O., Threatt, P. H., Seladi, T., Goldberg, J. B., da Silva, R. R., et al. (2020). Differences in cystic fibrosis-associated *Burkholderia* spp. bacteria metabolomes after exposure to the antibiotic trimethoprim. *ACS Infectious Diseases*, *6*(5), 1154–1168. https://doi.org/10.1021/acsinfecdis.9b00513.

Mohimani, H., Gurevich, A., Mikheenko, A., Garg, N., Nothias, L. F., Ninomiya, A., et al. (2017). Dereplication of peptidic natural products through database search of mass spectra. *Nature Chemical Biology*, *13*(1), 30–37. https://doi.org/10.1038/nchembio.2219.

Mohimani, H., Gurevich, A., Shlemov, A., Mikheenko, A., Korobeynikov, A., Cao, L., et al. (2018). Dereplication of microbial metabolites through database search of mass spectra. *Nature Communications*, *9*(1), 4035. https://doi.org/10.1038/s41467-018-06082-8.

Nothias, L. F., Petras, D., Schmid, R., Dührkop, K., Rainer, J., Sarvepalli, A., et al. (2020). Feature-based molecular networking in the GNPS analysis environment. *Nature Methods*, *17*(9), 905–908. https://doi.org/10.1038/s41592-020-0933-6.

Patiny, L., & Borel, A. (2013). ChemCalc: A building block for tomorrow's chemical infrastructure. *Journal of Chemical Information and Modeling*, *53*(5), 1223–1228. https://doi.org/10.1021/ci300563h.

Pluskal, T., Castillo, S., Villar-Briones, A., & Oresic, M. (2010). MZmine 2: Modular framework for processing, visualizing, and analyzing mass spectrometry-based molecular profile data. *BMC Bioinformatics*, *11*, 395. https://doi.org/10.1186/1471-2105-11-395.

Roepstorff, P., & Fohlman, J. (1984). Proposal for a common nomenclature for sequence ions in mass spectra of peptides. *Biomedical Mass Spectrometry*, *11*(11), 601. https://doi.org/10.1002/bms.1200111109.

Rost, H. L., Sachsenberg, T., Aiche, S., Bielow, C., Weisser, H., Aicheler, F., et al. (2016). OpenMS: A flexible open-source software platform for mass spectrometry data analysis. *Nature Methods*, *13*(9), 741–748. https://doi.org/10.1038/nmeth.3959.

Schlosser, A., & Lehmann, W. D. (2000). Five-membered ring formation in unimolecular reactions of peptides: A key structural element controlling low-energy collision-induced dissociation of peptides. *Journal of Mass Spectrometry*, *35*(12), 1382–1390. https://doi.org/10.1002/1096-9888(200012)35:12<1382::AID-JMS84>3.0.CO;2-6.

Smith, C. A., Want, E. J., O'Maille, G., Abagyan, R., & Siuzdak, G. (2006). XCMS: Processing mass spectrometry data for metabolite profiling using nonlinear peak alignment, matching, and identification. *Analytical Chemistry*, *78*(3), 779–787. https://doi.org/10.1021/ac051437y.

Sumner, L. W., Amberg, A., Barrett, D., Beale, M. H., Beger, R., Daykin, C. A., et al. (2007). Proposed minimum reporting standards for chemical analysis Chemical Analysis Working Group (CAWG) Metabolomics Standards Initiative (MSI). *Metabolomics*, *3*(3), 211–221. https://doi.org/10.1007/s11306-007-0082-2.

Tsugawa, H., Cajka, T., Kind, T., Ma, Y., Higgins, B., Ikeda, K., et al. (2015). MS-DIAL: Data-independent MS/MS deconvolution for comprehensive metabolome analysis. *Nature Methods*, *12*(6), 523–526. https://doi.org/10.1038/nmeth.3393.

van der Hooft, J. J., Wandy, J., Barrett, M. P., Burgess, K. E., & Rogers, S. (2016). Topic modeling for untargeted substructure exploration in metabolomics. *Proceedings of the National Academy of Sciences of the United States of America*, *113*(48), 13738–13743. https://doi.org/10.1073/pnas.1608041113.

Wang, M., Carver, J. J., Phelan, V. V., Sanchez, L. M., Garg, N., Peng, Y., et al. (2016). Sharing and community curation of mass spectrometry data with Global Natural Products Social Molecular Networking. *Nature Biotechnology*, *34*(8), 828–837. https://doi.org/10.1038/nbt.3597.

Yang, J. Y., Sanchez, L. M., Rath, C. M., Liu, X., Boudreau, P. D., Bruns, N., et al. (2013). Molecular networking as a dereplication strategy. *Journal of Natural Products*, *76*(9), 1686–1699. https://doi.org/10.1021/np400413s.

CHAPTER THIRTEEN

Methods for the design and characterization of peptide antibiotics

Angela Cesaro[a,b,c], Marcelo Der Torossian Torres[a,b,c], and Cesar de la Fuente-Nunez[a,b,c,*]

[a]Machine Biology Group, Departments of Psychiatry and Microbiology, Institute for Biomedical Informatics, Institute for Translational Medicine and Therapeutics, Perelman School of Medicine, University of Pennsylvania, Philadelphia, PA, United States
[b]Departments of Bioengineering and Chemical and Biomolecular Engineering, School of Engineering and Applied Science, University of Pennsylvania, Philadelphia, PA, United States
[c]Penn Institute for Computational Science, University of Pennsylvania, Philadelphia, PA, United States
[*]Corresponding author: e-mail address: cfuente@upenn.edu

Contents

Abstract

Multi-drug resistant infections cause the death of millions of people worldwide. Today, there is an urgent need to identify innovative and sustainable alternatives to conventional antibiotics and to develop outside the box strategies to counter drug resistance. Versatile molecules such as antimicrobial peptides (AMPs), which display multiple mechanisms of action, have been explored as templates constituting a new generation of antibiotics. Here, we review recent methodological advances for the design, structural and functional characterization of AMPs. The methodologies outlined here have been validated and well established and may be used as a guide for the discovery, design, development, and reprogramming of peptide antibiotics.

Methods in Enzymology, Volume 663
ISSN 0076-6879
https://doi.org/10.1016/bs.mie.2021.11.003

1. Introduction

Antibiotics are the most powerful medicines that we have to prevent and treat acute and chronic bacterial infections. Despite their relevance, the misuse of these drugs has accelerated the development of multi-drug resistant pathogens globally (Magana et al., 2020). Drug resistance may emerge upon exposure to antibiotics through genetically encoded enzymes that inactivate antibiotics, lack of specific targets, and reduced membrane permeability, among others (Magana et al., 2020). Antibiotic resistance may arise through horizontal gene transfer and/or mutations (Magana et al., 2020). Given the current limitations in terms of drug development, validation, and commercialization, the discovery of new antimicrobials that are less prone to select for bacterial resistance, in addition to effective methods for their design, are needed (de la Fuente-Nunez et al., 2017).

Antimicrobial peptides (AMPs) have attracted significant attention because of their multiple mechanisms of action that may occur concomitantly. Consequently, the evolution of resistance toward these compounds is less probable when compared with traditional antibiotics (Spohn et al., 2019). AMPs are small (4–50 amino acid residues), cationic, and amphipathic (generally 40%–60% of hydrophobic amino acid residues) molecules that play a role in the host defense system of essentially all organisms, having multiple mechanisms of action (Torres et al., 2021). Despite their diversity in terms of amino acid composition and cellular targets, AMPs typically present affinity for hydrophobic and/or negatively charged prokaryotic membranes with strong electrical potential gradients. Their multiple membrane-targeting mechanisms of action give AMPs an advantage over conventional antimicrobials that typically have a single target. Even minor changes in the amphipathic and cationic sequences of AMPs can modulate their selectivity toward bacterial membranes than eukaryotic membrane, further underscoring their potential as therapeutics for human health (Swetha, Priya, & Pandian, 2021). Interactions (e.g., electrostatic and hydrophobic) of peptides with prokaryotic cell membranes enable their entry into bacterial cells by (1) permeabilizing the phospholipid bilayer; and/or (2) disrupting the bacterial cell membrane by interfering with hydrophobic interactions between the phospholipids present in the bilayer (Glukhov et al., 2005; Magana et al., 2020; Umeyama et al., 2006).

Because of their vast sequence space, physicochemical properties, and antimicrobial potency, AMPs represent versatile scaffolds for the design of antibiotics. The first step for designing AMPs involves understanding the role of each amino acid residue contained within a given sequence in the overall secondary structure and biological function (i.e., antimicrobial activity). Computational and heuristic methods have been recently developed to predict the structure-activity relationship of these molecules (Torres et al., 2021). Generally, such approaches involve the use of descriptors, which are vectors that describe physicochemical and structural properties intrinsically correlated with the biological activity of a given molecule (Beltran, Aguilera-Mendoza, & Brizuela, 2018). Numerous descriptors may influence the structure and function of peptides, including net charge, hydrophobicity, length, topological properties, and energetic parameters. Selecting the most relevant descriptors to systematically explore a given peptide is crucial to identify active antimicrobial agents. Here, we provide an overview of the latest methods for the design and experimental characterization of AMPs.

2. Methods for the design of AMPs

Computational and heuristic approaches have been developed for AMP discovery and design (Torres & de la Fuente-Nunez, 2019a; Torres, Melo, Crescenzi, Notomista, & de la Fuente-Nunez, 2021). The performance of such methods depends on how the different physicochemical and structural descriptors are explored and understood (Torres & de la Fuente-Nunez, 2019b). The main difference between computational and heuristic methodologies are that the former is based on complex molecular descriptors, whereas the latter depends on well-known and empirically obtained descriptors. The most common approach for the heuristic design of AMPs are structure-activity relationship (SAR) systematic studies that enable the design of specific peptides based on standard physicochemical properties and simple structural observations (Torres et al., 2021). Next, we will briefly outline the most commonly used computational methods and provide details on the heuristic design and experimental validation of AMPs.

2.1 Computational methods

The recent increase in computational power and data processing capabilities have revolutionized drug discovery by exponentially accelerating the identification of new bioactive molecules (Lin, Li, & Lin, 2020). Underlying these advances is an increasingly deep understanding of the role of molecular descriptors on biological function, opening new avenues for rational design and activity prediction.

Traditionally, quantitative structure-activity relationship (QSAR) methods were developed to predict the activity of new peptides by quantitatively correlating their biological properties to their molecular descriptors, e.g., the number of rotatable bonds, electronegativities, polarizabilities, and topological features (Beltran et al., 2018). More recently, computational methods have enabled the exploration of descriptors that are not empirically obtained and generated through knowledge-based, graph-theoretical, molecular mechanical, or quantum-mechanical tools (Torres & de la Fuente-Nunez, 2019b). Because of their multiple levels of complexity, these molecular descriptors are often not ascribable to a single physicochemical or structural feature of a peptide. The main hurdle for obtaining computational methods with high accuracy is the lack of standardization in experimental procedures needed to validate the biological activities of the training sets used (Mercer et al., 2020; Meurer et al., 2021). For this reason, the most valuable generative and comparative models thus far have been obtained by validating the models and establishing a feedback loop with results obtained from standardized experiments (Jiménez-Luna et al., 2021; Torres & de la Fuente-Nunez, 2019b).

2.1.1 Comparative models

Comparative models seek to find patterns conferring similar biological activity to different peptide sequences (Torres et al., 2021). Such patterns can be based on physicochemical features intrinsic to amino acid residues or structural motifs relevant to biological activity. For instance, sequences with antimicrobial properties that are embedded within proteins (termed encrypted peptides) can be identified by means of pattern recognition algorithms that search for physicochemical features such as amphipathicity and net positive charge. Pane *et al.* reported a computational-experimental framework to discover novel encrypted AMPs (Pane et al., 2017) based on the linear correlation between the antimicrobial potency and the net positive charge, hydrophobicity, and length of peptide sequences by considering the relative contribution of each of these parameters (Pane et al., 2017). The scoring function employed by this method was generated based on known antimicrobial activity profiles of previously reported natural and synthetic peptides. This method successfully identified peptides encrypted within the human pepsinogen A3 that possessed *in vitro* and *in vivo* anti-infective properties (Pane et al., 2018). A similar computational approach was taken to explore the entire human proteome as a source of peptide antibiotics (Torres et al., 2021). The peptides identified were highly active against pathogens, displayed synergistic activity when co-administered at very low concentrations both *in vitro* and *in vivo* and did not readily select for bacterial resistance. The results obtained revealed that the human proteome could be mined to identify previously undisclosed encrypted peptides with antimicrobial properties, which are encoded in proteins from body systems other than the immune system (Torres et al., 2021) (Fig. 1A).

2.1.2 Generative models

Generative models may enable the design of new peptide antibiotics based on their biological functions (Torres & de la Fuente-Nunez, 2019b). For instance, genetic algorithms that "evolve" molecules toward a desired biological function follow the basic principles of Darwinian natural selection (de Magalhães et al., 2014). Despite the redundancy of sequences generated by a genetic algorithm, this model can identify novel artificial peptides with distinct functions and compositions. Porto *et al.*, for example, used a plant-derived template to generate synthetic peptides through a fitness function, driven by the exploration of combinatorial sequence space, which considers mostly hydrophobicity and hydrophobic moment for the generation of new derivatives (Porto et al., 2018) (Fig. 1B). Porto *et al.* demonstrated that guavanin 2, an *in silico* evolved peptide, displays increased antimicrobial and anti-infective activities than its parent peptide Pg-AMP1, a naturally occurring peptide from the guava fruit (Porto et al., 2018). A different methodology was described by Supady *et al.* who developed a framework leveraging a genetic algorithm to search for peptides within a potential energy surface window above the global minimum instead of finding the global minimum potential energy surface (Supady, Blum, & Baldauf, 2015). Furthermore, Yoshida *et al.* presented a closed-loop combining a genetic algorithm and machine learning, which improved the antimicrobial activity of a natural peptide template by 160-fold after only three cycles of predictions (Yoshida et al., 2018) (Fig. 1C). Muller *et al.* described an approach for computer-guided peptide design based on a long short-term memory (LSTM) recurrent neural network. This method captured patterns present in α-helical AMPs and, starting from the learned representation, generated novel active antimicrobial peptides (Müller, Hiss, & Schneider, 2018) (Fig. 1D).

2.2 Non-computational methods

Heuristic approaches are design methodologies able to generate highly active peptide antibiotics by introducing changes within the AMPs sequence

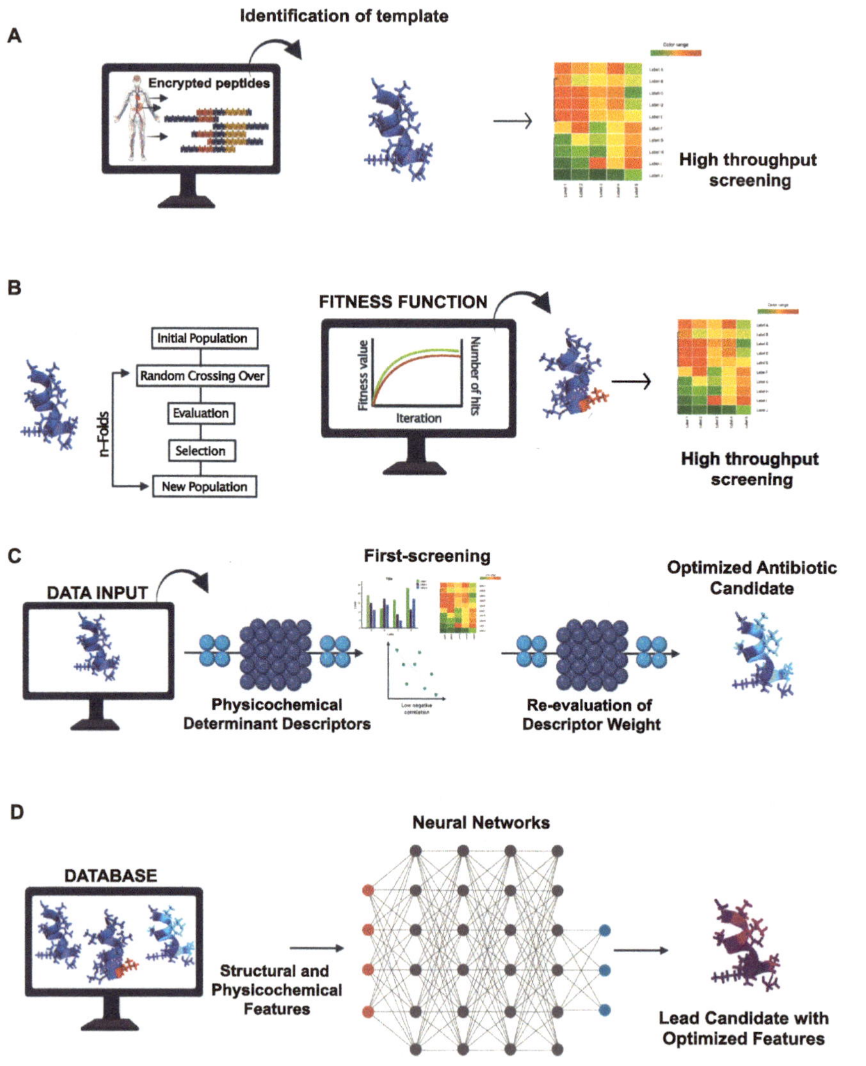

Fig. 1 Computational design methods for peptide antibiotics. (A) Comparative modeling enables the discovery of bioactive templates encrypted within protein precursors by comparing them to known molecules. (B) Generative models such as genetic algorithms may be used for creating novel antimicrobial candidates using a particular fitness function. (C) Machine learning provides tools to improve and predict the biological activity of peptides from physicochemical and biological activity data. (D) Neural networks may be employed to explore the structural and physicochemical descriptors of peptide activity and can be used to predict lead peptide candidates with optimized features.

and empirically observing their effect on physicochemical and structural properties. Such a method has been applied to rationally design peptides based on specific features, such as net positive charge, hydrophobicity, and length (Torres et al., 2018). Three main steps are involved in AMP design:

1. Selection of an appropriate parent peptide scaffold.
2. Understanding the effect on antimicrobial activity of each residue from the parent peptide sequence.
3. Systematic structure-function-guided design studies.

The selection of a scaffold is contingent on the biological function of interest. Significantly active AMPs are commonly selected for this purpose since the properties conferring antimicrobial activity are relatively well understood. These studies typically involve modifying peptide sequences through mutations, deletions, insertions, and chemical modifications (e.g., by capping functional groups and adding functional groups). High-throughput mutational screens have been the most used and successful method for understanding the antimicrobial activity of peptides. Mutations have the advantage of preserving the length of a molecule while identifying the role of individual amino acids on antimicrobial activity. Combinatorial mutational libraries also constitute valuable tools for understanding peptide function (Bionda et al., 2016).

An alternative strategy to combinatorial mutational libraries is represented by systematic single residue substitutions, which entails introducing a desired residue into each position within a given peptide sequence, thus revealing the biological function of each amino acid. For example, Torres *et al.* generated an Ala-scan screening library of derivatives from polybia-CP, a toxic wasp venom peptide. Ala is the most used amino acid residue in systematic screens because it has the simplest side chain (methyl group), preserving the backbone size and allowing the evaluation of the effects of the native side chains groups on both structure and activity. The information obtained from the first cycle of mutagenesis led to the design of a second generation of polybia-CP derivatives with mutations that served to fine tune the antimicrobial and cytotoxic properties of the peptide (Torres et al., 2018). Another strategy reported recently involved substituting specific amino acid residues with positively charged side chain groups known to influence the initial electrostatic interactions of peptides with biological membranes, and consequently, their antimicrobial potency (Arias et al., 2018). Arias *et al.* reported a library of derivatives from Trp-Rich peptides obtained via substitution of Arg with Lys residues. Replacing Arg within the sequence of the lead peptide was shown to increase the antimicrobial activity and potentiated the peptide's ability to permeabilize the cytoplasmic membrane (Arias et al., 2018). Thus, by comparing differences in features such as hydrophobicity, amphipathicity, net charge, and topological surface-related properties, peptide function may be tuned, and sequences reprogrammed to achieve the desired activity (Fig. 2).

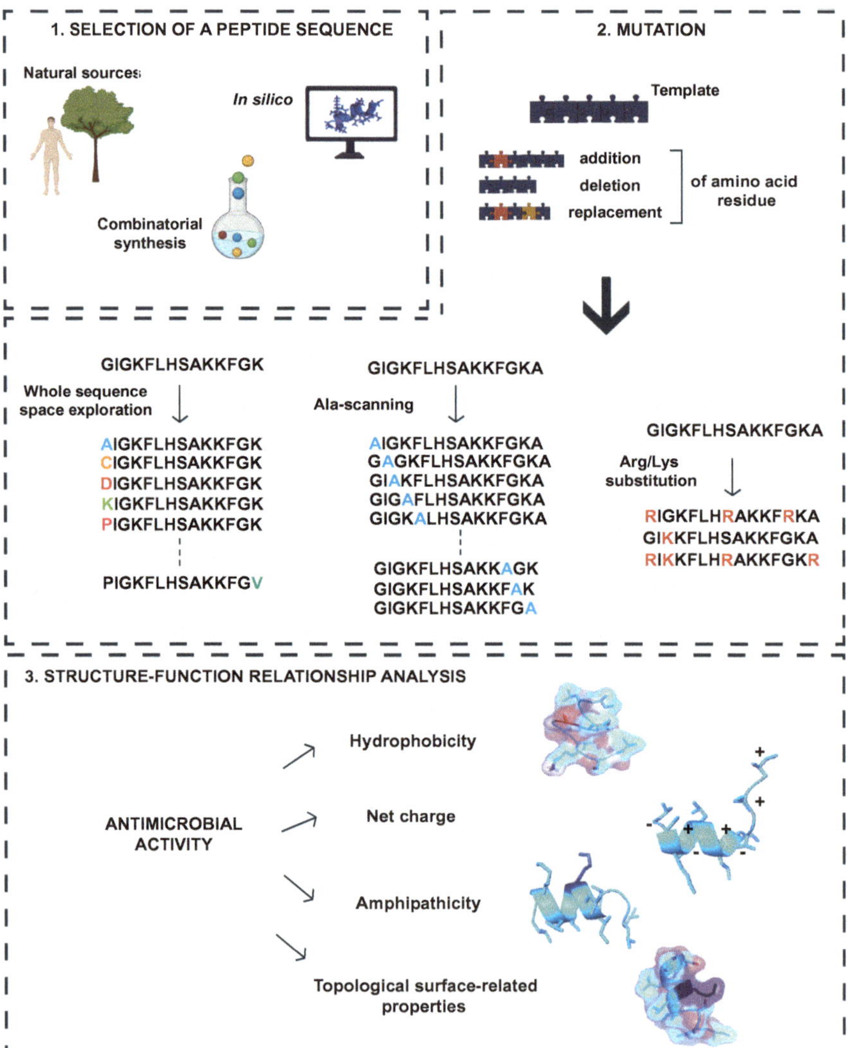

Fig. 2 Non-computational design methods for peptide antibiotics. Non-computational (heuristic) approaches mainly consist of three steps: (1) Selection of the lead peptide candidate; (2) Mutation generation (e.g., combinatorial libraries, Ala-scan, substitution of specific amino acid residues; (3) Structure – function relationship analyses.

3. Platforms for AMP screening

3.1 Physicochemical features

Simple descriptors such as hydrophobicity, net charge, and secondary structure can strongly influence the antimicrobial activity of a peptide and can be measured empirically. To avoid the need for substantial experimentation, physicochemical feature values can be obtained using online servers, which approximate structures and fixed parameters to fit peptides into widely used prediction models. Helical wheel projections are a well-known example of how features can be extracted within seconds and is applicable to most AMPs, which are α-helical (>90%) (Beltran et al., 2018; Torres et al., 2021). Helicity is a consequence of the amphipathic distribution of amino acid residues within the sequence. When amphipathic peptides are in hydrophobic/hydrophilic interfaces, such as the bacterial membrane/extracellular environment, they transition from random coiled to helical structures. This structural transition is caused by the intramolecular formation of hydrogen bonds and lead to optimal interactions with the phospholipids from the membrane leading to the lipid bilayer destabilization (Lifson & Roig, 1961). The software used to generate helical wheel projections predicts physicochemical features based on the peptide's primary sequence and approximates their secondary structure to idealized α-helices (100° turn). These approximations usually correlate with experimental observations in the case of amphipathic peptides with sequences sufficiently long to contain two helical steps and help to understand whether these simple molecular descriptors are the driving forces behind the interactions between AMPs and their biological membrane targets (Beltran et al., 2018) (Fig. 3A).

3.2 *In vitro* antimicrobial activity evaluation

Minimal inhibitory concentration (MIC) assays are used to evaluate the antimicrobial activity of peptides and other molecules (Wiegand, Hilpert, & Hancock, 2008).

First, the growth rate of each bacterial strain of interest is thoroughly evaluated. All experimental conditions must be standardized, including selecting the appropriate culture medium, bacterial growth phase, nutrient concentration, molecule solubility, optimal temperature, aerobic or anaerobic conditions, and pH, among others (Mercer et al., 2020; Meurer et al., 2021) (Fig. 3B).

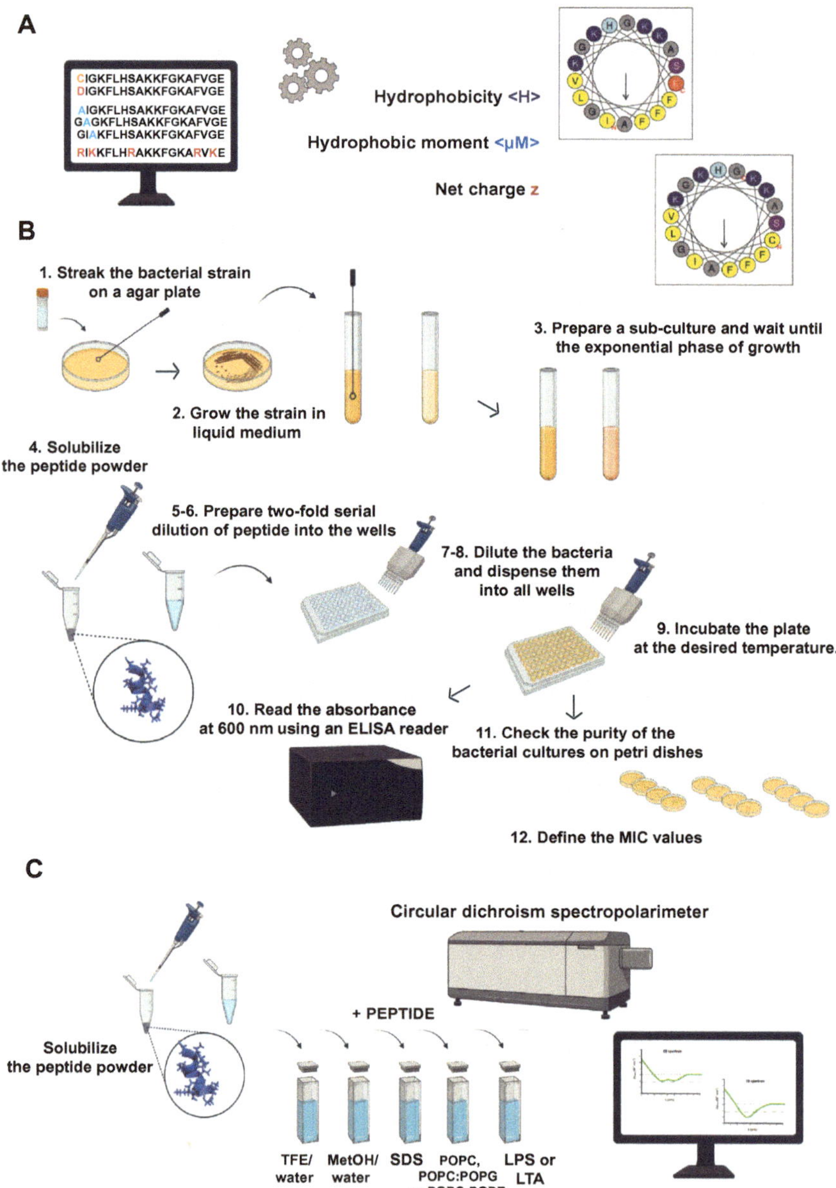

Fig. 3 Platforms for preliminary antimicrobial screening of therapeutic peptide candidates. (A) Empirical value definition of physical-chemical features (e.g., hydrophobicity, charge, and secondary structure) (B) *In vitro* antimicrobial activity evaluation of peptides by establishing their minimal inhibitory concentration (MIC). (C) Circular dichroism spectroscopy analyses used to elucidate the secondary structure of peptides. In these assays, peptides are incubated with different secondary structure inducers (TFE, SDS, POPC, POPC:POPG, LPS and LTA).

Materials and reagents
- Sterile inoculating loops
- Bacterial strains
- Sterile liquid and solid culture media for bacteria
- Sterile Petri dishes
- Sterile 96-wells microtiter plates
- Peptide powder
- Sterile Milli-Q water
- Sterile culture tubes

Equipment
- Rotating shaker incubator for culture tubes
- Petri dish and plate incubator for bacterial cells
- Spectrophotometer plate reader for OD_{600} turbidity measurement

Methods
1. Streak the bacterial strain of interest on a nutrient rich agar plate [e.g., Luria Bertani (LB) agar for *E. coli*] by using a sterile inoculating loop and incubate the plate at the optimal temperature for the microbe under study (e.g., at 37 °C for *E. coli*) to facilitate colony growth.
2. Grow the strains of interest in 4–5 mL using the appropriate medium [e.g., Luria Bertani (LB) broth pH 7 for *E. coli*]. Pick 1–3 colonies from the agar plate to start a 5 mL liquid culture (at 37 °C for 16 h in the case of *E. coli*).
3. Prepare a sub-culture (pre-inoculum) diluting the saturated bacterial sample 100-fold in 4–5 mL of fresh medium. Then, incubate the pre-inoculum at the optimal temperature with shaking (e.g., 37 °C at 200 r.p.m. for *E. coli*) and wait until bacteria reach the exponential phase of growth (e.g., approximately 2–3 h for *E. coli*).
4. Solubilize the peptide powder in sterile water and then dilute the antimicrobial in the test medium to 2 × the highest concentration used for testing.
5. Pipette 200 μL of 2 × concentrated peptide solutions into column 1 of the 96-well plate and 100 μL of test medium (e.g., LB broth pH 7 for *E. coli*) in the other wells.
6. Using a multi-pipettor set at 100 μL, prepare two-fold serial dilutions by mixing the peptide into the wells in column 1 with 100 μL of the medium in the other columns by pipetting up and down a couple of times. Stop the serial dilution at column 10 and discard the extra 100 μL from column 10.

7. Dilute the bacteria in the test medium before mixing with the peptides. The appropriate bacterial dilution for a standard MIC assay must account for a final concentration of 10^5–10^6 CFU mL^{-1}.
8. Dispense bacteria into all wells except for column 12, which will be the sterility control and serve as blank for the plate scanner.
9. Incubate the plate at the desired temperature (e.g., at 37 °C for *E. coli*).
10. When a visible turbidity is obtained, which is indicative of bacteria growth (e.g., approximately 20–24 h of incubation for *E. coli*), read the absorbance at 600 nm using a plate reader.
11. To count the number of bacterial cells, expressed as colony-forming units (CFUs), in all conditions tested, dilute and plate each well on agar plates (e.g., LB agar for *E. coli*).
12. Define the MIC as the lowest drug concentration that reduces bacterial growth by 100%.

Note: It is suggested to include a positive control in all assays, such as a broad-spectrum antibiotic (e.g., gentamicin, ciprofloxacin, chloramphenicol, and gramicidin S), to verify the quality of the method and to compare the antimicrobial potency of the peptide candidate with a conventional antimicrobial drug.

3.3 Circular dichroism spectroscopy

Circular dichroism (CD) spectroscopy is broadly used to analyze the secondary structure of peptides in solution. This technique allows measuring differences in absorption of left-handed polarized light with respect to right-handed polarized light that is indicative of the structural asymmetry of these molecules (Tromelin, Andriot, & Guichard, 2006). In such a way, CD analyses discriminate between disordered (random coil) and ordered (e.g., α-helix or β-sheet) structures of optically active molecules like peptides (Yeaman & Yount, 2003). The structural profile of peptides with antimicrobial features are usually obtained in water, helical inducer media (e.g., trifluoroethanol/water mixtures, lipid lamellar vesicles, and biological component of bacterial membranes) or β-like structures inducer medium (methanol/water mixtures). Most AMPs (~90%) undergo a helix-coil transition when at the interface between the microorganisms' membranes and the extracellular environment, (Gaglione et al., 2017, 2021; Torres et al., 2017, 2021) (Fig. 3C). However, there are constrained AMPs, specially defensins and peptides from bacteria that present stable β-like structures because of their cyclic structure (Huan et al., 2020). Thus, validating the structural tendency of peptides in a given environment helps understand how mutations within the sequence affect topological features and antimicrobial activity with respect to the parent molecule.

Materials and reagents

- Peptide
- Sterile Milli-Q water
- PBS buffer pH 7.4: 137 mmol L^{-1} NaCl, 2.7 mmol L^{-1} KCl, 10 mmol L^{-1} Na_2HPO_4, 1.8 mmol L^{-1} KH_2PO_4
- Sodium dodecyl sulfate (SDS) 15–20 mmol L^{-1}
- Trifluoroethanol (TFE)
- Methanol (MeOH)
- 2-oleoyl-1-pamlitoyl-sn-glyecro-3-phosphocholine (POPC) 10 mmol L^{-1}
- 2-oleoyl-1-pamlitoyl-sn-glyecro-3-phosphocholine (POPC) and 2-oleoyl-1-pamlitoyl-sn-glyecro-3-glycerol (POPG) (3:1, v/v) 10 mmol L^{-1}
- 2-oleoyl-1-pamlitoyl-sn-glyecro-3-phosphocholine (POPC) and 2-Oleoyl-1-palmitoyl-sn-glycero-3-phosphoethanolamine (POPE) (3:1, v/v) 10 mmol L^{-1}
- Lipopolysaccharide (LPS) 5 mg mL^{-1}
- Lipoteichoic acid (LTA) 5 mg mL^{-1}
- Quartz cuvette with 0.1 cm path length

Equipment

- Circular dichroism spectropolarimeter

Methods

1. Solubilize the peptide powder in sterile water to a final concentration of 250 μmol L^{-1}.
2. Run a blank sample of the medium of interest by adding the medium solution (250 μL) in the Quartz cuvette. The most used media for studies with peptides are: water, PBS buffer, TFE/water and MeOH/water mixtures, SDS micelles, unilamellar vesicles (e.g., POPC, POPC:POPG, POPC:POPE), biological component of bacterial membranes (e.g., LPS for Gram-negative bacteria or LTA for Gram-positive bacteria).
3. Collect the first spectrum at 25 °C in the 190–260 nm (far-UV). Follow 0.2 nm intervals, with a 50 nm/min scan rate, 1.0 nm bandwidth and a 4 s response.

   Note: The CD spectra of the blank solutions are subtracted from the spectra of the samples to decrease the background effects during the measurement.
4. Remove 50 μL of the control buffer from the cells and add 50 μL of peptide at 250 μmol L^{-1}.

5. Collect the second spectrum at 25 °C in the 190–260 nm (far-UV) at 0.2 nm intervals, with a 50 nm/min scan rate, 1.0 nm bandwidth and a 4 s response.

The CD spectrum of unstructured peptides usually displays a single negative band at 200 nm, whereas α-helical structures are characterized by two negative bands at 208 and 222 nm and one positive band at 192 nm. β-like structures, on the other hand, present a negative band at 217 nm and a positive band at 195 nm.

3.4 Biocompatibility evaluation

Peptide candidates with promising antimicrobial activity must show low toxicity toward eukaryotic cells to be considered preclinical candidates (Oddo et al., 2016). Although most AMPs interact selectively with the negatively charged lipid components of the bacterial membrane and display low affinity toward the zwitterionic phospholipids present in mammalian cells, they may exert cytotoxic effects by targeting the eukaryotic membrane or intracellular targets (Greco et al., 2020).

3.5 Hemolytic activity assay

In the initial stages of peptide characterization, hemolysis represents the most common assessment to define biocompatibility (Greco et al., 2020). The ability of AMPs to lyse mammalian red blood cells (RBCs) is usually evaluated on human blood samples, however blood from sheep, rat, pig, mouse, or rabbit are also routinely utilized (Oddo et al., 2016) (Fig. 4A).

Materials and reagents

- EDTA anti-coagulated blood (3–4 mL at ~1.8 $mg\,mL^{-1}$)
- PBS buffer pH 7.4: 137 $mmol\,L^{-1}$ NaCl, 2.7 $mmol\,L^{-1}$ KCl, 10 $mmol\,L^{-1}$ Na_2HPO_4, 1.8 $mmol\,L^{-1}$ KH_2PO_4
- Sterile 96-well microtiter plates (clear, round bottom)
- Sterile 96-well microtiter plates (clear, flat bottom)
- Sodium dodecyl sulfate (SDS)
- Peptide
- Sterile Milli Q water

Equipment

- Spectrophotometer plate reader for absorbance measurements at 405 nm ($A_{405\ nm}$)
- Plate and tubes centrifuge

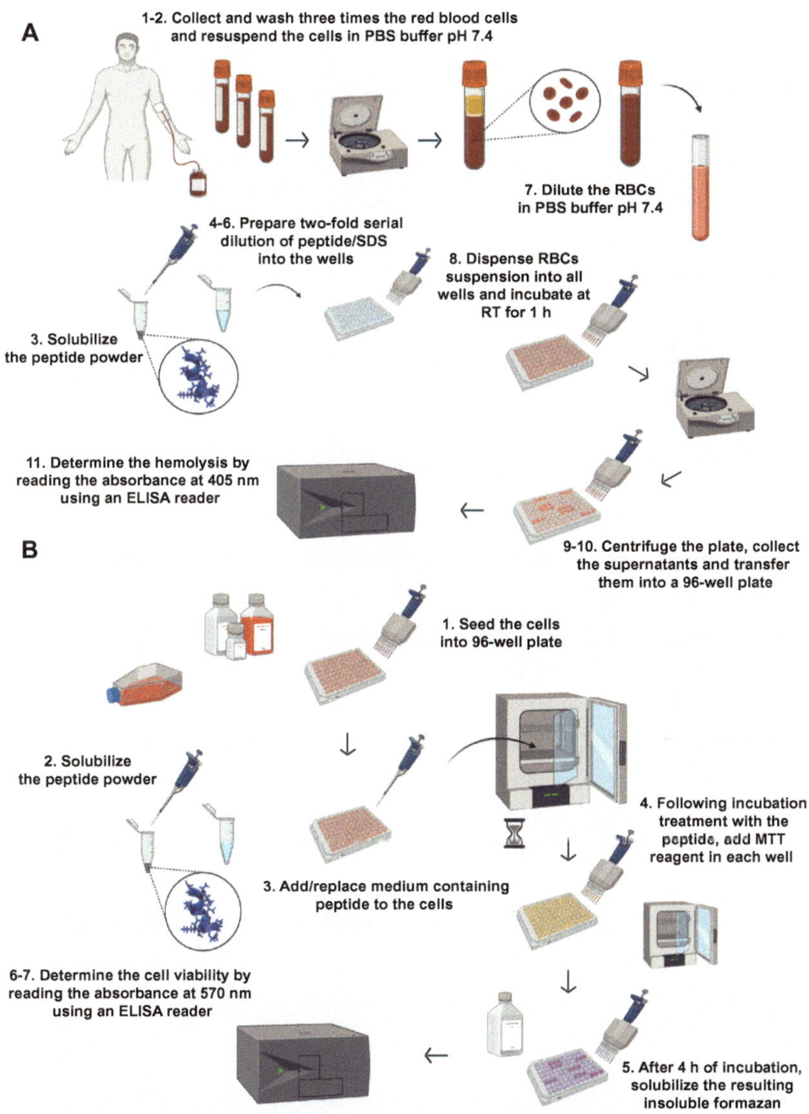

Fig. 4 Platforms for preliminary biocompatibility screening of peptide candidates. (A) Hemolytic activity evaluation by defining the percentage of hemolysis of each peptide toward red blood cells (RBCs). (B) Quantification of cellular metabolic activity, expressed as cell viability percentage, by performing MTT assays using mammalian cell cultures previously incubated with peptides.

Methods

1. Collect red blood cells (RBCs) from 3 to 4 mL of an anti-coagulated blood sample and wash three times by centrifugation at 800 × *g* for 10 min with PBS buffer pH 7.4.
2. Resuspend the RBCs in PBS buffer pH 7.4 by restoring the initial volume of the blood sample (3–4 mL).
3. Solubilize the peptide powder in sterile water and then dilute the antimicrobial in PBS buffer to 2 × the highest concentration required for testing.
4. Using a 96-well microtiter plate with rounded bottoms, pipette 150 μL of 2 × peptide solutions into the wells in column 1 and 75 μL of PBS buffer in the wells from columns 2 to 10.
5. Using a multi-pipettor set at 75 μL, prepare two-fold serial dilution by mixing the peptide into the wells in column 1 with 75 μL of PBS in the other columns by pipetting up and down a couple of times. Stop the serial dilution at column 10 and discard the extra 75 μL from column 10.
6. Add 75 μL of SDS diluted in PBS buffer to a final concentration of 2% in column 11 and 75 μL of PBS buffer in column 12 to prepare the positive and negative control samples, respectively, for the assay.
7. Dilute the RBCs in PBS buffer to a final concentration of 0.04% before mixing with the peptides.
8. Add 75 μL of RBCs suspension into all wells and incubate the plate at room temperature for 1 h.
9. Next, centrifuge the plate for 10 min at 1300 × *g* at room temperature.
10. Collect 100 μL of supernatant from each well and transfer into a 96-well flat bottom plate being careful not to generate bubbles.
11. Hemolysis is determined by reading the absorbance at 405 nm using an ELISA reader.
12. The percentage of hemolysis is expressed by comparing the results obtained with those from control samples containing PBS (negative control) or 1% (v/v) SDS in PBS solution (positive control, complete lysis):

$$\text{Hemolysis}\ (\%) = \frac{(\text{Abs}_{405\ \text{nm}}\ \text{peptide} - \text{Abs}_{405\ \text{nm}}\ \text{negative control})}{(\text{Abs}_{405\ \text{nm}}\ \text{positive control} - \text{Abs}_{405\ \text{nm}}\ \text{negative control})} \times 100.$$

Note: Triton-X at a final concentration of 0.2% is a viable and inexpensive alternative as a positive control for these assays. However, similar to SDS, triton-X favors foaming affecting the absorbance readouts (Oddo et al., 2016). Alternatively, melittin, an AMP from *Apis mellifera* venom, is recommended as a positive control because of its broad spectrum cytolytic activities at sub-micromolar concentrations.

3.6 MTT (3-(4,5-dimethylthiazol-2-yl)-2,5-diphenyltetrazolium bromide) tetrazolium reduction assay

The MTT assay is a colorimetric method used to quantify cellular metabolic activity as an indicator of cell viability, proliferation, and indirect cytotoxicity (Gaglione, Cesaro, Dell'Olmo, et al., 2020; Ibrahim et al., 2021). Several cell lines are used to investigate the cytotoxicity of AMPs. Many studies select both normal and tumoral cells as generic eukaryotic models [e.g., Human embryonic kidney 293 cells (HEK-293), human liver cancer cells (Hep G2), human colorectal adenocarcinoma cells (Caco-2)] (Greco et al., 2020; Torres et al., 2018), whereas others prefer to test the peptides on cell lines from specific tissue or organs on the basis of the future application [e.g., normal human keratinocytes (NHKs), immortalized human keratinocytes (HaCaT), human lung epithelial cells (Beas-2B), murine fibroblasts (BALBc 3T3), human dermal fibroblasts (HDFs), human intestinal epithelial cells (HIEC-6), human retinal pigmented epithelial cells (ARPE-19), murine macrophages (RAW 264.7), or human monocytic cells (THP-1)] (Di Grazia et al., 2015; Do et al., 2014; Gaglione et al., 2017, 2020; Ko et al., 2020; Vishweshwaraiah et al., 2021). The yellow tetrazolium salt (3-(4,5-dimethylthiazol-2-yl)-2,5-diphenyltetrazolium bromide or MTT) is reduced to purple formazan crystals by NAD(P)H-dependent oxidoreductase enzymes from metabolically active cells. The insoluble formazan crystals can be dissolved with an acid anhydrous solution and the resulting purple colored solution correlates proportionally with cell viability (Awati et al., 2020) (Fig. 4B).

Materials and reagents

- Cell culture medium without phenol red
- MTT: (3-(4,5-dimethylthiazol-2-yl)-2,5-diphenyltetrazolium bromide) tetrazolium
- 0.04 M HCl in anhydrous isopropanol buffer
- Sterile 96-well microtiter plates (clear, flat bottom)
- Peptide
- Sterile Milli-Q water

Equipment

- Cell culture dedicated incubator at 37 °C and 5% CO_2
- Spectrophotometer plate reader for absorbance measurements at 405 nm ($A_{405\ nm}$)

Methods for attached cells

1. Seed the cells into a 96-wells plate at a density of 1×10^3 to 10^6 cells per well in 100 μL of cell culture medium [e.g., Human embryonic kidney 293 (HEK 293) at a density of 1×10^3 cells per well in 100 μL of Dulbecco's Modified Eagle Medium (DMEM) supplemented with 10% Fetal Bovine Serum, 1% penicillin-streptomycin and 1% L-glutamine] and incubate from 6 to 24 h at 37 °C and 5% CO_2 before treatment to allow cell attachment at the bottom of the plate (e.g., an incubation time of 20 h is sufficient to ensure attachment of HEK 293 cells to the bottom of the plate).
2. Solubilize the peptide in sterile water and dilute an aliquot of the stock solution in the cell culture medium that will be used in the assay (e.g., DMEM supplemented with 10% Fetal Bovine Serum, 1% penicillin-streptomycin and 1% L-glutamine for HEK 293) to the highest concentration desired.
3. Prepare twofold serial dilutions of the peptide in the same medium buffer and gently replace the cell culture supernatants with fresh medium for the negative control or medium containing the peptide at different concentrations.
4. Following incubation treatment with the peptide, replace the cell culture supernatants with 0.5 mg mL^{-1} MTT (3-(4,5-dimethylthiazol-2-yl)-2,5-diphenyltetrazolium bromide) reagent dissolved in medium buffer without red phenol (100 μL/well).
5. After 4 h of incubation at 37 °C and 5% CO_2, solubilize the resulting insoluble formazan salts with 0.04 M HCl in anhydrous isopropanol.
6. Cell viability is determined by reading absorbance at 570 nm using an ELISA reader.
7. The percentage of viable cells is expressed as the percentage values compared to those of the untreated control cells.

Methods for cells in suspension

1. Seed into a 96-well plate the eukaryotic cells at a density of 2×10^3 to 10^6 cells/well in 50 μL of complete cell culture medium without phenol red [e.g., human monocytes (THP-1) at a density of 1×10^3 cells/well in 100 μL of RPMI 1640 medium supplemented with 10% Fetal Bovine Serum, 1% penicillin-streptomycin and 1% L-glutamine].
2. Solubilize the peptide powder in sterile water and then dilute the antimicrobial in the cell culture medium without phenol red (e.g., RPMI 1640 supplemented with 10% Fetal Bovine Serum, 1% penicillin-streptomycin and 1% L-glutamine for THP-1) to 2× the highest concentration to assay. Prepare two-fold serial dilutions of the peptide in the same buffer.

3. Add 50 μL of fresh medium for the negative control or medium containing peptide at different concentrations to the cells seeded into 96-well plate.
4. Following treatment with the peptide, add 5 mg mL^{-1} MTT (3-(4,-5-dimethylthiazol-2-yl)-2,5-diphenyltetrazolium bromide) reagent dissolved in medium buffer without red phenol in each well (10 μL/well).
5. After 4 h of incubation at 37 °C and 5% CO_2, solubilize the resulting insoluble formazan salts with 0.04 M HCl in anhydrous isopropanol.
6. Cell viability is determined by reading absorbance at 570 nm using an ELISA reader.
7. Cell viability % is expressed as the normalized values compared to the untreated control cells.

Note: Store the MTT reagent in the dark as it is light sensitive.

3.7 *In vivo* antimicrobial activity evaluation

Peptide candidates with promising antimicrobial activity *in vitro* and excellent biocompatibility profiles can then be tested in animal clinical models to ensure their potential translatability. The preclinical anti-infective efficacy of peptides can be assessed, for example, by using a mouse thigh infection model (Silveira et al., 2021). The main clinically relevant bacterial strains that may be used either in isolation or growing together in a consortium include *Staphylococcus aureus*, *Staphylococcus epidermidis*, and *Pseudomonas aeruginosa* (Fig. 5).

Materials and reagents

- Sterile inoculating loops
- Bacterial strains
- Sterile liquid and solid culture media for bacteria
- Sterile Petri dishes
- Sterile 96-wells microtiter plates
- Peptide
- Sterile Milli-Q water
- Sterile culture tubes
- PBS buffer pH 7.4: 137 mmol L^{-1} NaCl, 2.7 mmol L^{-1} KCl, 10 mmol L^{-1} Na_2HPO_4, 1.8 mmol L^{-1} KH_2PO_4
- Female CD-1 mice (5–6-weeks old)
- Cyclophosphamide
- Sterile syringes
- Sterile scalpel

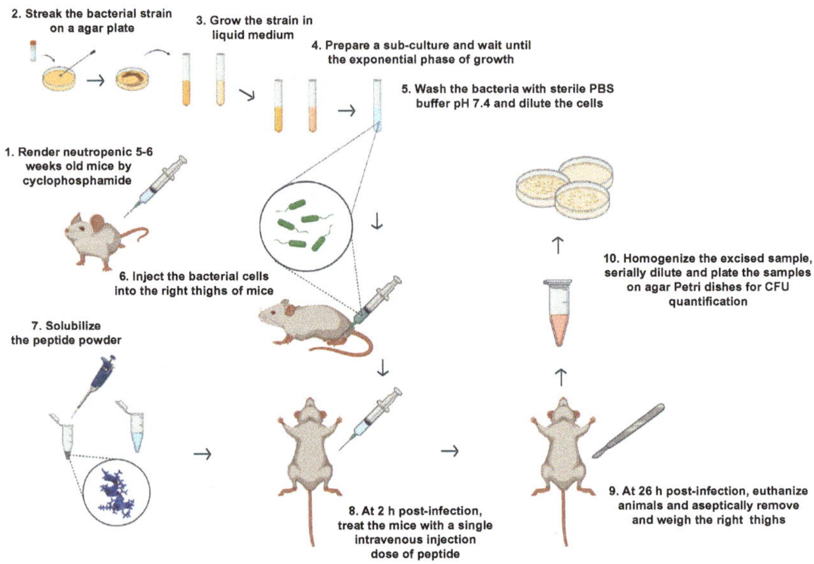

Fig. 5 *In vivo* antimicrobial activity evaluation of peptide therapeutics. The murine thigh infection model constitutes a preclinical assay to verify the anti-infective efficacy of promising peptide antibiotic candidates.

Equipment

- Rotating shaker incubator for culture tubes
- Tissue lyser
- Dedicated incubator for bacterial cells

Methods

1. Render female CD-1 mice (5–6 weeks old) neutropenic by injecting cyclophosphamide with two consecutive doses of 150 and 100 mg kg^{-1} administered 4 and 1 days before the infection.
2. Streak the bacterial strain of interest on an agar plate by using a sterile inoculating loop and incubate the samples at the optimum temperature to allow the colonies growth.
3. Grow the strains of interest in the appropriate medium from 1 to 3 colonies to start a liquid culture (overnight culture).
4. Prepare a sub-culture (pre-inoculum) diluting by 100-fold the saturated bacterial sample in fresh medium and wait until the inoculum reaches the exponential phase of growth
5. Wash the bacteria twice with sterile PBS buffer (13,000 rpm for 1 min) and resuspend the cells to a final concentration of 10^5 CFU/100 μL in PBS buffer.
6. Inject 100 μL of 10^5 CFU into the right thigh of the mouse.

7. Solubilize the peptide in sterile water and then dilute the antimicrobial in sterile PBS buffer to the concentration to assay. Keep it refrigerated until use.
8. At 2h post-infection, treat the mice with a single intravenous injection dose of peptide at 1, 2.5, 5, 10 or 20 $mg Kg^{-1}$.

   Note: peptides can also be administered intravenously (Schmidt et al., 2016).
9. At 26h post-infection, euthanize animals and aseptically remove the infected right thigh. The weight of the thigh is used to ensure the same amount of tissue for one of the each samples collected.
10. Homogenize the excised sample using a bead beater for 20 min (25 Hz), serially dilute and plate the samples on agar Petri dishes for CFU quantification.

4. Conclusion

The unique structural and physicochemical properties of peptide antibiotics make them versatile scaffolds. Recent advances in computational tools and experimental assays now enable the effective design of peptide therapeutics. In this chapter, we have described an array of methods for characterizing the structure and biological function of antimicrobial peptides. The experimental tools described here can be used to evaluate the biological properties of designed peptides and to understand the influence of structural and physicochemical properties on peptide bioactivity. Peptide antibiotics can be modified to act selectively against specific pathogens, decrease cytotoxicity, or demonstrate increased efficacy *in vivo*. Reprogramming peptides represents a promising approach for antibiotic discovery.

References

Arias, M., et al. (2018). Improving the activity of trp-rich antimicrobial peptides by Arg/Lys substitutions and changing the length of cationic residues. *Biomolecules*, *8*(2). https://doi.org/10.3390/biom8020019.

Awati, S. S., et al. (2020). In vitro antioxidant potential and cytotoxicity study of asparagus aethiopicus l. Extracts on ht-29 human colon cancer cell line. *Indian Journal of Pharmaceutical Education and Research*, *54*(3). https://doi.org/10.5530/ijper.54.3s.156.

Beltran, J. A., Aguilera-Mendoza, L., & Brizuela, C. A. (2018). Optimal selection of molecular descriptors for antimicrobial peptides classification: An evolutionary feature weighting approach. *BMC Genomics*, *19*. https://doi.org/10.1186/s12864-018-5030-1.

Bionda, N., et al. (2016). Identification of novel cyclic lipopeptides from a positional scanning combinatorial library with enhanced antibacterial and antibiofilm activities. *European Journal of Medicinal Chemistry*, *108*. https://doi.org/10.1016/j.ejmech.2015.11.032.

de la Fuente-Nunez, C., et al. (2017). Next-generation precision antimicrobials: Towards personalized treatment of infectious diseases. *Current Opinion in Microbiology*. https://doi.org/10.1016/j.mib.2017.05.014.
de Magalhães, C. S., et al. (2014). A dynamic niching genetic algorithm strategy for docking highly flexible ligands. *Information Sciences*, *289*(1). https://doi.org/10.1016/j.ins.2014.08.002.
Di Grazia, A., et al. (2015). The frog skin-derived antimicrobial peptide esculentin-1α(1-21) NH2 promotes the migration of human HaCaT keratinocytes in an EGF receptor-dependent manner: A novel promoter of human skin wound healing? *PLoS One*, *10*(6). https://doi.org/10.1371/journal.pone.0128663.
Do, N., et al. (2014). Cationic membrane-active peptides—Anticancer and antifungal activity as well as penetration into human skin. *Experimental Dermatology*, *23*(5). https://doi.org/10.1111/exd.12384.
Gaglione, R., Cesaro, A., Dell'Olmo, E., et al. (2020). Cryptides identified in human apolipoprotein B as new weapons to fight antibiotic resistance in cystic fibrosis disease. *International Journal of Molecular Sciences*, *21*(6), 2049. https://doi.org/10.3390/ijms21062049.
Gaglione, R., et al. (2017). Novel human bioactive peptides identified in Apolipoprotein B: Evaluation of their therapeutic potential. *Biochemical Pharmacology*, *130*, 34–50. https://doi.org/10.1016/j.bcp.2017.01.009.
Gaglione, R., et al. (2021). Impact of a single point mutation on the antimicrobial and fibrillogenic properties of cryptides from human apolipoprotein B. *Pharmaceuticals*, *14*-(7). https://doi.org/10.3390/ph14070631.
Glukhov, E., et al. (2005). Basis for selectivity of cationic antimicrobial peptides for bacterial versus mammalian membranes. *The Journal of Biological Chemistry*, *280*(40), 33960–33967. https://doi.org/10.1074/jbc.M507042200.
Greco, I., et al. (2020). Correlation between hemolytic activity, cytotoxicity and systemic in vivo toxicity of synthetic antimicrobial peptides. *Scientific Reports*, *10*(1). https://doi.org/10.1038/s41598-020-69995-9.
Huan, Y., et al. (2020). Antimicrobial peptides: Classification, design, application and research progress in multiple fields. *Frontiers in Microbiology*. https://doi.org/10.3389/fmicb.2020.582779.
Ibrahim, A. M. H., et al. (2021). The effect of surface modification of TiMg composite on the in-vitro degradation response, cell survival, adhesion, and proliferation. *Materials Science and Engineering: C*, *127*. https://doi.org/10.1016/j.msec.2021.112259.
Jiménez-Luna, J., et al. (2021). Artificial intelligence in drug discovery: Recent advances and future perspectives. *Expert Opinion on Drug Discovery*, *16*(9). https://doi.org/10.1080/17460441.2021.1909567.
Ko, S. J., et al. (2020). Bee venom-derived antimicrobial peptide melectin has broad-spectrum potency, cell selectivity, and salt-resistant properties. *Scientific Reports*, *10*(1). https://doi.org/10.1038/s41598-020-66995-7.
Lifson, S., & Roig, A. (1961). On the theory of helix-coil transition in polypeptides. *The Journal of Chemical Physics*, *34*(6). https://doi.org/10.1063/1.1731802.
Lin, X., Li, X., & Lin, X. (2020). A review on applications of computational methods in drug screening and design. *Molecules*. https://doi.org/10.3390/molecules25061375.
Magana, M., et al. (2020). The value of antimicrobial peptides in the age of resistance. *The Lancet Infectious Diseases*, *20*(9), e216–e230. https://doi.org/10.1016/S1473-3099(20)30327-3.
Mercer, D. K., et al. (2020). Antimicrobial susceptibility testing of antimicrobial peptides to better predict efficacy. *Frontiers in Cellular and Infection Microbiology*. https://doi.org/10.3389/fcimb.2020.00326.

Meurer, M., et al. (2021). Antimicrobial susceptibility testing of antimicrobial peptides requires new and standardized testing structures. *ACS Infectious Diseases*. https://doi.org/10.1021/acsinfecdis.1c00210.

Müller, A. T., Hiss, J. A., & Schneider, G. (2018). Recurrent neural network model for constructive peptide design. *Journal of Chemical Information and Modeling*, *58*(2). https://doi.org/10.1021/acs.jcim.7b00414.

Oddo, A., et al. (2016). An amphipathic undecapeptide with all D-amino acids shows promising activity against colistin-resistant strains of *Acinetobacter baumannii* and a dual mode of action. *Antimicrobial Agents and Chemotherapy*, *60*(1). https://doi.org/10.1128/AAC.01966-15.

Pane, K., et al. (2017). Antimicrobial potency of cationic antimicrobial peptides can be predicted from their amino acid composition: Application to the detection of 'cryptic' antimicrobial peptides. *Journal of Theoretical Biology*, *419*, 254–265. https://doi.org/10.1016/j.jtbi.2017.02.012.

Pane, K., et al. (2018). Identification of novel cryptic multifunctional antimicrobial peptides from the human stomach enabled by a computational–experimental platform. *ACS Synthetic Biology*, 7(9), 2105–2115. https://doi.org/10.1021/acssynbio.8b00084.

Porto, W. F., et al. (2018). In silico optimization of a guava antimicrobial peptide enables combinatorial exploration for peptide design. *Nature Communications*, *9*(1). https://doi.org/10.1038/s41467-018-03746-3.

Schmidt, R., et al. (2016). Pharmacokinetics and in vivo efficacy of optimized oncocin derivatives. *Journal of Antimicrobial Chemotherapy*, *71*(4). https://doi.org/10.1093/jac/dkv454.

Silveira, G. G. O. S., et al. (2021). Antibiofilm peptides: Relevant preclinical animal infection models and translational potential. *ACS Pharmacology and Translational Science*. https://doi.org/10.1021/acsptsci.0c00191.

Spohn, R., et al. (2019). Integrated evolutionary analysis reveals antimicrobial peptides with limited resistance. *Nature Communications*, *10*(1), 4538. https://doi.org/10.1038/s41467-019-12364-6.

Supady, A., Blum, V., & Baldauf, C. (2015). First-principles molecular structure search with a genetic algorithm. *Journal of Chemical Information and Modeling*, *55*(11). https://doi.org/10.1021/acs.jcim.5b00243.

Swetha, T. K., Priya, A., & Pandian, S. K. (2021). Natural molecules against QS-associated biofilm formation of pathogens. In *Microbial and natural macromolecules*. https://doi.org/10.1016/b978-0-12-820084-1.00014-4.

Torres, M. D. T., et al. (2017). Antimicrobial activity of leucine-substituted decoralin analogs with lower hemolytic activity. *Journal of Peptide Science*, *23*(11). https://doi.org/10.1002/psc.3029.

Torres, M. D. T., et al. (2018). Structure-function-guided exploration of the antimicrobial peptide polybia-CP identifies activity determinants and generates synthetic therapeutic candidates. *Communications Biology*, *1*(1), 221. https://doi.org/10.1038/s42003-018-0224-2.

Torres, M. D. T., & de la Fuente-Nunez, C. (2019a). Reprogramming biological peptides to combat infectious diseases. *Chemical Communications*, *55*(100), 15020–15032. https://doi.org/10.1039/c9cc07898c.

Torres, M. D. T., & de la Fuente-Nunez, C. (2019b). Toward computer-made artificial antibiotics. *Current Opinion in Microbiology*. https://doi.org/10.1016/j.mib.2019.03.004.

Torres, M. D. T., Melo, M. C. R., Crescenzi, O., Notomista, E., & de la Fuente-Nunez, C. (2021). Mining for encrypted peptide antibiotics in the human proteome. *Nature Biomedical Engineering*. https://doi.org/10.1038/s41551-021-00801-1.

Torres, M. D. T., et al. (2021). Synthetic biology and computer-based frameworks for antimicrobial peptide discovery. *ACS Nano*, *15*(2), 2143–2164. https://doi.org/10.1021/acsnano.0c09509.

Tromelin, A., Andriot, I., & Guichard, E. (2006). Protein-flavour interactions. In *Flavour in food*. https://doi.org/10.1533/9781845691400.2.172.
Umeyama, M., et al. (2006). Interactions of bovine lactoferricin with acidic phospholipid bilayers and its antimicrobial activity as studied by solid-state NMR. *Biochimica et Biophysica Acta (BBA)—Biomembranes*, *1758*(9), 1523–1528. https://doi.org/10.1016/j.bbamem.2006.06.014.
Vishweshwaraiah, Y. L., et al. (2021). Rational design of hyperstable antibacterial peptides for food preservation. *NPJ Science of Food*, *5*(1). https://doi.org/10.1038/s41538-021-00109-z.
Wiegand, I., Hilpert, K., & Hancock, R. E. W. (2008). Agar and broth dilution methods to determine the minimal inhibitory concentration (MIC) of antimicrobial substances. *Nature Protocols*, *3*(2), 163–175. https://doi.org/10.1038/nprot.2007.521.
Yeaman, M. R., & Yount, N. Y. (2003). Mechanisms of antimicrobial peptide action and resistance. *Pharmacological Reviews*. https://doi.org/10.1124/pr.55.1.2.
Yoshida, M., et al. (2018). Using evolutionary algorithms and machine learning to explore sequence space for the discovery of antimicrobial peptides. *Chem*, *4*(3). https://doi.org/10.1016/j.chempr.2018.01.005.

CHAPTER FOURTEEN

Solid-phase synthesis of novel antimicrobial peptoids with α- and β-chiral side chains

Biljana Mojsoska*
Department of Science and Environment, Roskilde University, Denmark
*Corresponding author: e-mail address: biljana@ruc.dk

Contents

Abstract

Bacterial resistance to conventional antibiotics poses an immense threat to human health and consequently many bacterial infections arise from multi-drug resistant pathogens. It is, therefore, necessary to continue to develop novel antimicrobials. Peptoids are a novel class of antimicrobial agents that mimic the structures of peptides. These *N*-substituted glycines differ from peptides in the side chain attachment site where in peptoids, the side chain functionality is introduced at the nitrogen rather than the *α*-carbon in peptides. During the process of design and development of antimicrobial peptoids, several key elements such as charge and amphiphilicity appear to guide peptoid bioactivity against bacterial pathogens. In this study, we report a quick synthesis of three *novel* cationic antimicrobial peptoids that contain charged, hydrophobic, and chiral monomers. This study is the first to use a monomer that contains both charge and chiral center for the design of the novel peptoid sequences.

Methods in Enzymology, Volume 663
ISSN 0076-6879
https://doi.org/10.1016/bs.mie.2021.10.010

1. Introduction

Peptoids are peptide-like synthetic molecules that were originally developed at the Chiron Corporation as a simple way to make large libraries of structurally diverse chemical compounds (Zuckermann, Kerr, Kent, & Moos, 1992). Peptoids differ from peptides in the chemical structure where the side chain functionality is introduced in the peptoid backbone nitrogen rather than the α-carbon in peptides.

In the last 3 decades, many research studies have demonstrated diverse biological activities of peptoids (Sun & Li, 2018). Some work includes peptoid activity as molecular modulators, e.g. ligands for GPCRs (Zuckermann et al., 2002) and protease inhibitors (Mazaleyrat et al., 1994), novel bio-inspired nanomaterials (Sun & Zuckermann, 2013; Tran, Gael, Connolly, & Zuckermann, 2011) and other work emphasizes on the potent antiviral (Diamond et al., 2021), anti-cancer (Huang et al., 2014), and most importantly, antimicrobial activities (Chongsiriwatana et al., 2008; Kapoor et al., 2011; Mojsoska et al., 2017; Mojsoska, Zuckermann, & Jenssen, 2015). Antimicrobial peptoid structures often mimic the side chain functionality of known antimicrobial peptides which act as host-defense molecules with diverse microbial activity ranging from Gram-negative to Gram-positive bacteria. Other active peptoids are generated *de novo* (current study) (Huang, Benson, Sung, Shin, & Torres, and Kent Kirshenbaum., 2013; Kapoor et al., 2011; Khara et al., 2020; Ryge & Hansen, 2005), where they can be either linear or cyclic. The main physicochemical properties that govern antimicrobial activities in peptides and peptoids are related to the overall amphiphilic structure, especially the ratio between the hydrophobic side chains and those that carry positive charges (Mojsoska et al., 2015). Antimicrobial properties of cationic peptoids originate from initial electrostatic interactions with the negatively charged microbial membranes, where a sufficient charge causes displacement of positively charged counter ions at the outer surface layers and induces changes in the bacterial surface. Peptoid activity requires an average net charge of +3 for initial electrostatic interaction with the negatively charged molecules found on microbial membranes (Mojsoska & Jenssen, 2015) such as bacterial lipopolysaccharides (LPS), teichoic acids etc. Additionally, hydrophobic residues can further destabilize the microbial membrane composition via three major mechanisms: barrel-stave, toroidal and carpet. Each mechanism relies on the threshold concentration of the antimicrobial peptoid/peptide on the surface of the membrane, and the different integration and spanning across the microbial membrane than leads to membrane disruption and bacterial death (Mojsoska et al., 2015). Peptoid hydrophobicity has also been associated with cell toxicity (Frederiksen, Hansen, Björkling, & Franzyk, 2019; Mojsoska et al., 2015). Peptoids offer enzymatic stability, increased bioavailability, and highly variable side-chain chemistry (Seo, Lee, & Zuckermann, 2011) thereby allowing this novel class of molecules to join other clinically important classes of antimicrobials. There are several peptide and peptoid modifications that have been implemented while designing potent antimicrobials and along with these traditional modification strategies (e.g. beta-peptides, peptide/peptoid hybrids), sequence rearrangement have been applied to many peptide and peptoid structures to improve their biological activities (Fischer, 2005; Mojsoska et al., 2015).

For the synthesis of biologically active peptoid structures, four methods have been reported, one of which, the submonomer solid-phase methodology (Proulx, Yoo, Connolly, & Zuckermann, 2015; Zuckermann et al., 1992) has been by far the most used. The other three methods include (1) solid-phase monomer peptoid synthesis, where monomers are synthesized before they are coupled to the growing peptoid chain (Frederiksen et al., 2019), (2) aldehyde or ketone assisted reductive alkylation of glycines (Tal-Gan, Freeman, Klein, Levitzki, & Gilon, 2010) and photolithographic synthesis of peptoids (Li et al., 2004).

The backbone in peptoids is rather rigid and alone it does not allow great secondary structure flexibility. However, secondary structure can be influenced by introducing monomers that contain side chain chirality. For example, several studies have reported that the use of *N*spe alpha-chiral aromatic (S)-*N*-(1-phenethyl) glycine (Fig. 1) does assist helical conformation in peptoids while adapting all cis-amide helix (Armand et al., 1997; Sanborn, Wu, Zuckermann, & Barron, 2002). The R variant of *N*spe has also been shown to contribute to secondary structure in peptoids where the requirement for full helical confirmation is 50 % of (R)-*N*spe in the sequence and also a *N*spe placed at the C-terminus (Wu, 2001). Blackwell and coworkers (Stringer, Aaron Crapster, Guzei, & Blackwell, 2011) have further demonstrated that introducing (S)-*N*-(1-napthylethyl) glycine-*Ns1*npe in a peptoid sequence can introduce stable helical structure in peptoids.

Our primary intention in this study was to design and present a quick synthesis of peptoids with specific physicochemical characteristics that will allow them to interact with bacterial membranes (G+ and G-) and serve as

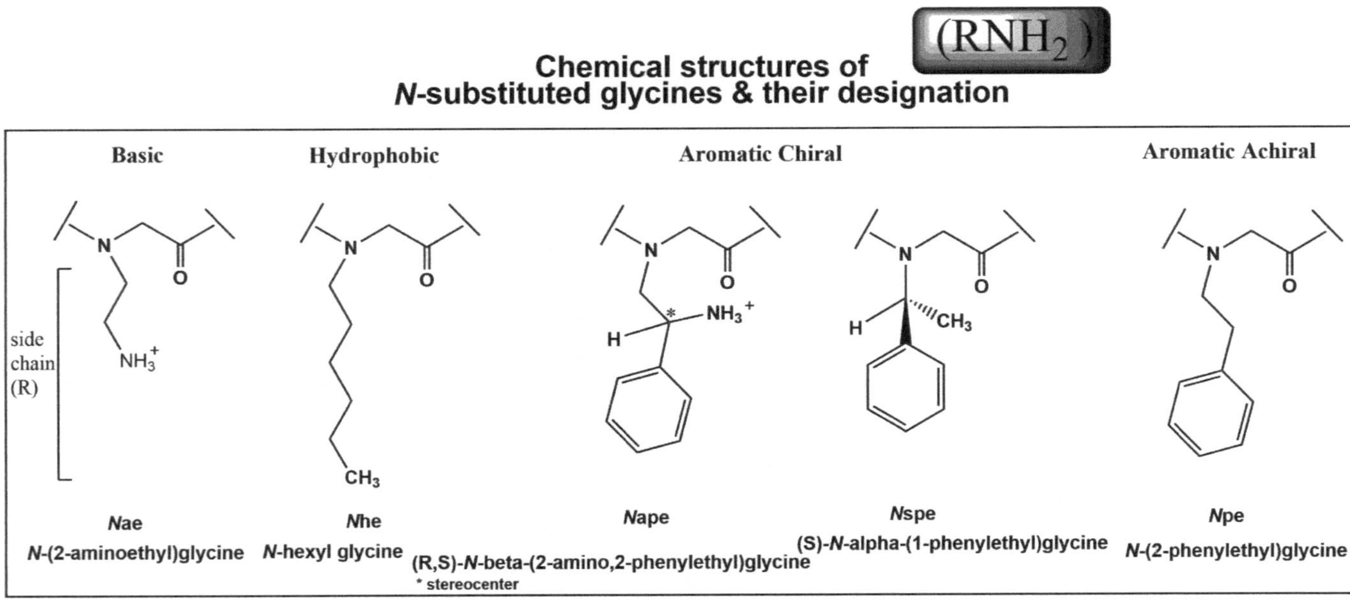

Fig. 1 Chemical structure of monomers used in the synthesis of peptoid BM4-BM6. Different side-chain functionalities and monomer abbreviations are shown.

antimicrobial agents. We synthesized peptoids composed of five different monomers (Fig. 1) arranged in different positions along the heptamer backbone. Incorporating monomers with alpha and beta side chains allows synthesis of peptoids that can be readily used for many structure-activity relationships (SAR) studies where one could evaluate the relationship between secondary structures and biological activities. Circular dichroism of peptoid sequences containing such building blocks can be used to deduce the secondary structures of peptoids.

2. Materials

1. Reaction vessels: Plastic 5-mL Syringe with stopper and PTFE filter. Use new reaction vessel for every new synthesis. Syringe pressure caps (Sigma, non-sterile) were used as stoppers and removed each time the syringe is attached on the liquid collection reservoir.
2. Rink amide MHBA resin (loading = 0.3–0.4 mmol/g, Novablochem CAS# 431041-83-7), 100–200 mesh. Note: the use of this resin gives peptoids with amidated C-terminal.
3. Vacuum pump connected with liquid collection reservoir.
4. Dimethylformamide (DMF)-Solvent and washing organic liquid.
5. *N*-methyl-2 pyrrolidone (NMP, Sigma), organic solvent for amines.
6. 20% 4-methyl piperidine in DMF, deprotection.
7. Bromoacetic acid (BrAcOH), acylation.
8. *N*, *N*-diisopropylcarbodiimide (DIC, Acros Organics), acylation.
9. Acylation mixture: prepare fresh solution (daily) of 0.6 M bromoacetic acid. For example, dissolve 1.67 g BrAcOH in 20 mL DMF. Before each acylation step for one peptoid mix 86 μL DIC (0.56 mmol) and 1 mL BrAcOH (0.6 mmol).
10. Amine 1: (*S*)-(-)-alpha-methylbenzylamine (Sigma, CAS #115568), 0.5 M in NMP.
    Note: 0.5–2 M amine concentration can be used. For longer sequences higher concentrations are recommended.
11. Amine 2: α-(Boc-amino) phenethylamine (Merck, CAS#:142910-85-8).
12. Amine 3: Hexylamine (Sigma, CAS #111-26-2).
13. Amine 4: Benzylamine (TGI, CAS# B0406).
14. Amine 5: *N*-Boc-ethylenediamine (Merck, CAS# 57260-73-8).
15. Dichloromethane (DCM).
16. Trifluoroacetic acid (TFA), cleavage.

17. Triisopropylsilane (TIPS, Merck, CAS#6485-79-6), cleavage.
18. Cleavage mixture (prepare fresh): 49 % TFA: 49 % DCM: 2 %TIPS.
19. Acetonitrile (MeCN), purification, and HPLC-MS analysis.

3. Equipment

- Incubator orbital shaker (VWR).
- Solvent evaporator (Biotage V10).
- Lyophylizer, freeze drying of peptoid products.
- Analytical HPLC-MS with UV detection: Dionex Ultimate 3000 system interfaced with a HESII electrospray ion source to Thermo Finnigan LTQ-XL linear ion trap mass spectrometer.
- AKTA low-pressure chromatography system with UV detection and fraction collection.

4. Submonomer peptoid synthesis (manual)

Here we describe the design and synthesis of three peptoids (BM4, BM5, and BM6, for sequence, see Fig. 2), which display antimicrobial activities against Gram-negative and Gram-positive bacteria. We designed the 7mer peptoids to have cationic, hydrophobic, and chiral side-chain functionalities as most antimicrobial peptoids are amphiphilic and in some cases, the observed antimicrobial activity has been attributed to the secondary structures. Secondary structures in peptoids rely on having chiral side chains. In our design, we have used *N*ae to introduce positive charge, *N*he lipid-like property, *N*pe hydrophobic bulk side chain, and *N*spe for alpha chirality. In addition, this study is the first to introduce a novel *N*-substituted glycine side chain, *N*ape, in the design of antimicrobial peptoids, which introduces beta chirality and charge in the peptoid chain. We used a solid-phase sub-monomeric approach to synthesize the desired peptoids. This synthesis approach uses resin activation followed by acylation and displacement steps. Several solid supports (resins) can be used depending on the desired C-terminal functionality. Rink amide MHBA and Tenta Gel S are commonly used resins that generate C-terminal amide, while the use of 2-chlorotrityl chloride resins give C-terminal acid. For the acylation steps, bromoacetic acid and chloroacetic acids have been used (Burkoth, Fafarman, Charych, Connolly, & Zuckermann, 2003). DIC-assisted activation of the haloacetic acids generates amide bonds via S_N2 nucleophilic displacement by

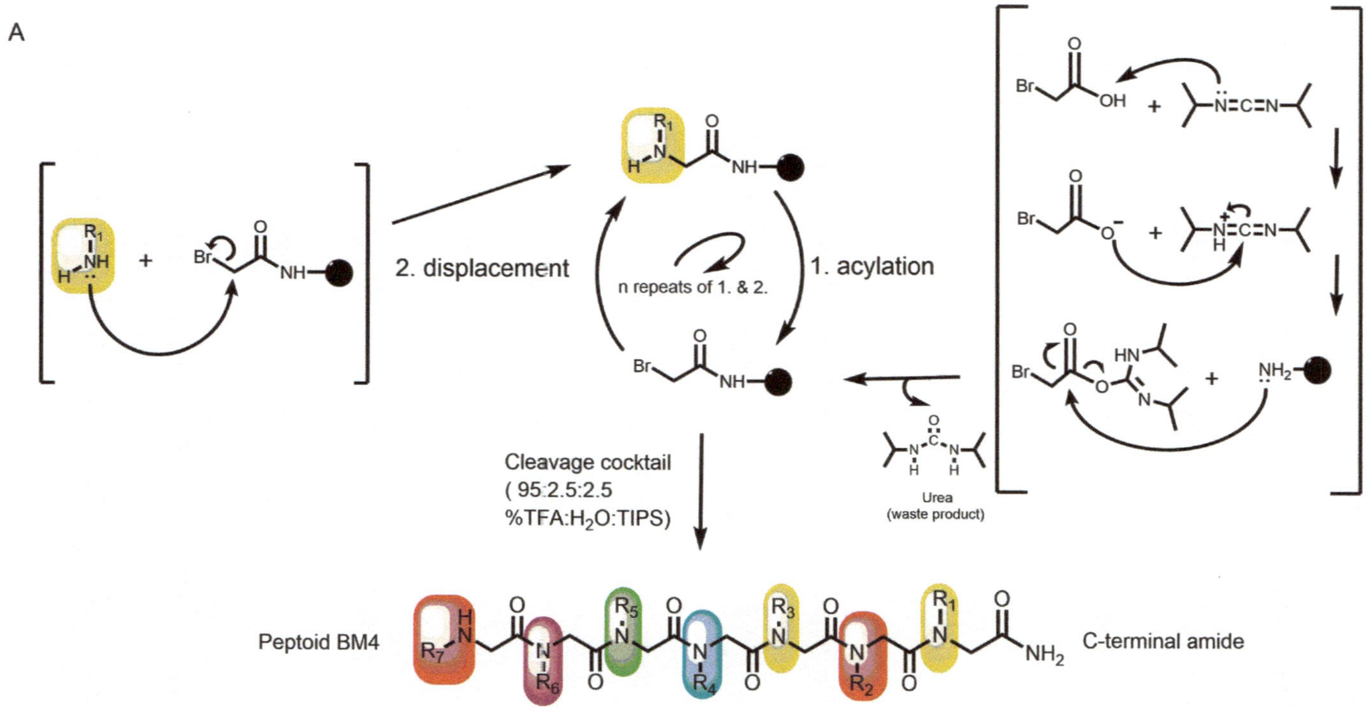

B

Peptoid BM 4	R_1,R_3=*N*ae,	R_2,R_7=*N*spe,	R_4=*N*ape,	R_5=*N*pe,	R_6=*N*he
Peptoid BM 5	R_1,R_6=*N*spe,	R_2=*N*he,	R_3=*N*pe,	R_4=*N*ape,	R_5,R_7=*N*ae
Peptoid BM 6	R_1,R_7=*N*spe,	R_2=*N*pe,	R_3=*N*ape,	$R_{4,5}$=*N*ae,	R_6=*N*he

Fig. 2 Solid-phase submonomer peptoid synthesis scheme. (A) Steps in peptoid synthesis, (1) acylation using BrAcOH and DIC, and (2) Displacement step where the desired amine is added. The sequence of peptoid BM4 is shown. (B) Side chain substitutions for peptoids BM (4-6) are shown.

primary amines, followed by production of urea as a waste product that is easily washed away from the solid support (Fig. 2A). During peptoid synthesis multiple side-chain reactions have been reported and some of these largely rely on the chemical composition of the amine side chains used for the synthesis. The most common reactions that have been characterized include the formation of *N*-substituted diketopiperazine (Culf & Ouellette, 2010; Lone, Arnous, Hansen, Mojsoska, & Jenssen, 2020), and nucleophilic attack of a side chain free secondary amine with the bromoacetyl group on the growing main chain (intramolecular alkylation) (Lone et al., 2020).

4.1 Resin activation and Fmoc removal

Time: 30–40 min at RT

1. Weigh 200 mg resin and place it in a reaction vessel.
2. Resin swelling: Add 4 mL DMF and close the reaction vessel with a silicon stopper or syringe plunger. Leave for at least 30–40 min with shaking.
3. Remove the cap, place the reaction vessel (syringe) in the liquid reservoir connected with the vacuum device pump and drain the DMF.
4. Wash the resin with 3 mL DMF. Repeat this step 3 times.
5. Remove the syringe, cap the lower end, and add 1 mL of 20% 4-methylpiperidine to remove the Fmoc group from the resin. Close the syringe and place it in the shaker at RT. Wait 2 min.
6. After 2 min drain the piperidine solution by placing the syringe on the liquid reservoir as in step 3 and wash the resin 2x2 mL DFM.
7. Remove the syringe, cap the bottom, and add 2 mL 20% 4-methypiperidine. Close the syringe and place it in the shaker for 12 min.
8. Drain the solution as in step 6.
9. Wash thoroughly with 3x2 mL and 5x3 mL DMF under vacuum.

Notes:

Step 2: It is possible to leave the beads to swell overnight. DMF can be replaced with NMP.

Step 4: In each washing step, it is optional to remove the syringe, cap the bottom, add DMF, close the top and shake the syringe. Then open the top, take the cap off, and place the syringe on the liquid reservoir connected with the pump to drain the washing liquid.

The resin should be completely soaked when adding reaction solutions!

4.2 Manual sub-monomer coupling

Time: 22–25 min

1. Remove the syringe from the liquid reservoir and cap the bottom.
2. Acylation: add 1.86 mL of the acylation mixture (BrAcOH + DIC). Incubate by gentle shaking in the shaking incubator for 5 min at 35 °C.
3. Drain and rinse with 2 mL DMF (repeat 4x)
4. Displacement: Add 1 mL of 0.5 M amine prepared in NMP. For the synthesis of BM4, the first amine used to displace the bromine is R^1 = *N*ae. Make sure the resin is covered by liquid and if necessary, adjust by adding a minimum volume of NMP.
5. Close the syringe and place it in the incubator shaker for a minimum of 20 min at 35 °C.
6. Remove the plunger and then the cap and place the syringe on the liquid reservoir connected with a vacuum pump. Drain the amine solution.
7. Wash the resin with 2 mL DMF (5 x 1 min).
8. Repeat steps 2–7 to grow the peptoid chain.
9. After the final displacement is done, rinse with 2 mL DMF (repeat 4x) and then 2 mL DCM (repeat 3x).
10. Leave the syringe on the liquid contained connected with the vacuum pump for 10–30 min for the resin to dry.
11. Close and cap the syringe until cleavage.

Notes:

Step 2: If the product yield is low, depending on the sequence length and the monomers used in the design, the acylation step can be repeated.

Step 5: Addition of bulky, hydrophobic amines benefits from increased displacement time to a minimum of 30 min.

Step 11: dried resin can be stored at -20 °C until cleavage.

Amine 1, *N*spe, displacement time 30 min

Amine 2, *N*ape, displacement time 30 min

Amine 3, *N*he, displacement time 20 min

Amine 4, *Npe*, displacement time 30 min

Amine 5, *N*ae, displacement time 20 min

Pause point: To pause during synthesis, ensure that you have performed steps 2–11. To continue growing the peptoid chain, restart the synthesis by acclimating the resin at RT followed by re-swelling the resin as in step 1.2 and repeat the submonomer synthesis cycle as in steps 2–8. Avoid pausing your synthesis after the addition of the second amine as there is a risk of formation of diketopiperazine side products.

4.3 Product cleavage from the solid support and side-chain deprotection

Time: 20–30 min at RT with shaking

1. Prepare 4 mL fresh cleavage cocktail (49% aq. TFA, 49% DCM, 2 % TIPS)
2. Remove the top of the syringe and add the cleavage cocktail mixture. Close the syringe, and close it tightly using parafilm. Shake for 60 min at RT.
3. Take the syringe and remove the top. Place the syringe on a stationary lab holder and place a 20 mL scintillation glass vial below. Remove the cap and collect the cleavage cocktail that now contains the peptoid product. *Optional*: Add 500 μL–1 mL of fresh cleavage cocktail to gather any residual product.
4. Evaporate the TFA using Biotage V10 evaporator or by blowing a stream of nitrogen.
5. Redissolve the crude oil in 6 mL of MeCN: water (1:1). For hydrophobic peptoids, try dissolving the oil in small amounts of MeCN and then gradually add water until you reach complete dissolution (clear solution).
6. Freeze and lyophilize.
7. Repeat steps 5–6 2x to obtain fine white powder.
8. Measure the weight of the crude product and store the peptoid crude at -20 °C.

Optional: Test cleavage: take a small amount of resin beads and perform cleavage to determine whether you have the correct sequence and the purity of the peptoid product. You can also use this procedure to try different cleavage conditions and/or monitor the progress of the synthesis. Test cleavage can be performed in a 1.5 mL microcentrifuge tube or HPLC glass vials. Filtering of the small amount of resin can be performed using mini-scale columns or any small, fritted syringe.

Optional: Take 50 μL at step 5 and run the peptoid crude on HPLC-MS to check for correct product synthesis and purity as in Section 4.4.

Caution: Wear proper eyewear and adequate gloves when handling highly concentrated TFA. In case of any spills, make sure you clean and handle the spilled liquid appropriately.

4.4 Purification and product analysis

1. Analytical HPLC-ESI-MS can be used to determine the purity of the crude peptoid product as well as whether the desired sequences have been successfully synthesized.
2. Prepare small amount (200–500 μL) of 100 μg/mL–500 μg/mL of peptoid crude in mobile phase (MeCN: water: 0.01% formic acid (FA)) or water. Make sure the solution is completely dissolved. Filter 200 μL of the solution using a 0.45 μm syringe filter to remove particles that can clog the column into an HPLC vial.
3. Add the sample vial to the auto sample holder and inject 2–5 μL sample.
4. Analytical HPLC-ESI-MS was performed on Dionex Ultimate 3000 system interfaced with a HESII electrospray ion source to Thermo Finnigan LTQ-XL linear ion trap mass spectrometer.
5. The sample was run on a C_{18} reverse-phase column (2.6 μm/100 Å, Phenomenex at 40 °C, with a flow rate of 0.5 mL/min using an increasing gradient. The gradient started with 5% buffer A (MeCN: water: formic acid, 9:1:0.1) and increased over 15 min to 65% and finally to 100% over 3 min with an additional isocratic gradient at 100% A for 2 min at 0.7 mL/min. UV-vis was recorded at 214, 220, and 280 nm.
6. FreeStyle software (ThermoFisher) was used to inspect the HPLC chromatogram and the electrospray ionization spectra of the peptoids. BM4-6 peptoids have a molecular weight of 1018.32 g/mol and molecular ion $[M+H]^+$ at m/z: 1018.58 and $[M+Na]^+$ at m/z: 1040.67 ware observed in the ESI spectra. FreeStyle also allows determination of the purity of the crude peptoid product.
7. Preparative LC (AKTA) was used to purify the product to the desired purity of >95%. The total crude peptoid product was dissolved in 10% MeCN in water with 0.1% TFA to a concentration below the maximum loading capacity of the instrument column. The peptoid crude was purified using linear gradient of 10–65% MeCN in water (0.1% TFA) with flow rate of 2 mL/min, on C_{18} reverse-phase column (250 x 10 x 10 μm/100 Å, Kinetex, Phenomenex). All fractions were collected and those corresponding to the desired product were checked for purity using analytical HPLC (step 5) and after complete purity was confirmed, the fractions were combined, frozen at -80 °C and lyophilized.
8. The final pure peptoid product weight was noted.

4.5 TFA-free peptoids for biological testing

After cleavage and purification, residual TFA is not desired when preparing peptoid products that will be taken for biological testing. Peptoid TFA salts can be removed however, during this process peptoid product loss is inevitable. TFA salt exchange using 100 mM hydrochloric acid (HCl) is a commonly practiced methodology. The following protocol has been used to remove TFA from peptoids BM4-BM6.

1. Add 320 μL of 100 mM HCl to 5 mg BM4-BM6 peptoids.
2. Mix the solution for 3 min.
3. Freeze the solution at -80 °C.
4. Lyophilize.
5. Repeat steps 2–4 (min 2 x times) for a complete TFA/HCl salt exchange.

5. Safety considerations and standards

1. Many of the chemicals are hazardous and need to be handled with special care.
2. All the reactions should be performed in a fume hood and appropriate eyewear, gloves, and lab coat should always be worn.
3. It is suggested to print chemical safety data sheets (e.g. https://chemicalsafety.com/sds-search/) for the following chemicals and be aware of the toxicity/hazard they cause so that you are careful when handling these chemicals.
4. TFA is highly toxic and corrosive. Wear appropriate eye protection as well as gloves and lab coat when handling solutions containing high concentrations of TFA.
5. NPM and DMF are toxic and readily absorbed by the skin, inhaled, or ingested. Wear proper lab gear and work in the fume hood. Replace gloves immediately if you notice a spill of these solvents on your gloves.
6. DIC is fatal if inhaled, causes serious eye damage and is very toxic to aquatic life. Work in the fume hood and wear protective lab gear.
7. Dispose all chemicals in the proper chemical waste containers.

References

Armand, P., Kirshenbaum, K., Falicov, A., Dunbrack, R. L., Dill, K. A., Zuckermann, R. N., et al. (1997). Chiral N-substituted glycines can form stable helical conformations. *Folding & Design*, *2*(6), 369–375. https://doi.org/10.1016/S1359-0278(97)00051-5.

Burkoth, T. S., Fafarman, A. T., Charych, D. H., Connolly, M. D., & Zuckermann, R. N. (2003). Incorporation of unprotected heterocyclic side chains into peptoid oligomers via solid-phase submonomer synthesis. *Journal of the American Chemical Society*, *125*(29), 8841–8845. https://doi.org/10.1021/ja0352101.

Chongsiriwatana, N. P., Patch, J. A., Czyzewski, A. M., Dohm, M. T., Ivankin, A., Gidalevitz, D., et al. (2008). Peptoids that mimic the structure, function, and mechanism of helical antimicrobial peptides. *Proceedings of the National Academy of Sciences of the United States of America*, *105*(8), 2794–2799. https://doi.org/10.1073/pnas.0708254105.

Culf, A. S., & Ouellette, R. J. (2010). Solid-phase synthesis of N-substituted glycine oligomers (α-peptoids) and derivatives. *Molecules*, *15*(8), 5282. https://doi.org/10.3390/MOLECULES15085282.

Diamond, G., Molchanova, N., Herlan, C., Fortkort, J. A., Lin, J. S., Figgins, E., et al. (2021). Potent antiviral activity against HSV-1 and SARS-CoV-2 by antimicrobial peptoids. *Pharmaceuticals*, *14*, 304. https://doi.org/10.3390/PH14040304.

Fischer, P. (2005). The design, synthesis and application of stereochemical and directional peptide isomers: A critical review. *Current Protein & Peptide Science*, *4*(5), 339–356. https://doi.org/10.2174/1389203033487054.

Frederiksen, N., Hansen, P. R., Björkling, F., & Franzyk, H. (2019). Peptide/peptoid hybrid oligomers: The influence of hydrophobicity and relative side-chain length on antibacterial activity and cell selectivity. *Molecules*, *24*, 4429. https://doi.org/10.3390/MOLECULES24244429.

Huang, M. L., Benson, M. A., Sung, B. Y., Shin, V. J., & Torres, and Kent Kirshenbaum. (2013). Amphiphilic cyclic peptoids that exhibit antimicrobial activity by disrupting staphylococcus aureus membranes. *European Journal of Organic Chemistry*, *17*, 3560–3566. https://doi.org/10.1002/ejoc.201300077.

Huang, W., Seo, J., Willingham, S. B., Czyzewski, A. M., Gonzalgo, M. L., Weissman, I. L., et al. (2014). Learning from host-defense peptides: Cationic, amphipathic peptoids with potent anticancer activity. *PLoS One*, *9*(2), e90397. https://doi.org/10.1371/JOURNAL.PONE.0090397.

Kapoor, R., Wadman, M. W., Dohm, M. T., Czyzewski, A. M., Spormann, A. M., & Barron, A. E. (2011). Antimicrobial peptoids are effective against Pseudomonas Aeruginosa biofilms. *Antimicrobial Agents and Chemotherapy*, *55*(6), 3054–3057. https://doi.org/10.1128/AAC.01516-10.

Khara, J. S., Mojsoska, B., Mukherjee, D., Langford, P. R., Robertson, B. D., Jenssen, H., et al. (2020). Ultra-short antimicrobial peptoids show propensity for membrane activity against multi-drug resistant Mycobacterium tuberculosis. *Frontiers in Microbiology*. https://doi.org/10.3389/fmicb.2020.00417.

Li, S., Bowerman, D., Marthandan, N., Klyza, S., Luebke, K. J., Garner, H. R., et al. (2004). Photolithographic synthesis of peptoids. *Journal of the American Chemical Society*, *126*(13), 4088–4089. https://doi.org/10.1021/JA039565W.

Lone, A., Arnous, A., Hansen, P. R., Mojsoska, B., & Jenssen, H. (2020). Synthesis of peptoids containing multiple Nhtrp and Ntrp residues: A comparative study of resin, cleavage conditions and submonomer protection. *Frontiers in Chemistry*, *2020*(8), 370. https://doi.org/10.3389/FCHEM.2020.00370.

Mazaleyrat, J. P., Rage, I., Mouna, A. M., Šavrda, J., Wakselman, M., Boulay, R., et al. (1994). Peptoid mimics of a C2-symmetric inhibitor of the HIV-1 protease. *Bioorganic and Medicinal Chemistry Letters*, *4*(11), 1281–1284. https://doi.org/10.1016/S0960-894X(01)80345-0.

Mojsoska, B., Carretero, G., Larsen, S., Mateiu, R. V., Jenssen, H., et al. (2017). Peptoids Successfully Inhibit the Growth of Gram Negative E. Coli Causing Substantial Membrane Damage. *Scientific Reports*. https://doi.org/10.1038/srep42332. In press.

Mojsoska, B., & Jenssen, H. (2015). Peptides and peptidomimetics for antimicrobial drug design. *Pharmaceuticals*, *8*(3), 366–415. https://doi.org/10.3390/ph8030366.

Mojsoska, B., Zuckermann, R. N., & Jenssen, H. (2015). Structure-activity relationship study of novel peptoids that mimic the structure of antimicrobial peptides. *Antimicrobial Agents and Chemotherapy*, *59*(7), 4112–4120. https://doi.org/10.1128/AAC.00237-15.

Proulx, C., Yoo, S., Connolly, M. D., & Zuckermann, R. N. (2015). accelerated submonomer solid-phase synthesis of peptoids incorporating multiple substituted N-aryl glycine monomers. *Journal of Organic Chemistry*, *80*(21), 10490–10497. https://doi.org/10.1021/acs.joc.5b01449.

Ryge, T. S., & Hansen, P. R. (2005). Novel lysine-peptoid hybrids with antibacterial properties. *Journal of Peptide Science*, *11*(11), 727–734. https://doi.org/10.1002/PSC.705.

Sanborn, T. J., Wu, C. W., Zuckermann, R. N., & Barron, A. E. (2002). Extreme stability of helices formed by water-soluble poly-N-substituted glycines (polypeptoids) with alpha-chiral side chains. *Biopolymers*, *63*(1), 12–20. https://doi.org/10.1002/BIP.1058.

Seo, J., Lee, B. C., & Zuckermann, R. N. (2011). Peptoids: Synthesis, characterization and nanostructures. *Comprehensive Biomaterials*, *2*, 53–76. http://www.ronznet.com/RZ_paper_pdfs/Seo_Comp_Biomat_2011.pdf.

Stringer, J. R., Aaron Crapster, J., Guzei, I. A., & Blackwell, H. E. (2011). Extraordinarily robust polyproline type I peptoid helices generated via the incorporation of α-chiral aromatic N-1-naphthylethyl side chains. *Journal of the American Chemical Society*, *133*(39), 15559–15567. https://doi.org/10.1021/JA204755P.

Sun, J., & Li, Z. (2018). Peptoid applications in biomedicine and nanotechnology. *Peptide Applications in Biomedicine, Biotechnology and Bioengineering*, 183–213. https://doi.org/10.1016/B978-0-08-100736-5.00007-7.

Sun, J., & Zuckermann, R. N. (2013). Peptoid polymers: A highly designable bioinspired material. *ACS Nano*. https://doi.org/10.1021/nn4015714.

Tal-Gan, Y., Freeman, N. S., Klein, S., Levitzki, A., & Gilon, C. (2010). Synthesis and structure-activity relationship studies of peptidomimetic PKB/Akt inhibitors: The significance of backbone interactions. *Bioorganic and Medicinal Chemistry*, *18*(8), 2976–2985. https://doi.org/10.1016/J.BMC.2010.02.031.

Tran, H., Gael, S. L., Connolly, M. D., & Zuckermann, R. N. (2011). Solid-phase submonomer synthesis of peptoid polymers and their self-assembly into highly-ordered nanosheets. *Journal of Visualized Experiments*, *57*, 1–7. https://doi.org/10.3791/3373.

Wu, C. W., Sanborn, T. J., Huang, K., Zuckermann, R. N., Barron, A. E., Wu, C. W., et al. (2001). Peptoid oligomers with α-chiral, aromatic side chains: Sequence requirements for the formation of stable peptoid helices. *Journal of the American Chemical Society*, *123*(28), 6778–6784. https://pubs.acs.org/doi/full/10.1021/ja003154n.

Zuckermann, R. N., Kerr, J. M., Kent, S. B. H., & Moos, W. H. (1992). Efficient method for the preparation of peptoids [oligo(N-substituted glycines)] by submonomer solid-phase synthesis. *Journal of the American Chemical Society*, *114*, 10646–10647.

Zuckermann, R. N., Martin, E. J., Spellmeyer, D. C., Stauber, G. B., Shoemaker, K. R., Kerr, J. M., et al. (2002). Discovery of nanomolar ligands for 7-transmembrane G-protein-coupled receptors from a diverse N-(substituted)glycine peptoid library. *Journal of Medicinal Chemistry*, *37*(17), 2678–2685. https://doi.org/10.1021/JM00043A007.

CHAPTER FIFTEEN

Molecular networking in infectious disease models

Morgan B. Harris[a,c,†], Mahbobeh Lesani[b,c,†], Zongyuan Liu[a,c,†], and Laura-Isobel McCall[a,b,c,*]

[a]Department of Chemistry and Biochemistry, University of Oklahoma, Norman, OK, United States
[b]Department of Microbiology and Plant Biology, University of Oklahoma, Norman, OK, United States
[c]Laboratories of Molecular Anthropology and Microbiome Research, University of Oklahoma, Norman, OK, United States
*Corresponding author: e-mail address: lmccall@ou.edu

Contents

Abstract

Small molecule metabolites are the product of many enzymatic reactions. Metabolomics thus opens a window into enzyme activity and function, integrating effects at the post-translational, proteome, transcriptome and genome level. In addition, small molecules can themselves regulate enzyme activity, expression and function both via substrate

† Authors contributed equally. Author order determined randomly.

Methods in Enzymology, Volume 663
ISSN 0076-6879
https://doi.org/10.1016/bs.mie.2021.09.018

availability mechanisms and through allosteric regulation. Metabolites are therefore at the nexus of infectious diseases, regulating nutrient availability to the pathogen, immune responses, tropism, and host disease tolerance and resilience. Analysis of metabolomics data is however complex, particularly in terms of metabolite annotation. An emerging valuable approach to extend metabolite annotations beyond existing compound libraries and to identify infection-induced chemical changes is molecular networking. In this chapter, we discuss the applications of molecular networking in the context of infectious diseases specifically, with a focus on considerations relevant to these biological systems.

1. Introduction

Metabolomics allows insight into systems by systematically detecting and quantifying the metabolites present in samples (Idle & Gonzalez, 2007). Metabolites are small molecules under 1500 Da with diverse chemical structures. Traditionally, metabolites include substrates, products, or intermediates of central metabolic pathways such as glycolysis, TCA cycle, or fatty acid oxidation. However, metabolites can also be involved in many critical systems such as regulation of gene expression, enzymatic inhibition, or signaling pathways (Newsom & McCall, 2018). In the broadest sense, metabolites can also include small molecules that were encountered and absorbed from the environment such as food or personal care product-derived molecules. The metabolome is thus the profile of metabolites found in a system and represents the intersection between cellular pathways, organ functions, behaviors, and habitat (Lewis, McCall, Sharp, & Spicer, 2020). The metabolome can be investigated through mass spectrometry (MS) or nuclear magnetic resonance spectroscopy (NMR)-based detection and identification methods. Mass spectrometry measures the mass-to-charge (m/z) ratio of analytes, both intact (MS1 level) or following fragmentation (MS2 level). Metabolomic approaches can be targeted, whereby a specific list of molecules of interest are quantified, or untargeted, whereby MS data is acquired across a broad mass range, with no pre-existing target m/z list. In the latter case, the goal is to study the broadest range of metabolites, for a given sample preparation protocol and detection system, without being pre-constrained by a priori notions. Major challenges in untargeted metabolomics include the complexity of the resulting data and the annotation of the resulting metabolite features (Wang et al., 2016).

Global Natural Product Social Molecular Networking (GNPS) provides a high throughput infrastructure to address these issues. Specifically, users can annotate their mass spectrometry data by comparing experimental

MS2 spectra to collected reference MS2 spectra from public libraries of natural and synthetic small molecules, including lipids, central metabolism intermediates, plant and microbial natural products, drugs, personal care products, etc. These libraries are extended regularly by community addition of new reference spectra. In addition, GNPS enables users to search for analogs of specific compounds in their spectral data (Wang et al., 2016). For example, phosphatidylcholine 36:1 and phosphatidylcholine 36:3 have similar structures and thus similar MS2 fragmentation patterns. If only phosphatidylcholine 36:3 is present in the reference libraries, phosphatidylcholine 36:1 would remain unannotated by direct matching of spectra and precursor mass. In contrast, analog searches consider their MS2 spectral similarities, enabling phosphatidylcholine 36:1 annotation by analog matching to phosphatidylcholine 36:3 reference spectra.

The other advantage of GNPS is the ability to connect metabolite features with similar fragmentation spectra, indicating structural similarity, and thus generate molecular networks. These networks facilitate extended metabolite annotations ("annotation propagation"): although not all nodes in a molecular network match reference library spectra, connected nodes can be annotated by using the network and considering common chemical modifications such as oxidation or methylation. Molecular networks also enable analysis of the frequency of specific chemical modifications under a given experimental condition. Molecular networks can be visualized directly on the GNPS website or exported to Cytoscape software (see Section 3) (Shannon et al., 2003; Wang et al., 2016).

GNPS offers two types of molecular networking, classical and feature-based molecular networking (FBMN). Classical molecular networking is generated from unprocessed MS2 spectral data (from open-format .mzXML or .mzML files). It automatically clusters similar MS2 spectra into network nodes. Sometimes, however, MS2 fragmentation spectra from one single metabolite are variable, and in those cases classical molecular networking generates multiple nodes for this metabolite instead of only one node. FBMN uses MS1 information, including chromatographic retention time and feature peak area to group and align features in the molecular network. FBMN can thus distinguish between isomers with similar MS2 spectra but different retention times. In addition, FBMN networks can be directly related to metabolite feature quantification, since FBMN node size can be set to reflect MS1 peak area. On the other hand, classical networking is easy to set up from raw data, whereas FBMN needs pre-processing that is affected by user expertise. Classical molecular networking is also better suited to very large sample numbers or to integrate data collected through different instrumental conditions (Jarmusch et al., 2020; Nothias et al., 2020; Wang et al., 2016).

GNPS and molecular networking were initially designed for natural product discovery (Wang et al., 2016; Watrous et al., 2012), and therefore have seen the greatest applications in that field (e.g., Heine et al., 2018; Nothias et al., 2018; Senges et al., 2018). However, recent work has demonstrated their utility in the field of infectious diseases. Many infectious diseases can be difficult to detect and diagnose. In this context, GNPS was used to annotate metabolites linked to Chagas disease severity (Hoffman et al., 2021; McCall et al., 2017) or to discover metabolites predictive of *Staphylococcus aureus* bacteremia mortality (Wozniak et al., 2020). Metabolomics and molecular networking have also been used to identify molecules associated with pathogen or pathobiont colonization in mammalian or insect systems (Eberhard, Klimpel, Guarneri, & Tobias, 2021; Garg et al., 2017; Melnik et al., 2019). Chronic infectious diseases can also present a challenge in understanding effects on organ systems or unforeseen complications due to infection. Molecular networking applied to acute and chronic *Trypanosoma cruzi* infection identified changes in specific metabolites based on tissue location and helped discover a potential treatment to prevent damage associated with *Trypanosoma cruzi* infection (Hossain et al., 2020). To study communication between fungal pathogens and host, molecular networking was applied to the study of *Cryptococcus gattii* extracellular vesicles (Reis et al., 2021). Notably, these examples are all very recent. Thus, to build on these promising expansions of GNPS and molecular networking implementation, this protocol focuses specifically on molecular networking applications in the context of infectious diseases.

2. Practical considerations

All material should be handled using the appropriate biosafety protocol approved by your institution. For samples collected from animal models of infection, all approved Institutional Animal Care and Use Committee protocols should be followed.

There are multiple strategies for metabolite extraction, chromatography and mass spectrometry data acquisition. We will not discuss them here in detail, and instead refer the reader to the multiple excellent reviews on this topic (e.g., Dunn et al., 2011; Ivanisevic & Want, 2019; Want et al., 2013). In general, the metabolite extraction method should be chosen based on sample type and the properties of the metabolites of interest (for example polar or non-polar metabolites; lipids; microbial signaling molecules).

We focus here on molecular networking applied to liquid chromatography-tandem mass spectrometry data (Nothias et al., 2020; Wang et al., 2016), though it has recently been extended to gas chromatography–mass spectrometry data (Aksenov et al., 2021). Chromatography and mass spectrometry conditions should be optimized for the available instrument and metabolites of interest. Irrespective of these choices, it is however critical to acquire tandem mass spectrometry (MS/MS, also known as MS2) data to be able to perform molecular networking analysis of liquid chromatography-mass spectrometry data. In our experience, the best molecular network quality is obtained when MS/MS data is acquired for all samples under consideration, rather than only for a pooled sample. In all cases, MS2 should be acquired in data-dependent mode. High quality data will determine the success of this protocol, while poor quality data will lead to poor, uninterpretable or incorrect findings.

Additional important considerations include selection of sampling site and biofluid, which should be relevant to the pathogenesis of the disease of interest. Though more challenging to access, tissue samples are in our opinion preferable because they directly reflect local metabolic alterations caused by infection. Multiple adjacent samples will enable reconstruction of the spatial impact of infection on the metabolome ("chemical cartography") (see for example, Dean et al., 2021; Hossain et al., 2020; McCall et al., 2017; Parab et al., 2021). However, should this not be possible or desired, tissue sampling site should always be consistent across all experimental samples.

Molecular networking is performed through the Global Natural Product Social Molecular Networking (GNPS) web interface. The GNPS website has excellent detailed instructions on how to perform molecular networking (https://ccms-ucsd.github.io/GNPSDocumentation/). Our goal here is not to supersede these instructions, but rather to highlight practical considerations for molecular networking specifically in the context of infectious diseases. Molecular networking can be performed straight from instrumental data files in open .mzXML or .mzML format ("classical molecular networking") or by integrating feature finding using software with molecular networking (FBMN) (Nothias et al., 2020; Wang et al., 2016). In this protocol, we focus on the downstream data analyses, once a network has been generated.

3. Protocol

3.1 Before you begin

Timing: 4 h-weeks (depending on number of samples, data size and prior familiarity with molecular networking).

1. Collect samples from infection model. Extract metabolites and collect data-dependent LC-MS2 data
2. Install Cytoscape software from https://cytoscape.org/
3. Read the instructions on the GNPS website on how to set up a molecular networking job and watch the videos
   - (a) https://ccms-ucsd.github.io/GNPSDocumentation/networking/
   - (b) Published protocols are also available from Phelan (2020) and Aron et al. (2020)
4. Set up a classical or feature-based molecular network following these instructions
5. The following example networks are available to the reader to practice this protocol and were used to generate the figures herein:
   - (a) Classical molecular network example: https://gnps.ucsd.edu/ProteoSAFe/status.jsp?task=4f706aaba6ed4bc1b56791b12bf98777 (re-run from Hossain et al., 2020)
   - (b) Feature-based molecular network example: https://gnps.ucsd.edu/ProteoSAFe/status.jsp?task=a2ae5d8bb4dd440ba885f7d0b3ebbfac (re-run with from Hossain et al., 2020)
6. Follow instructions below, Visualizing and interpreting your molecular network, on steps to download the data

Critical: These approaches are only compatible with data-dependent MS2 data. They cannot be applied to data-independent MS2 data, or to MS1-only data. Refer to the GNPS documentation for suitable file types (https://ccms-ucsd.github.io/GNPSDocumentation/isgnpsright/). Although molecular networking can now be performed using GC–MS data (Aksenov et al., 2021), we focus here on LC–MS/MS data analysis.

Critical: To be able to perform analyses of the impact of infection, ensure that your metadata contains categories pertaining to infection status, and that file names in the metadata table perfectly match names of the data files. To avoid mis-filtering, we recommend using "infected" and "no_infection," or similar non-overlapping terms, rather than "infected" and "uninfected."

3.2 Key resources table

Reagent or resource	Source
Deposited data	
"Classical" molecular networking job	Processed in GNPS, from mzXML or mzML file formats
Example (re-run from Hossain et al. 2020)	https://gnps.ucsd.edu/ProteoSAFe/status.jsp?task=4f706aaba6ed4bc1b56791b12bf98777
Feature-based molecular networking job	Processed in GNPS, from mass spectrometry data processing programs (e.g., MZmine (Pluskal, Castillo, Villar-Briones, & Orešič, 2010), XCMS (Forsberg et al., 2018; Smith, Want, O'Maille, Abagyan, & Siuzdak, 2006; Tautenhahn, Böttcher, & Neumann, 2008), MS-DIAL (Lai et al., 2017; Tsugawa et al., 2015), OpenMS (Röst et al., 2016; Sturm et al., 2008))
Example (re-run from Hossain et al. 2020)	https://gnps.ucsd.edu/ProteoSAFe/status.jsp?task=a2ae5d8bb4dd440ba885f7d0b3ebbfac
Software and algorithms	
GNPS	https://gnps.ucsd.edu/
Cytoscape	https://cytoscape.org/

3.3 Materials and equipment

- Computer. The analyses described in this protocol can be performed on any standard internet-connected desktop computer or laptop. Software is compatible with Windows, Mac and Linux.
- Molecular network (from feature-based or classical networking job).
- Cytoscape software.

3.4 Step-by-step method details

3.4.1 Visualizing and interpreting your molecular network

Timing: <40 min

Detailed steps for visualizing and interpreting molecular networks in general may be found in the GNPS documentation, in Aron et al. (2020), and in Phelan (2020). For reader convenience, we provide a brief outline of these steps here, as a necessary prerequisite for subsequent analysis in the context of infectious diseases. These steps apply to both feature-based and classical molecular networks. As an example, we used the feature-based molecular network from: https://gnps.ucsd.edu/ProteoSAFe/status.jsp?task=a2ae5d8bb4dd440ba885f7d0b3ebbfac.

1. Access your GNPS results page.
2. Some visualization is available via the GNPS browser interface (View Spectral Families (In Browser Network Visualizer)). Refer to the GNPS documentation and (Aron et al., 2020) for detailed instructions.
3. For best visualization and flexibility, and to enable the subsequent analyses described in this protocol, download the data to Cytoscape through either of the following options:
   (a) For large datasets:
   - On your GNPS results page, select: Export/Download Network Files (Download Cytoscape Data) or Export/Download Network Files (Download GraphML for Cytoscape) (Fig. 1A (yellow box)).
   - Click Download in the upper right. After downloading, unzip the folder.
   - Open Cytoscape.
   - Select the "Import Network from File" button in Cytoscape (Fig. 1C) and import the graphml format file from Fig. 1B.

   (b) For small datasets:
   - On your GNPS results page, select: Advanced Views-External Visualization (Direct Cytoscape Preview/Download, Fig. 1A (red box)).
   - The file opens in Cytoscape directly.
4. After importing the network into Cytoscape, the network displays in the right upper panel (Fig. 1D). To see more details on the network, zoom in with the tool bar magnifying glass icon (yellow box, Fig. 1D).
5. Each node represents one metabolite feature (Fig. 2A). Nodes with similar MS2 spectra are connected to each other and form a sub-network of features (chemical family, Fig. 2B). The lines that connect the nodes (features) are called edges (Fig. 2C). Some nodes (self-looped nodes) are not connected to other nodes and are displayed in the lowest part of the network (Fig. 2D). These self-looped nodes represent metabolite features whose MS2 spectra did not have sufficient similarity to any other MS2 spectra in the dataset (based on the cutoffs established when setting up the molecular networking job).

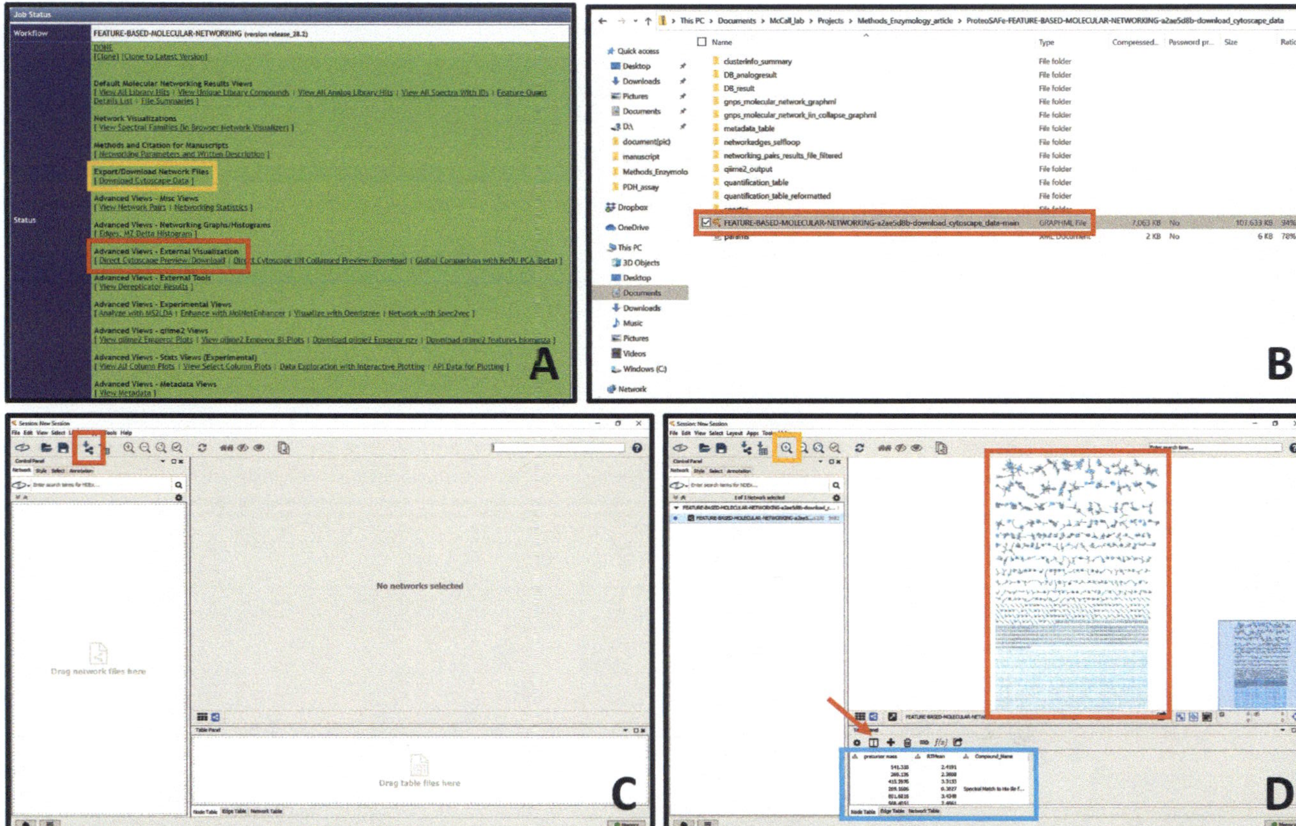

Fig. 1 Importing and visualizing the molecular network. (A) Downloading GNPS job file with "Direct Cytoscape preview/download" link (red box) for small datasets and "Download Cytoscape data" link (yellow box) for large datasets. (B) "Graphml" format file in the downloaded folder. (C) "Import network from file system" button in Cytoscape to import graphml format file (red box). (D) Example network displayed in Cytoscape (red box). Yellow box, zoom icon. Blue box, node table. Arrow, Show Columns button.

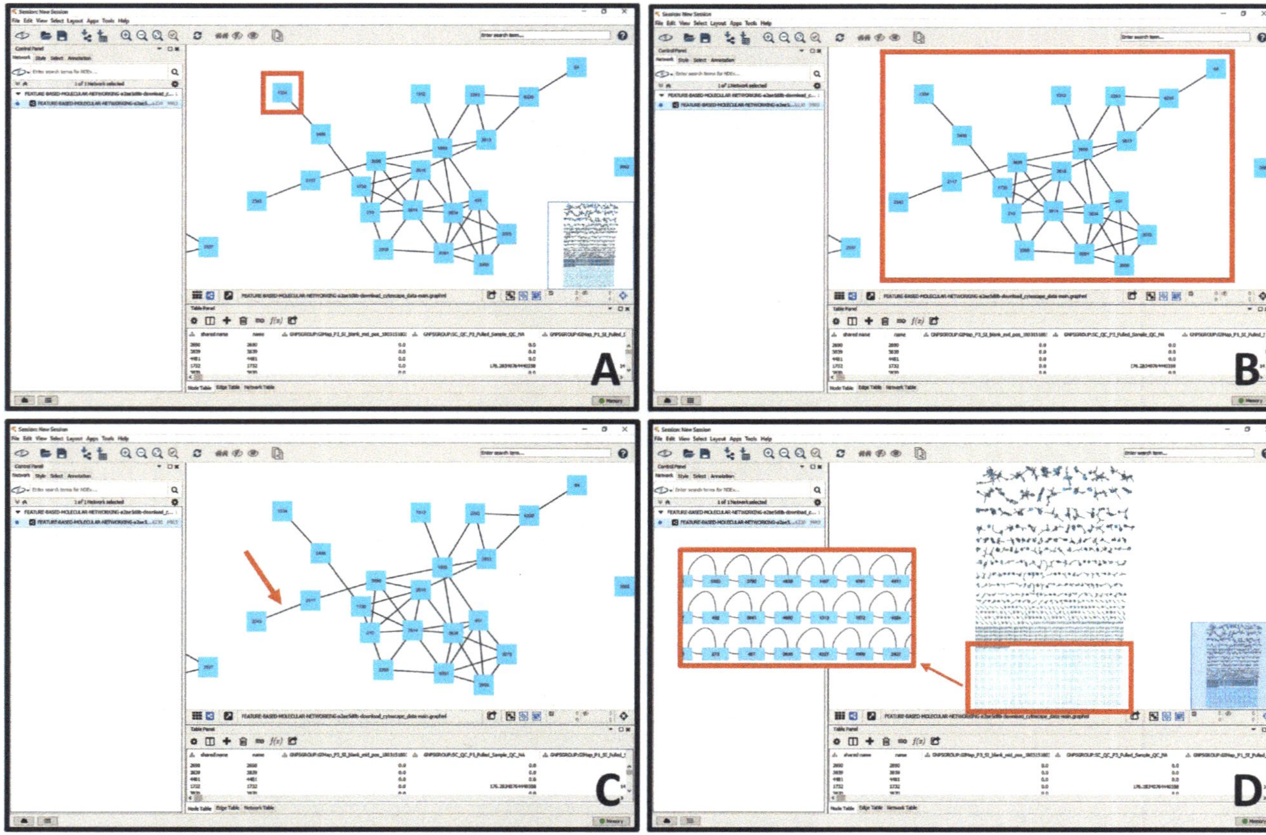

Fig. 2 Network and network features. (A) A node in a network (red box). (B) A sub-network of related, connected nodes (red box). (C) An edge that connects two nodes (arrow). (D) Self-looped nodes (singletons) with no connection with any other nodes (red box).

6. The total number of nodes and edges can be found on the left panel of the network tab (Fig. 3).
7. To determine the number of subnetworks:
   (a) Export network table using File→Export→Table. In the dropdown menu, select the name of your network default node table, then press OK.
   (b) In Excel, determine the number of unique values in the "componentindex" column. In this example, we have 810 sub-families in our full molecular network. −1 values represent self-looped nodes.
8. Node, edge, and network styles can be changed through the Style tab on the left panel. The Style tab has three sub-tabs for node, edge, and network.
   (a) Nodes can be labeled with the annotation obtained via GNPS spectral matching.
      - Select the Node tab.
      - Select the Label drop-down menu by clicking the down arrow. Choose Compound_name for Column using the dropdown menu and Passthrough Mapping for Mapping Type (Fig. 4).
   (b) Node color can be changed in a similar fashion using the Fill Color drop-down menu.

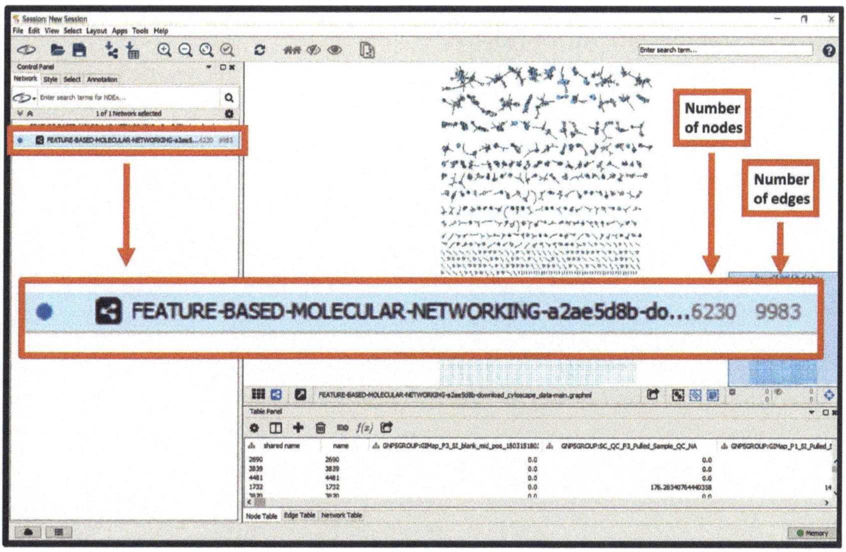

Fig. 3 Total number of nodes and edges.

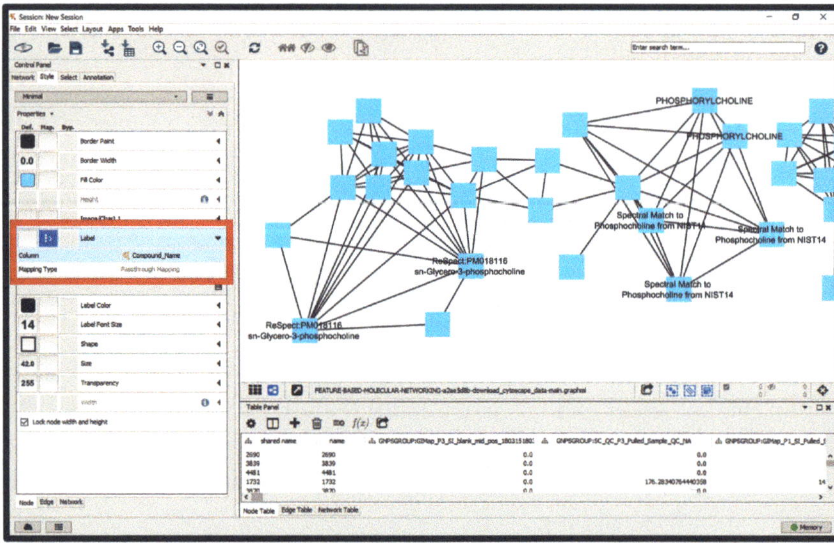

Fig. 4 Cosmetic changes (node, edge, and network styles). This figure shows how to customize the node labels with the compound name (red box).

- **(c)** Node shape can be changed in a similar fashion using the Shape drop-down menu.
- **(d)** Additional detail is available here: (Aron et al., 2020; Phelan, 2020; Shannon et al., 2003) and https://ccms-ucsd.github.io/GNPS Documentation/cytoscape/.

9. The node table (bottom right, Fig. 1D) contains information on each of the nodes within your network. To simplify the displayed table, use the "Show Columns" button (Fig. 1D, arrow).

- **(a)** Pick Select None.
- **(b)** Then manually select columns of interest. We recommend selecting at least: cluster index, Compound_Name, Analog: Compound_Name (Feature-based molecular networking), precursor mass, RTMean (Classical molecular networking) and RTConsensus (Feature-based molecular networking).

10. Annotated features can be verified by inspecting the mirror plot displaying the MS2 spectral concordance between your experimental spectra and the library reference spectra, and assessing the match cosine score, number of matched peaks, reference library quality, and your instrumental parameters. Plausibility in the context of your infection model should also be considered. Refer to the GNPS documentation and (Aron et al., 2020) for more details.

11. A quick way to visualize the mirror plot for a specific annotation is as follows:
    (a) In the Node table under the network view in Cytoscape (bottom right pane, Fig. 5A), scroll to the column named "cluster index" (Fig. 5A, red box).
    (b) Find the cluster index value for the feature/node with the annotation you would like to verify.
    (c) From your GNPS results page, select View All Library Hits (Fig. 5B).
    (d) Search by cluster index.
    (e) Visualize the mirror plot by clicking the View Mirror Match button (Fig. 5C, red box).
    (f) A higher cosine score indicates a higher similarity between the experimental and reference MS2 spectra, for a maximum of 1 (identical MS2 spectra) and a minimum of 0 (highly dissimilar MS2 spectra).
12. A putative annotation for unannotated features can be obtained through network annotation propagation. Refer to Aron et al. (2020) and the GNPS documentation for more details. Briefly:
    (a) For the node of interest, locate its neighbors using the First Neighbors of Selected Nodes (undirected) icon (Fig. 6A, blue box).
    (b) Determine the mass difference between the unannotated node and any connected annotated nodes. Assess for chemical plausibility while also considering instrumental parameters.
        - This information can also be found in the network Edge table, in column "EdgeAnnotation."
    (c) If first neighbors are not annotated, these steps can be repeated iteratively until no connected neighbors remain.
    (d) See Optimizations and troubleshooting for additional complementary analysis tools that can be implemented if no nodes in the sub-network are annotated.
13. To select specific nodes based for example on precursor mass or compound name:
    (a) Click on the Select tab (PC) or Filter tab (Mac) on the left panel.
    (b) Select "+," then Column filter (Fig. 6A).
    (c) Use the dropdown Choose column menu to select the parameter of interest (e.g., Node: precursor mass or Node: Compound_Name).
    (d) Hit Apply icon at the bottom of the Select tab (Fig. 6B).
    (e) Selected nodes are highlighted in yellow in your right-hand network view.

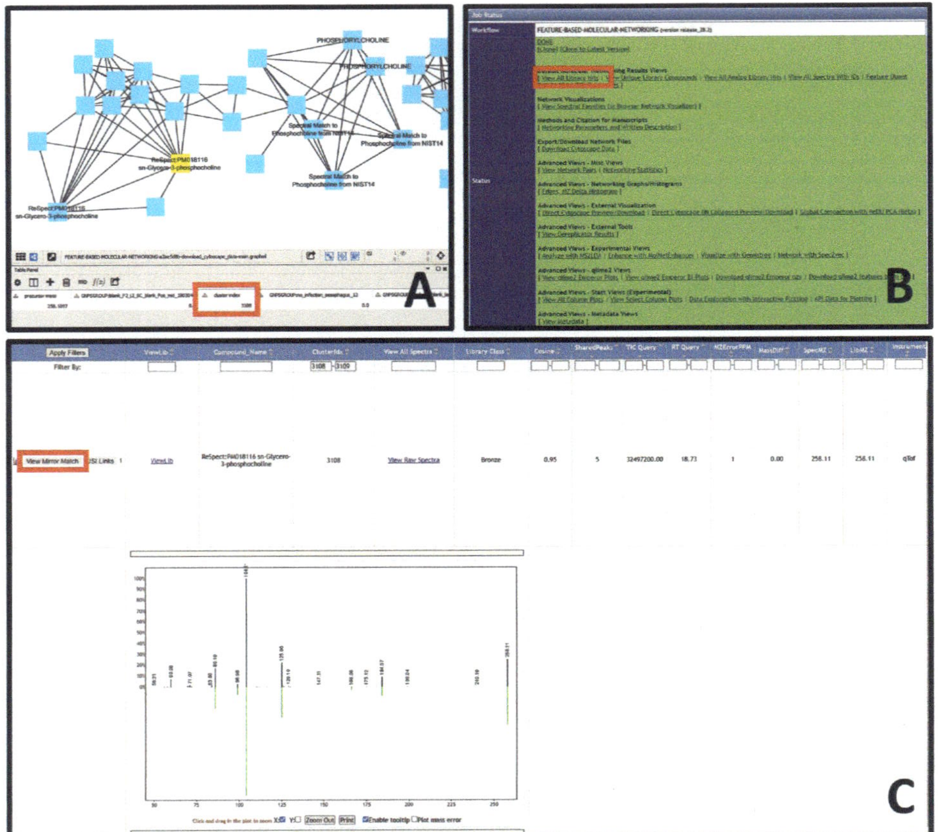

Fig. 5 Visualizing the mirror plot for a specific annotation. (A) Locating the node cluster index on the bottom right pane in Cytoscape (red box). (B) GNPS results page with "View All Library Hits" link (red box). (C) Example of mirror plot obtained by clicking the View Mirror Match link (red box) for the specified cluster index.

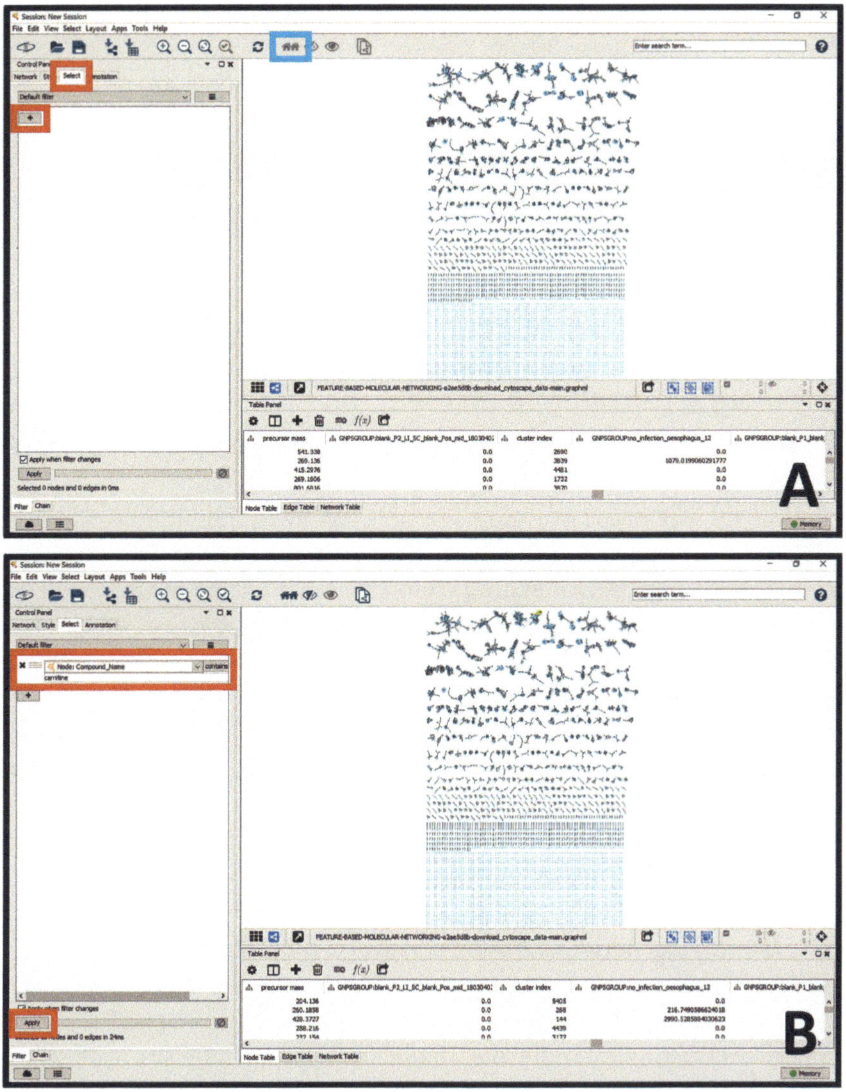

Fig. 6 Selecting a specific node. (A) "Select" tab and "+" bottom on Cytoscape (red boxes). First Neighbors of Selected Nodes button (blue box). (B) Filtering the specific node with "Node: Compound_Name" as an example for parameter of interest and "Apply" icon.

14. To subset networks, for example to create a new network that only includes nodes annotated as acylcarnitines and members of the acylcarnitine family of metabolites (Hossain et al., 2020):

(a) Select nodes as described above:

- In the dropdown Choose column menu, select Node: Compound_Name.
- In the adjacent dropdown menu, select contains.
- In the search bar, type: carnitine.

(b) Use the First Neighbors of Selected Nodes (undirected) icon to select all neighbors (Fig. 7A).

(c) Continue expanding neighbors until all neighbors have been selected.

(d) Use the top "New Network From Selection (all edges)" icon (Fig. 7B, red box) to generate the new subnetwork.

(e) The new subnetwork can be visualized via the left panel Network tab.

(f) Rename the new subnetwork by right-clicking on the current network name and selecting Rename Network (Fig. 7C).

- The edge table and node table related to this subnetwork can be exported as a csv file as described above (step 7a).

Pause point: Saving. When exiting Cytoscape, you will be prompted to save your session. This will preserve the formatting you established, so that it will be restored next time you open the saved Cytoscape session in Cytoscape software.

Pause point: Saving network view for figures. A publication-quality figure of the network can be saved through File → Export → Network Views as Graphics, and setting the export file format to pdf.

Note: Molecular network link should be shared in publications in the Data Availability section or appropriate equivalent.

3.4.2 Identifying network nodes perturbed by infection

Timing: 10–15 min

1. This protocol will work for both feature-based and classical molecular networks. However, the pie charts are a better indication of relative metabolite feature abundance in feature-based molecular networks and thus the use of feature-based molecular networks is recommended for these analyses.
2. Make the pie charts.
   (a) Select the "Style" tab.
   (b) Select the box in the Def. column by "Image/Chart 1" (Fig. 8A, red box).
   (c) A new window will pop up, named "Graphics."
   (d) Switch to Pie chart tab (Fig. 8B, arrow).

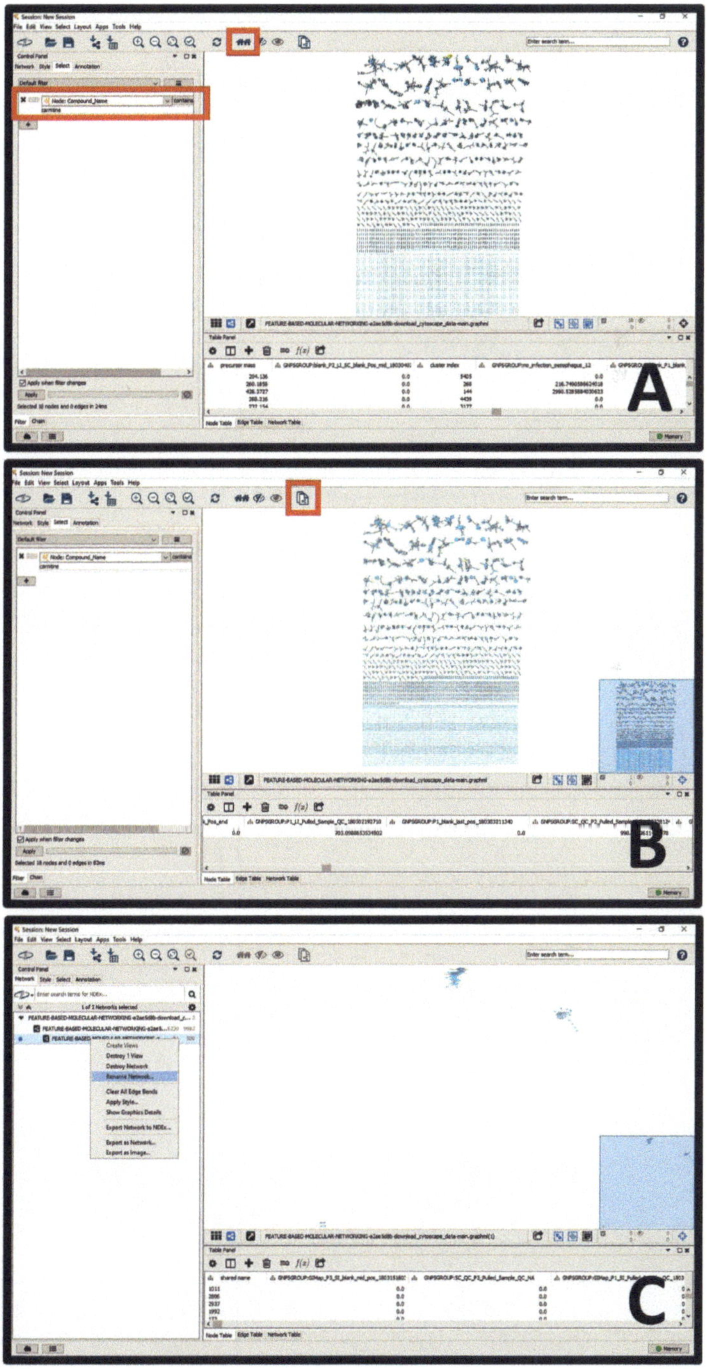

Fig. 7 Steps to subset network. (A) Selecting specific nodes based on parameters of interest and hitting the First Neighbors of Selected Nodes (red boxes). (B) "New Network From Selection (all edges)" icon to generate new subnetwork (red box). (C) Renaming the new network.

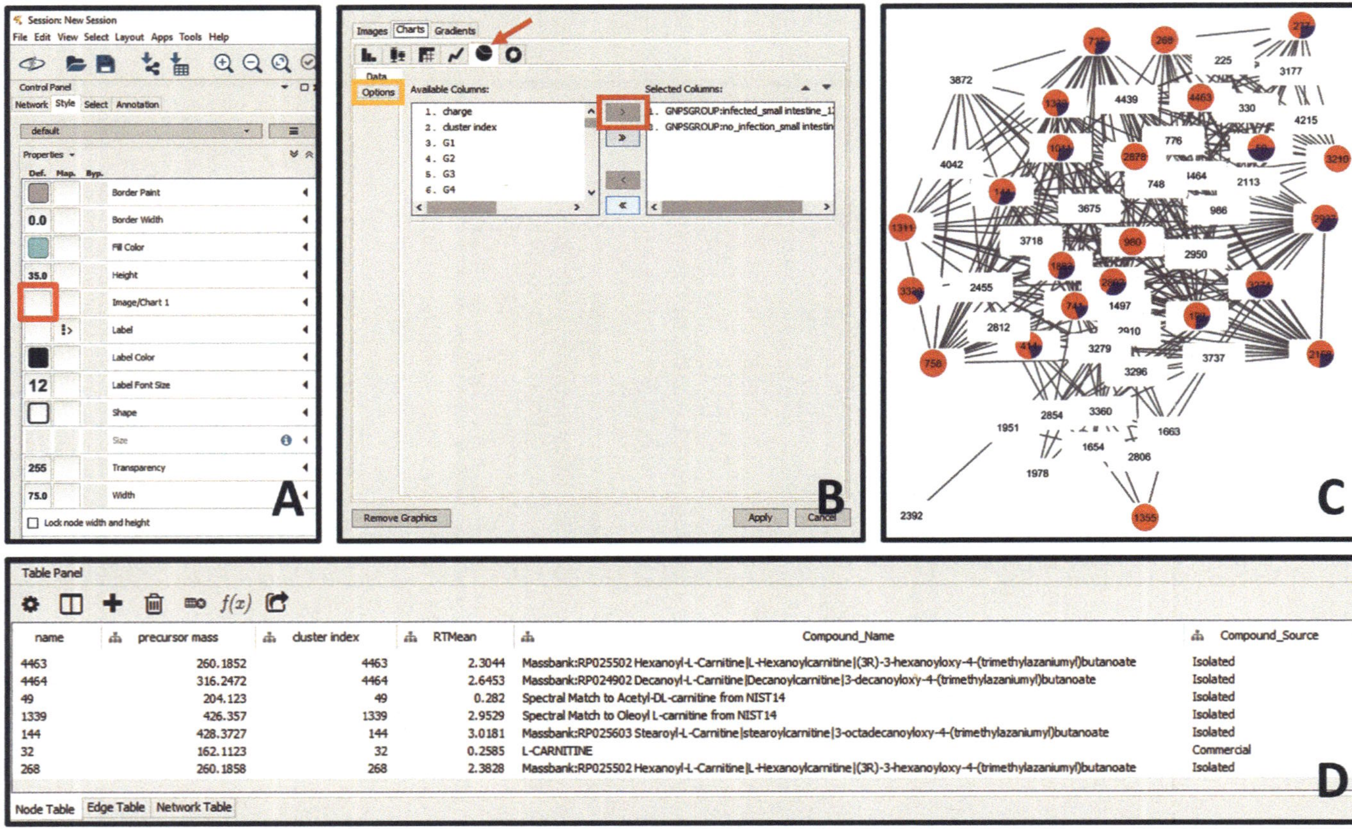

Fig. 8 Making feature abundance pie charts. (A) Select the "Image/chart" box (red box). (B) Choose pie chart tab (arrow) and select the groups to plot (red box). The options tab (yellow box) can be used to set pie chart colors. (C) Visualize the molecular network. (D) Network table with metabolite feature information.

(e) In the "Data" tab, choose the columns you are interested in and click the Add Selected arrow (Fig. 8B, red box) to put them into "Selected Columns". In this example, we choose "GNPSGROUP: infected_small intestine_12" and "GNPSGROUP: no_infection_small intestine_12."

(f) In the "Options" tab (Fig. 8B, yellow box), you can customize the pie chart colors. In our example, we choose red for "GNPSGROUP: infected_small intestine_12" and blue for "GNPSGROUP:no_infection_small intestine_12."

(g) Click Apply in the bottom right.

(h) You will have a pie chart for each node, displaying the relative ratio of peak areas between selected groups (Fig. 8C).

(i) (Optional) For clearer pie chart visualization, change from the default blue node background to a white background. Refer to Section 3.4.1 section above for instructions.

3. Visually explore the network for nodes showing differential abundance between infected and uninfected samples (see *Note* for statistical analyses of these differences).
4. If you wish to export information on a select infection-perturbed sub-family, the best way to do so is to create a new subnetwork from this family. Refer to Section 3.4.1 for instructions.
5. Inspect node annotations. Refer to Section 3.4.1 for instructions.
6. Expected results: Fig. 8C.

Note: When molecular network is large or to provide statistical support for step 3, instead of visual network exploration, begin with statistical tools such as random forest analysis, fold change comparison between infected and uninfected samples or statistical significance testing, to identify metabolite features differing between infected and uninfected samples. Then, search for these features in the network as described in Section 3.4.1, step 13.

Note: Pie chart colors should be selected keeping in mind accessibility to color-blind readers. Avoid red (infected) to green (uninfected) visualizations in particular.

Pause point: You may save your Cytoscape session as described in Section 3.4.1.

Critical: Annotations should always be checked for quality of the chemical match (mirror plot, cosine score, number of matched peaks). For analog matches, chemical plausibility of the mass difference should always be considered. A final filter should always be the biological plausibility of the annotation. Refer to the GNPS documentation and (Aron et al., 2020), for more details.

3.4.3 Identifying infection-correlated chemical modifications

Timing: 50 min

The goal of this protocol is to determine whether different chemical modifications occur in infected and uninfected samples. For example, you may want to determine whether there is increased oxidative modifications in infected samples, as a potential indicator of oxidative stress (Spickett, 2020), or whether degree of saturation differs between infection conditions. Higher degree of lipid saturation has been correlated with induction of pro-inflammatory pathways (Monson, Trenerry, Laws, Mackenzie, & Helbig, 2021). This protocol takes advantage of the molecular network structure, where related chemicals are connected by edges, which are annotated with the mass difference between nodes, representing a chemical group differing between the two nodes (Fig. 9A). This protocol can be used with classical or feature-based molecular networking.

1. Open your network in Cytoscape (as described in Section 3.4.1).
2. Subset your network to only include infection-associated nodes in one subnetwork and nodes found in uninfected samples in a second subnetwork.
   - **(a)** First, use the Select (Filter) tab to select uninfected samples.
     - In the example feature-based molecular network, uninfected samples can be found by selecting Node: ATTRIBUTE_condition and selecting "contains" in the drop-down menus, followed by typing in the search bar: no_infection.
   - **(b)** Create a new sub-network from selected samples using the New Network From Selection (all edges) button.
   - **(c)** Rename this network under the network tab to "uninfected."
   - **(d)** Last, remove singletons (unconnected) nodes (Fig. 2D). These are unsuitable for edge analysis, since they are not connected to any other nodes. This can be performed by reformatting your subnetwork using the Layout top menu and selecting "Apply preferred layout." Singleton nodes can then be visualized, highlighted using shift + mouse drag and then deleted using standard keyboard shortcuts.
     - if you do not wish to delete nodes, a new subnetwork can instead be created by selecting manually all non-singleton nodes using shift + mouse drag and then creating a new subnetwork from the selection.

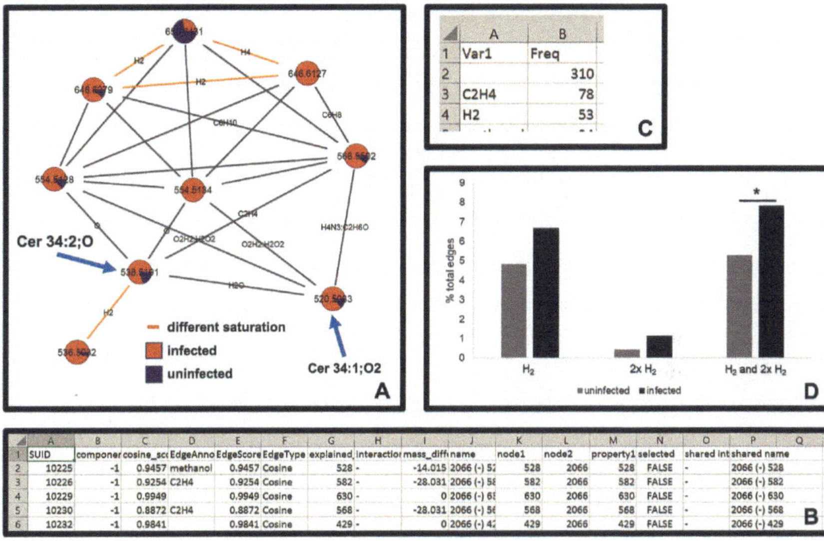

Fig. 9 Expected output of edge analysis. Feature-based networking results were subset to only include samples from the esophagus at 12 days post-infection and processed as described in the protocol. (A) Sub-network of ceramide-related metabolite features in infected and uninfected esophagus at 12 day timepoint. Annotated nodes indicated by arrows. Edges connecting metabolites differing by degree of saturation are highlighted in orange. (B) Partial screenshot of edge table export for all metabolite features detected in infected esophagus samples at 12 days post-infection. (C) Partial screenshot of aggregated abundance of each edge annotation, obtained from exported edge table. (D) Relative proportion of H_2 and $2 \times H_2$ (single bond differing in saturation between related metabolites or two bonds differing in saturation between related metabolites) between infected and uninfected esophagus samples at 12 day timepoint. $^*P < 0.05$, Fisher's exact test.

(e) Repeat these steps to select and create a new sub-network for infected samples called "infected," without singletons.
   - In the example feature-based molecular network, these can be found by selecting Node: ATTRIBUTE_condition contains infected.

3. Use the Export Table to File button to select and export "infected default edge" and "uninfected default edge" tables as .csv files to the location of your choice.
4. In the edge tables, the column of interest for this protocol is EdgeAnnotation.

5. Determine the number of occurrences of each annotation, in your "infected default edge" and "uninfected default edge" tables. This can be performed manually in Excel, but best practices would be to use reproducible code such as in Python or R, in Jupyter Notebooks or similar.
6. Combine these results for infected and uninfected samples into a single worksheet.
7. From this file, you can calculate the frequency of each edge annotation in infected and in uninfected samples and generate plots using excel or any standard code (see Fig. 9D for example). Statistical analysis can be performed using Fisher's exact test.

Critical: Automated edge annotations should always be considered in the context of chemical and biochemical plausibility in the infectious disease system under study.

Note: This method does not consider relative abundance of the connected nodes and chemical modifications, only presence/absence.

Note: This method could also be applied to a particular chemical family. For example, if interested in oxidative modification of phospholipids, the steps above should be followed, with additional subsetting to the chemical family of interest.

3.4.4 Optional steps: Demonstrating that specific network nodes of interest may be pathogen-derived

Timing: 40 min

This step requires data from axenically-cultured pathogen samples, acquired along with data from tissue samples. This data should be included when generating the molecular network. As an example, we use here a classical molecular network. However, these steps are suitable for both classical and feature-based molecular networks. Selection of the type of network to use will depend on the user's needs in terms of data filtering vs available culture samples and data clean-up processes.

1. Before you start, download the Venn and Euler Diagrams app from Cytoscape as follows:
   (a) In Cytoscape, select Apps and then App Manager (Fig. 10A).
   (b) Under Install Apps, find Venn and Euler Diagrams and select install (Fig. 10B).
2. Make groups in Cytoscape using the filtering tool. In our example we are looking to group the nodes from infected tissue, cultured parasites, and uninfected tissue.

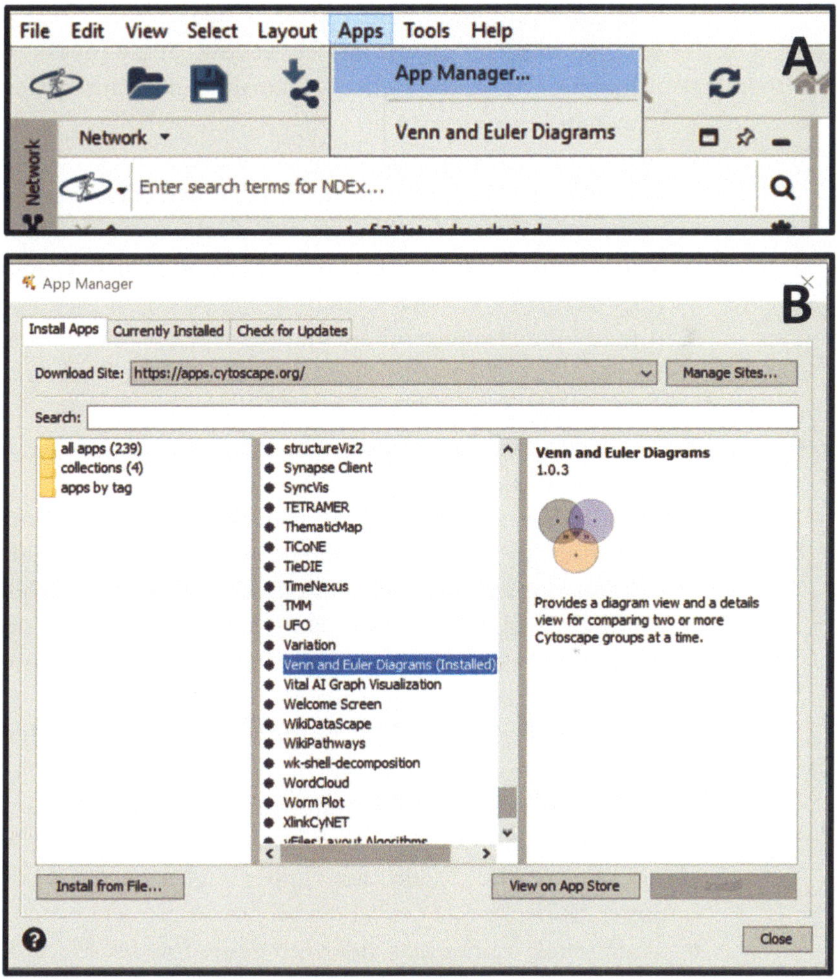

Fig. 10 Installing the Venn and Euler Diagrams app. (A) App manager menu. (B) Choosing the Venn and Euler Diagrams app.

- **(a)** Use the Select (or Filter) button on the left side of Cytoscape
- **(b)** Click the plus button and then Column Filter in the dropdown menu
- **(c)** Select the column you would like to filter by based on the groups you are trying to create
  - Infected tissue:
    1. To group all nodes present in infected tissue, select Node: ATTRIBUTE_condition as column filter.

2. In the text box underneath, type: infected.
   a. The selected nodes will be yellow. If no nodes are highlighted yellow, check to make sure your spelling matches metadata.
3. While the nodes are selected, move the mouse over a blank space between nodes on the network. Right click to pull up options for the selected nodes (Fig. 11B).
4. Select the option: Group and then Group Selected Nodes (Fig. 11B).
5. A textbox will appear where you can type the desired name for the group. In this example, we will call these "Infected."

- Cultured parasites:
  1. To group all nodes present in cultured parasites, select Node: ATTRIBUTE_study as column filter.
  2. For the cultured parasites, we had two types in our study, labeled tryps (*Trypanosoma cruzi* trypomastigotes) and ama (*T. cruzi* amastigotes) in our metadata. To include both of these types in this group, two column filters are needed.
     a. To add a second column filter, click the plus button and select column filter again.
  3. Select Node: ATTRIBUTE_study for this column filter as well.
  4. In the first box, type: tryps and in the second box type: ama.
  5. We want this group to include all nodes in trypomastigotes or amastigotes, so in the upper right select: Match any (OR).
  6. Right click on the network with desired nodes selected.
  7. Select the option: Group and then Group Selected Nodes.
  8. A textbox will appear to type the desired name for the group. We will call these "Cultured Parasites."
- Uninfected samples:
  1. This analysis included both uninfected host cells and uninfected tissue samples.
  2. To group all nodes present in the uninfected samples, select Node: ATTRIBUTE_study as the first column filter and Node: ATTRIBUTE_condition for the second column filter.
  3. Type C2C12 in the Node: ATTRUBUTE_study search box. These are uninfected cells of the cell type that were used to culture the *T. cruzi* parasites.

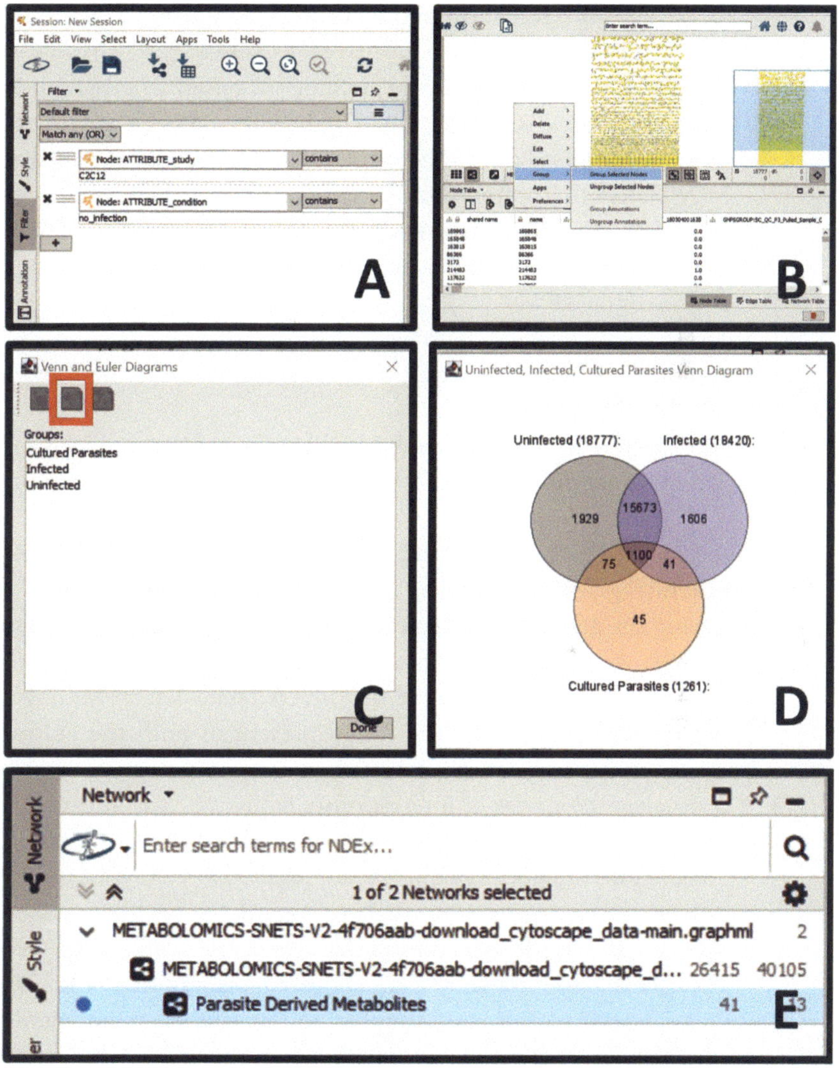

Fig. 11 Expected output from identifying candidate pathogen-derived nodes. (A) Screenshot of the filter setting used to make groups. "Match any" is used in the protocol, however different analyses may require "Match all." (B) Creation of a group from selected nodes. Selected nodes are highlighted in yellow. (C) Pop up screen for the Venn and Euler diagram app. All groups created will be shown in the box labeled "Groups." Multiple groups must be selected to create a Venn diagram (red box). (D) The expected Venn diagram. Nodes of interest are found at the intersection of Cultured Parasites and Infected. Clicking on that section will select the 41 nodes of interest on the network. (E) Screenshot of the subset network name (Parasite Derived Metabolites) created from the nodes of interest and containing 41 metabolite features.

4. Type no_infection in the Node: ATTRIBUTE_condition search box.
5. We want this group to include all nodes in either of the categories, so in the upper right select: Match any (OR).
6. Right click on the network with desired nodes selected.
7. Select the option: Group and then Group Selected Nodes.
8. A textbox will appear where you can type the desired name for the group. We will call these "Uninfected."

3. Make Venn diagrams
   (a) Select Apps and select Venn and Euler diagrams.
   (b) A pop up should appear with the groups delineated in the previous steps (Fig. 11C).
   (c) Select all three groups: Infected, Uninfected, and Cultured Parasites and then select the picture of the three-way Venn diagram (red box, Fig. 11C).
   (d) A pop up will appear with the Venn diagram using the groups selected (Fig. 11D).
   (e) Clicking on the section of interest in the Venn diagram will select the nodes in that group within in the network. In our case we are looking for the nodes present in the infected state that are pathogen-derived. These would be the nodes present in both the cultured parasites and the infected groups.
   (f) To visualize the nodes of interest more easily, you can subset those nodes into their own network.
      - Make sure the nodes of interest are selected in yellow by clicking on the intersection in the Venn diagram.
      - While the nodes of interest are selected, click the "New Network From Selection (all edges)" icon (red box, Fig. 7B).
   (g) The new network name will appear underneath the main network (Fig. 11E). To rename the new network, right click on the current network name and select: Rename Network.
   (h) To improve visibility of the nodes on this subset network, refer to Visualizing and interpreting your molecular network.
   (i) Annotations can be investigated following standard GNPS instructions, available here: https://ccms-ucsd.github.io/GNPSDocumentation/networkingviews/#view-all-library-hits.
   (j) Expected results: Venn diagram (Fig. 11D) and subnetwork containing 41 nodes shared between infected tissues and cultured parasites (Fig. 11E).

Note: Possibilities that the nodes are present in all sample types but below the limit of detection in uninfected controls should always be considered, as well as instrumental effects such as peak over-splitting or ion suppression.

Note: Absence of a node in axenic pathogen cultures that was present in infected samples should not automatically be interpreted as confirmation that the molecule is host-derived. The possibility that it is indeed a pathogen-derived molecule but only induced in infection settings and not in axenic culture should always be considered.

Note: Overlap between axenic pathogen cultures and uninfected samples is also possible. These metabolite features may represent metabolites common across host and pathogen, but decreased by infection in tissues (with the pathogen-derived metabolite within the tissues below the limit of detection).

Critical: Biological validation should always be performed as a follow-up, to address all the alternative possibilities discussed in Notes.

4. Advantages

The greatest strength of molecular networking is the ability to annotate metabolites that are not currently in reference metabolite libraries, but are structurally-related to these reference metabolites ("analog matches" and annotation propagation). This is particularly important when trying to discover novel pathogen-derived metabolites that are likely to be missing from existing databases. Parasitic infections in particular are under-studied even compared to other pathogens and thus novel parasite metabolites would otherwise be under-annotated in the absence of such methods. Furthermore, GNPS houses "living data": additional reference spectra are continually being added and datasets re-analyzed, so that annotations increase over time (Wang et al., 2016). By grouping metabolite features into families with shared fragmentation patterns (and thus shared chemical substructures), molecular networking also enables analyses at the chemical family-level and rapid visualization of family-level changes in metabolite abundance between infected and uninfected samples (Dean et al., 2021; Eberhard et al., 2021; Hoffman et al., 2021; Hossain et al., 2020; McCall et al., 2017; Parab et al., 2021). We have demonstrated that this can provide additional insight into metabolic changes across infection conditions that would have been missed if analyses had been performed exclusively at the single metabolite level (Dean et al., 2021; Hoffman et al., 2021; Hossain et al., 2020; McCall et al., 2017). These findings can even be used to guide drug development for infectious diseases (Hossain et al., 2020). Molecular network connectivity further facilitates the determination of the presence and frequency of certain chemical modifications such as oxidation, methylation, etc. This may have strong utility in understanding the oxidative environment present under many infection conditions, though applications so far have often focused on environmental samples (e.g., Hartmann et al., 2017; Roach et al., 2021). Lastly, molecular networking is embedded within the GNPS infrastructure, enabling straightforward integration with multiple other data analysis tools such as in silico annotation tools (NAP, MS2LDA, DEREPLICATOR (da Silva et al., 2018; Mohimani et al., 2017; van der Hooft, Wandy, Barrett, Burgess, & Rogers, 2016)), meta-analysis tools (ReDU-MS2 (Jarmusch et al., 2020)), 3D visualization ('ili (Protsyuk et al., 2018)), statistical analyses using QIIME2 (Bolyen et al., 2019; Caporaso et al., 2010), etc. Multiple other tools are being developed to facilitate and enrich analyses (e.g., GNPS Dashboard, MolNetEnhancer (Ernst et al., 2019; Petras et al., 2021)).

5. Limitations

As with all metabolomic analyses, there is still considerable "dark matter" (da Silva, Dorrestein, & Quinn, 2015): metabolite features that are unannotated and that cannot benefit from annotation propagation, because none of the connected nodes are annotated. While some or even many of these features may represent redundancy in the data (multiple adducts of the same metabolite, in-source fragmentation…) or artifacts (e.g., chromatographic effects, peak over-splitting during data processing) (Mahieu, Huang, Chen, & Patti, 2014; Sindelar & Patti, 2020), there is nevertheless still significant scope to discover new chemical matter (Cohen et al., 2017; Motley et al., 2017; Quinn et al., 2020). The continuous expansion of reference libraries, supplemented with new in silico tools integrated into GNPS (e.g., NAP, SIRIUS, COSMIC, CSI-FingerID (Böcker, Letzel, Lipták, & Pervukhin, 2009; da Silva et al., 2018; Dührkop et al., 2019; Duhrkop, Shen, Meusel, Rousu, & Bocker, 2015; Hoffmann et al., 2021)) can assist with annotating novel chemistry, while modified GNPS workflows (ion identity molecular networking) can reduce redundancy from the presence of multiple adducts with different MS2 fragmentation patterns (Schmid et al., 2021). In silico annotations should be confirmed with pure standards, where possible, according to the Metabolomics Standards Initiative (Sumner et al., 2007).

Molecular networking, and metabolomics data processing in general, are highly complex. We hope that this protocol, other published protocols (Aron et al., 2020; Phelan, 2020), and the resources on the GNPS website will assist with de-mystifying this approach. Although considerable functionality is available within the molecular networking infrastructure, additional statistical tools outside of GNPS and Cytoscape may be required to select infection-perturbed metabolite features to annotate (Dean et al., 2021; Hoffman et al., 2021; Hossain et al., 2020; McCall et al., 2017, 2018; Parab et al., 2021; Wozniak et al., 2020). The applicability, strengths and limitations of each of these tools should be evaluated prior to implementation. Lastly, standard limitations of LC-MS should always be kept in mind when analyzing resulting data, such as limit of detection, ion suppression, etc.

6. Optimization and troubleshooting

6.1 General troubleshooting

The first resource to troubleshoot unexpected outcomes is the GNPS molecular networking documentation. A second step is to review currently-discussed issues in the GNPS forum (https://groups.google.com/g/molecular_networking_bug_reports/) and troubleshooting pages (https://ccms-ucsd.github.io/GNPSDocumentation/troubleshooting/). Below, we discuss some common problems and solutions encountered when performing the analyses covered by this protocol.

6.2 Problem: Molecular network is too small or contains few annotations

In the case of feature-based molecular networking, the number of network nodes should be identical to the number of features in the feature table. Expected numbers of annotations will vary depending on the biological system of interest; however, common metabolite annotations such as amino acids, nucleotides and phospholipids will generally be obtained. If an internal standard was included at the time of data acquisition, is present in GNPS libraries and was not filtered out during pre-processing, it should be observed in the resulting annotations.

6.2.1 Potential solution to optimize the procedure

Solutions include verifying that the files were correctly converted to open format (classical molecular networking) and correctly processed by your mass spectrometry data processing program of choice (feature-based molecular networking).

6.3 Problem: Unable to link network node to metadata in cytoscape

In some cases, a network may be obtained with annotations as expected. However, each node has not been linked to metadata, so that feature abundance pie charts based on metadata cannot be generated (see Section 3.4.2).

6.3.1 Potential solution to optimize the procedure

The most likely cause is discrepancies between file names in data and metadata. Special characters, diaereses and umlauts should be avoided here (e.g., naïve should be entered as naive).

7. Safety considerations and standards

All data acquisition should follow local institutional regulations, including Institutional Biosafety Committee and Institutional Animal Care and Use Committee-approved protocols. Publications using these methods should provide networking links in manuscript Methods. We also encourage users of these methods to also deposit their own data back into the databases they used (for example GNPS/MassIVE if using these protocols).

8. Alternative methods/procedures

Should data not have been acquired from cultured pathogen samples at the time of data acquisition, a possible pathogen source can nevertheless be hypothesized by co-networking with publicly-available datasets using ReDU (https://redu.ucsd.edu/) or by single-spectrum search (MASST) of publicly-available datasets (Jarmusch et al., 2020; Wang et al., 2020). MASST can be accessed at https://masst.ucsd.edu/ or by navigating to the MASST search on the GNPS home page (https://gnps.ucsd.edu/). MASST will search available datasets for the presence of a given MS2 spectrum. The resulting list can then be mined to determine whether this MS2 is indeed pathogen-specific or common to other organisms (or to blank samples and controls). One strength of the MASST approach is the ability to focus on a specific metabolic feature of interest, and to determine specificity of that feature to the pathogen of interest (Wang et al., 2020). In contrast, ReDU enables global analysis, by building molecular networks that combine the user's data with existing spectral data (Jarmusch et al., 2020). This approach could for example be used to determine the percentage of metabolic features in infected samples that are shared with the pathogen. In that case, once the co-network has been built, similar steps would be pursued as in Section 3.4.4. A limitation of these approaches is that they require publicly-available data on the pathogen of interest, which may be limiting in the case of neglected diseases. However, these two approaches use "living" databases, which are continually expanding.

Alternative networking approaches that do not rely on cosine scoring are being developed (for example, Spec2Vec or MS2LDA) (Huber et al., 2021; van der Hooft et al., 2016). Several of these are being implemented in the GNPS infrastructure. The resulting networks can then be visualized and analyzed using the approaches described in this manuscript.

9. Summary

Overall, this protocol covers procedures to use molecular networking approaches in the context of infectious diseases. Methods described in this protocol enable the visualization and identification of infection-associated changes in metabolite abundance and in chemical modifications. They also enable hypothesis-generating approaches such as the identification of candidate pathogen-derived metabolites. Metabolomics data analysis and the molecular networking infrastructure is rapidly evolving. We encourage interested researchers to build from these standard approaches to address their biological questions of interest.

Acknowledgments

We wish to thank the inventors and developers of molecular networking, as well as all the contributors who have been iteratively improving it. We have endeavored to cite their publications throughout this manuscript. We also wish to thank the members of the McCall lab who tested these protocols: Kate Wheeler, Camil Gosmanov and Michael Jimenez Sandoval. Research using molecular networking in the context of infectious diseases in the McCall laboratory is supported by NIH 1R21AI148886, NIH 1R21AI156669 and PhRMA foundation 45188. Laura-Isobel McCall, Ph.D. holds an Investigators in the Pathogenesis of Infectious Disease Award from the Burroughs Wellcome Fund. The content is solely the responsibility of the authors and does not necessarily represent the official views of the funders.

References

Aksenov, A. A., Laponogov, I., Zhang, Z., Doran, S. L. F., Belluomo, I., Veselkov, D., et al. (2021). Auto-deconvolution and molecular networking of gas chromatography–mass spectrometry data. *Nature Biotechnology*, *39*(2), 169–173. https://doi.org/10.1038/s41587-020-0700-3.

Aron, A. T., Gentry, E. C., McPhail, K. L., Nothias, L.-F., Nothias-Esposito, M., Bouslimani, A., et al. (2020). Reproducible molecular networking of untargeted mass spectrometry data using GNPS. *Nature Protocols*, *15*(6), 1954–1991. https://doi.org/10.1038/s41596-020-0317-5.

Böcker, S., Letzel, M. C., Lipták, Z., & Pervukhin, A. (2009). SIRIUS: Decomposing isotope patterns for metabolite identification†. *Bioinformatics*, *25*(2), 218–224. https://doi.org/10.1093/bioinformatics/btn603.

Bolyen, E., Rideout, J. R., Dillon, M. R., Bokulich, N. A., Abnet, C. C., Al-Ghalith, G. A., et al. (2019). Reproducible, interactive, scalable and extensible microbiome data science using QIIME 2. *Nature Biotechnology*, *37*(8), 852–857. https://doi.org/10.1038/s41587-019-0209-9.

Caporaso, J. G., Kuczynski, J., Stombaugh, J., Bittinger, K., Bushman, F. D., Costello, E. K., et al. (2010). QIIME allows analysis of high-throughput community sequencing data. *Nature Methods*, 7(5), 335–336. https://doi.org/10.1038/nmeth.f.303.

Cohen, L. J., Esterhazy, D., Kim, S.-H., Lemetre, C., Aguilar, R. R., Gordon, E. A., et al. (2017). Commensal bacteria make GPCR ligands that mimic human signalling molecules. *Nature*, *549*(7670), 48–53. https://doi.org/10.1038/nature23874.

da Silva, R. R., Dorrestein, P. C., & Quinn, R. A. (2015). Illuminating the dark matter in metabolomics. *Proceedings of the National Academy of Sciences of the United States of America*, *112*(41), 12549–12550. https://doi.org/10.1073/pnas.1516878112.

da Silva, R. R., Wang, M., Nothias, L.-F., van der Hooft, J. J. J., Caraballo-Rodríguez, A. M., Fox, E., et al. (2018). Propagating annotations of molecular networks using in silico fragmentation. *PLoS Computational Biology*, *14*(4), e1006089. https://doi.org/10.1371/journal.pcbi.1006089.

Dean, D. A., Gautham, Siqueira-Neto, J. L., McKerrow, J. H., Dorrestein, P. C., & McCall, L.-I. (2021). Spatial metabolomics identifies localized chemical changes in heart tissue during chronic cardiac Chagas disease. *PLos Neglected Tropical Diseases*, *15*(10), e0009819. https://doi.org/10.1371/journal.pntd.0009819.

Dührkop, K., Fleischauer, M., Ludwig, M., Aksenov, A. A., Melnik, A. V., Meusel, M., et al. (2019). SIRIUS 4: A rapid tool for turning tandem mass spectra into metabolite structure information. *Nature Methods*, *16*(4), 299–302. https://doi.org/10.1038/s41592-019-0344-8.

Duhrkop, K., Shen, H., Meusel, M., Rousu, J., & Bocker, S. (2015). Searching molecular structure databases with tandem mass spectra using CSI:FingerID. *Proceedings of the National Academy of Sciences of the United States of America*, *112*(41), 12580–12585. https://doi.org/10.1073/pnas.1509788112.

Dunn, W. B., Broadhurst, D., Begley, P., Zelena, E., Francis-McIntyre, S., Anderson, N., et al. (2011). Procedures for large-scale metabolic profiling of serum and plasma using gas chromatography and liquid chromatography coupled to mass spectrometry. *Nature Protocols*, *6*(7), 1060–1083. http://www.nature.com/nprot/journal/v6/n7/abs/nprot.2011.335.html#supplementary-information.

Eberhard, F. E., Klimpel, S., Guarneri, A. A., & Tobias, N. J. (2021). Metabolites as predictive biomarkers for Trypanosoma cruzi exposure in triatomine bugs. *Computational and Structural Biotechnology Journal*, *19*, 3051–3057. https://doi.org/10.1016/j.csbj.2021.05.027.

Ernst, M., Kang, K. B., Caraballo-Rodríguez, A. M., Nothias, L.-F., Wandy, J., Chen, C., et al. (2019). MolNetEnhancer: Enhanced molecular networks by integrating metabolome mining and annotation tools. *Metabolites*, *9*(7), 144. Retrieved from https://www.mdpi.com/2218-1989/9/7/144.

Forsberg, E. M., Huan, T., Rinehart, D., Benton, H. P., Warth, B., Hilmers, B., et al. (2018). Data processing, multi-omic pathway mapping, and metabolite activity analysis using XCMS online. *Nature Protocols*, *13*(4), 633–651. https://doi.org/10.1038/nprot.2017.151.

Garg, N., Wang, M., Hyde, E., da Silva, R. R., Melnik, A. V., Protsyuk, I., et al. (2017). Three-dimensional microbiome and metabolome cartography of a diseased human lung. *Cell Host Microbe*, *22*(5). https://doi.org/10.1016/j.chom.2017.10.001. 705-716.e704.

Hartmann, A. C., Petras, D., Quinn, R. A., Protsyuk, I., Archer, F. I., Ransome, E., et al. (2017). Meta-mass shift chemical profiling of metabolomes from coral reefs. *Proceedings of the National Academy of Sciences*, *114*(44), 11685–11690. https://doi.org/10.1073/pnas.1710248114.

Heine, D., Holmes, N. A., Worsley, S. F., Santos, A. C. A., Innocent, T. M., Scherlach, K., et al. (2018). Chemical warfare between leafcutter ant symbionts and a co-evolved pathogen. *Nature Communications*, *9*(1), 2208. https://doi.org/10.1038/s41467-018-04520-1.

Hoffman, K., Liu, Z., Hossain, E., Bottazzi, M. E., Hotez, P. J., Jones, K. M., et al. (2021). Alterations to the cardiac metabolome induced by Chronic T. cruzi infection relate to the degree of cardiac pathology. *ACS Infectious Diseases*, 7(6), 1638–1649. https://doi.org/10.1021/acsinfecdis.0c00816.

Hoffmann, M. A., Nothias, L.-F., Ludwig, M., Fleischauer, M., Gentry, E. C., Witting, M., et al. (2021). Assigning confidence to structural annotations from mass spectra with COSMIC. *bioRxiv*. https://doi.org/10.1101/2021.03.18.435634. 2021.2003.2018.435634.

Hossain, E., Khanam, S., Dean, D. A., Wu, C., Lostracco-Johnson, S., Thomas, D., et al. (2020). Mapping of host-parasite-microbiome interactions reveals metabolic determinants of tropism and tolerance in Chagas disease. *Science Advances*, *6*(30), eaaz2015. https://doi.org/10.1126/sciadv.aaz2015.

Huber, F., Ridder, L., Verhoeven, S., Spaaks, J. H., Diblen, F., Rogers, S., et al. (2021). Spec2Vec: Improved mass spectral similarity scoring through learning of structural relationships. *PLoS Computational Biology*, *17*(2), e1008724. https://doi.org/10.1371/journal.pcbi.1008724.

Idle, J. R., & Gonzalez, F. J. (2007). Metabolomics. *Cell Metabolism*, *6*(5), 348–351. https://doi.org/10.1016/j.cmet.2007.10.005.

Ivanisevic, J., & Want, E. J. (2019). From samples to insights into metabolism: Uncovering biologically relevant information in LC-HRMS metabolomics data. *Metabolites*, *9*(12), 308. Retrieved from https://www.mdpi.com/2218-1989/9/12/308.

Jarmusch, A. K., Wang, M., Aceves, C. M., Advani, R. S., Aguirre, S., Aksenov, A. A., et al. (2020). ReDU: A framework to find and reanalyze public mass spectrometry data. *Nature Methods*. https://doi.org/10.1038/s41592-020-0916-7.

Lai, Z., Tsugawa, H., Wohlgemuth, G., Mehta, S., Mueller, M., Zheng, Y., et al. (2017). Identifying metabolites by integrating metabolome databases with mass spectrometry cheminformatics. *Nature Methods*. https://doi.org/10.1038/nmeth.4512.

Lewis, C. M., Jr., McCall, L. I., Sharp, R. R., & Spicer, P. G. (2020). Ethical priority of the most actionable system of biomolecules: The metabolome. *American Journal of Physical Anthropology*, *171*(2), 177–181. https://doi.org/10.1002/ajpa.23943.

Mahieu, N. G., Huang, X., Chen, Y. J., & Patti, G. J. (2014). Credentialing features: A platform to benchmark and optimize untargeted metabolomic methods. *Analytical Chemistry*, *86*(19), 9583–9589. https://doi.org/10.1021/ac503092d.

McCall, L. I., Morton, J. T., Bernatchez, J. A., de Siqueira-Neto, J. L., Knight, R., Dorrestein, P. C., et al. (2017). Mass spectrometry-based chemical cartography of a cardiac parasitic infection. *Analytical Chemistry*, *89*(19), 10414–10421. https://doi.org/10.1021/acs.analchem.7b02423.

McCall, L. I., Tripathi, A., Vargas, F., Knight, R., Dorrestein, P. C., & Siqueira-Neto, J. L. (2018). Experimental Chagas disease-induced perturbations of the fecal microbiome and metabolome. *PLoS Neglected Tropical Diseases*, *12*(3), e0006344. https://doi.org/10.1371/journal.pntd.0006344.

Melnik, A. V., Vázquez-Baeza, Y., Aksenov, A. A., Hyde, E., McAvoy, A. C., Wang, M., et al. (2019). Molecular and microbial microenvironments in chronically diseased lungs associated with cystic fibrosis. *mSystems*, *4*(5). https://doi.org/10.1128/mSystems.00375-19. e00375-00319.

Mohimani, H., Gurevich, A., Mikheenko, A., Garg, N., Nothias, L. F., Ninomiya, A., et al. (2017). Dereplication of peptidic natural products through database search of mass spectra. *Nature Chemical Biology*, *13*(1), 30–37. https://doi.org/10.1038/nchembio.2219.

Monson, E. A., Trenerry, A. M., Laws, J. L., Mackenzie, J. M., & Helbig, K. J. (2021). Lipid droplets and lipid mediators in viral infection and immunity. *Fems Microbiology Reviews*, *45*(4), fuaa066. https://doi.org/10.1093/femsre/fuaa066.

Motley, J. L., Stamps, B. W., Mitchell, C. A., Thompson, A. T., Cross, J., You, J., et al. (2017). Opportunistic sampling of roadkill as an entry point to accessing natural products assembled by bacteria associated with non-anthropoidal mammalian microbiomes. *Journal of Natural Products*, *80*(3), 598–608. https://doi.org/10.1021/acs.jnatprod.6b00772.

Newsom, S. N., & McCall, L. I. (2018). Metabolomics: Eavesdropping on silent conversations between hosts and their unwelcome guests. *PLoS Pathogens*, *14*(4), e1006926. https://doi.org/10.1371/journal.ppat.1006926.

Nothias, L.-F., Nothias-Esposito, M., da Silva, R., Wang, M., Protsyuk, I., Zhang, Z., et al. (2018). Bioactivity-based molecular networking for the discovery of drug leads in natural product bioassay-guided fractionation. *Journal of Natural Products*, *81*(4), 758–767. https://doi.org/10.1021/acs.jnatprod.7b00737.

Nothias, L. F., Petras, D., Schmid, R., Dührkop, K., Rainer, J., Sarvepalli, A., et al. (2020). Feature-based molecular networking in the GNPS analysis environment. *Nature Methods*, *17*(9), 905–908. https://doi.org/10.1038/s41592-020-0933-6.

Parab, A. R., Thomas, D., Lostracco-Johnson, S., Siqueira-Neto, J. L., McKerrow, J. H., Dorrestein, P. C., et al. (2021). Dysregulation of glycerophosphocholines in the cutaneous lesion caused by leishmania major in experimental murine models. *Pathogens*, *10*(5). https://doi.org/10.3390/pathogens10050593.

Petras, D., Phelan, V. V., Acharya, D., Allen, A. E., Aron, A. T., Bandeira, N., et al. (2021). GNPS dashboard: Collaborative analysis of mass spectrometry data in the web browser. *bioRxiv*. https://doi.org/10.1101/2021.04.05.438475. 2021.2004.2005.438475.

Phelan, V. V. (2020). Feature-based molecular networking for metabolite annotation. *Methods in Molecular Biology*, *2104*, 227–243. https://doi.org/10.1007/978-1-0716-0239-3_13.

Pluskal, T., Castillo, S., Villar-Briones, A., & Orešič, M. (2010). MZmine 2: Modular framework for processing, visualizing, and analyzing mass spectrometry-based molecular profile data. *BMC Bioinformatics*, *11*(1), 395.

Protsyuk, I., Melnik, A. V., Nothias, L. F., Rappez, L., Phapale, P., Aksenov, A. A., et al. (2018). 3D molecular cartography using LC-MS facilitated by Optimus and 'ili software. *Nature Protocols*, *13*(1), 134–154. https://doi.org/10.1038/nprot.2017.122.

Quinn, R. A., Melnik, A. V., Vrbanac, A., Fu, T., Patras, K. A., Christy, M. P., et al. (2020). Global chemical effects of the microbiome include new bile-acid conjugations. *Nature*, *579*(7797), 123–129. https://doi.org/10.1038/s41586-020-2047-9.

Reis, F. C. G., Costa, J. H., Honorato, L., Nimrichter, L., Fill, T. P., & Rodrigues, M. L. (2021). Small molecule analysis of extracellular vesicles produced by Cryptococcus gattii: Identification of a tripeptide controlling Cryptococcal infection in an invertebrate host model. *Frontiers in Immunology*, *12*(652). https://doi.org/10.3389/fimmu.2021.654574.

Roach, T. N. F., Dilworth, J., Christian Martin, H., Jones, A. D., Quinn, R. A., & Drury, C. (2021). Metabolomic signatures of coral bleaching history. *Nature Ecology & Evolution*, *5*(4), 495–503. https://doi.org/10.1038/s41559-020-01388-7.

Röst, H. L., Sachsenberg, T., Aiche, S., Bielow, C., Weisser, H., Aicheler, F., et al. (2016). OpenMS: A flexible open-source software platform for mass spectrometry data analysis. *Nature Methods*, *13*(9), 741–748. https://doi.org/10.1038/nmeth.3959.

Schmid, R., Petras, D., Nothias, L.-F., Wang, M., Aron, A. T., Jagels, A., et al. (2021). Ion identity molecular networking for mass spectrometry-based metabolomics in the GNPS environment. *Nature Communications*, *12*(1), 3832. https://doi.org/10.1038/s41467-021-23953-9.

Senges, C. H. R., Al-Dilaimi, A., Marchbank, D. H., Wibberg, D., Winkler, A., Haltli, B., et al. (2018). The secreted metabolome of *Streptomyces chartreusis* and implications for bacterial chemistry. *Proceedings of the National Academy of Sciences, 115*(10), 2490–2495. https://doi.org/10.1073/pnas.1715713115.

Shannon, P., Markiel, A., Ozier, O., Baliga, N. S., Wang, J. T., Ramage, D., et al. (2003). Cytoscape: A software environment for integrated models of biomolecular interaction networks. *Genome Research, 13*(11), 2498–2504. https://doi.org/10.1101/gr.1239303.

Sindelar, M., & Patti, G. J. (2020). Chemical discovery in the era of metabolomics. *Journal of the American Chemical Society, 142*(20), 9097–9105. https://doi.org/10.1021/jacs.9b13198.

Smith, C. A., Want, E. J., O'Maille, G., Abagyan, R., & Siuzdak, G. (2006). XCMS: Processing mass spectrometry data for metabolite profiling using nonlinear peak alignment, matching, and identification. *Analytical Chemistry, 78*(3), 779–787. https://doi.org/10.1021/ac051437y.

Spickett, C. M. (2020). Formation of oxidatively modified lipids as the basis for a cellular epilipidome. *Frontiers in Endocrinology, 11*(974). https://doi.org/10.3389/fendo.2020.602771.

Sturm, M., Bertsch, A., Gropl, C., Hildebrandt, A., Hussong, R., Lange, E., et al. (2008). OpenMS—An open-source software framework for mass spectrometry. *BMC Bioinformatics, 9*, 163. https://doi.org/10.1186/1471-2105-9-163.

Sumner, L. W., Amberg, A., Barrett, D., Beale, M. H., Beger, R., Daykin, C. A., et al. (2007). Proposed minimum reporting standards for chemical analysis Chemical Analysis Working Group (CAWG) Metabolomics Standards Initiative (MSI). *Metabolomics: Official Journal of the Metabolomic Society, 3*(3), 211–221. https://doi.org/10.1007/s11306-007-0082-2.

Tautenhahn, R., Böttcher, C., & Neumann, S. (2008). Highly sensitive feature detection for high resolution LC/MS. *BMC Bioinformatics, 9*(1), 504. https://doi.org/10.1186/1471-2105-9-504.

Tsugawa, H., Cajka, T., Kind, T., Ma, Y., Higgins, B., Ikeda, K., et al. (2015). MS-DIAL: Data-independent MS/MS deconvolution for comprehensive metabolome analysis. *Nature Methods, 12*(6), 523–526. https://doi.org/10.1038/nmeth.3393.

van der Hooft, J. J. J., Wandy, J., Barrett, M. P., Burgess, K. E. V., & Rogers, S. (2016). Topic modeling for untargeted substructure exploration in metabolomics. *Proceedings of the National Academy of Sciences of the United States of America*. https://doi.org/10.1073/pnas.1608041113.

Wang, M., Carver, J. J., Phelan, V. V., Sanchez, L. M., Garg, N., Peng, Y., et al. (2016). Sharing and community curation of mass spectrometry data with Global Natural Products Social Molecular Networking. *Nature Biotechnology, 34*(8), 828–837. https://doi.org/10.1038/nbt.3597.

Wang, M., Jarmusch, A. K., Vargas, F., Aksenov, A. A., Gauglitz, J. M., Weldon, K., et al. (2020). Mass spectrometry searches using MASST. *Nature Biotechnology, 38*(1), 23–26. https://doi.org/10.1038/s41587-019-0375-9.

Want, E. J., Masson, P., Michopoulos, F., Wilson, I. D., Theodoridis, G., Plumb, R. S., et al. (2013). Global metabolic profiling of animal and human tissues via UPLC-MS. *Nature Protocols, 8*(1), 17–32. https://doi.org/10.1038/nprot.2012.135.

Watrous, J., Roach, P., Alexandrov, T., Heath, B. S., Yang, J. Y., Kersten, R. D., et al. (2012). Mass spectral molecular networking of living microbial colonies. *Proceedings of the National Academy of Sciences of the United States of America, 109*(26), E1743–E1752. https://doi.org/10.1073/pnas.1203689109.

Wozniak, J. M., Mills, R. H., Olson, J., Caldera, J. R., Sepich-Poore, G. D., Carrillo-Terrazas, M., et al. (2020). Mortality risk profiling of *Staphylococcus aureus* bacteremia by multi-omic serum analysis reveals early predictive and pathogenic signatures. *Cell, 182*(5). https://doi.org/10.1016/j.cell.2020.07.040. 1311-1327.e1314.

MIX
Papier | Fördert
gute Waldnutzung
FSC® C083411

Kolibri 360 GmbH
Ferdinand-Jühlke-Straße 7
99095 Erfurt, Deutschland
produktsicherheit@kolibri360.de